International Conference on Structural, Mechanical and Materials Engineering (ICSMME 2017)

AA004779

IOP Conference Series: Materials Science and Engineering
Volume 248

Seoul, South Korea
13 – 15 July 2017

ISBN: 978-1-5108-5028-6
ISSN: 1757-8981

Printed from e-media with permission by:

Curran Associates, Inc.
57 Morehouse Lane
Red Hook, NY 12571

Some format issues inherent in the e-media version may also appear in this print version.

Copyright© (2017) by the Institute of Physics
All rights reserved. The material featured in this book is subject to
IOP copyright protection, unless otherwise indicated.

Printed by Curran Associates, Inc. (2017)

For permission requests, please contact the Institute of Physics
at the address below.

Institute of Physics
Dirac House, Temple Back
Bristol BS1 6BE UK

Phone: 44 1 17 929 7481
Fax: 44 1 17 920 0979

techtracking@iop.org

Additional copies of this publication are available from:

Curran Associates, Inc.
57 Morehouse Lane
Red Hook, NY 12571 USA
Phone: 845-758-0400
Fax: 845-758-2633
Email: curran@proceedings.com
Web: www.proceedings.com

Table of contents

Volume 248

2017 International Conference on Structural, Mechanical and Materials Engineering (ICSMME 2017)

13–15 July 2017, Seoul, South Korea

Accepted papers received: 27 September 2017
Published online: 20 October 2017

Preface

Advanced material properties and applications

Evaluation theores and models for the mechanical and structural engineering

IOP Conf. Series: Materials Science and Engineering **248** (2017) 011001 doi:10.1088/1757-899X/248/1/011001

Preface

The International Conference on Structural, Mechanical and Materials Engineering (ICSMME 2017) was held in Seoul, South Korea on July 13-15, 2017 and this book proceeding contains some of the contributions presented at the conference.

The primary objective of ICSMME 2017 is to provide a world class forum for exchange original ideas and new information, latest research and discussing scientific. Also it to bring together academics, scientists, engineers, postgraduates and other professionals in the area of material science and engineering technology from all over the world.

On behalf of the conference organizing committee, we would like to express our sincere appreciation to the keynote speakers, for their excellent and inspirational speeches, the reviewers, for their valuable inputs for improving the scientific content of these papers. We also would like to thank every technical committee member and author who has contributed to the success of the conference.

The Organizing Committee of ICSMME 2017

Committee Chair

Prof. Jongwan Hu | Incheon National University

Department of Civil and Environmental Engineering

Conference Photograph

Content from this work may be used under the terms of the Creative Commons Attribution 3.0 licence. Any further distribution of this work must maintain attribution to the author(s) and the title of the work, journal citation and DOI.
Published under licence by IOP Publishing Ltd

IOP Conf. Series: Materials Science and Engineering **248** (2017) 011001 doi:10.1088/1757-899X/248/1/011001

Sponsors

Incheon Disaster Prevention Research Center (IDPRC)

Recently, various efforts to prevent and prepare are vitally needed for prevention of Disasters and calamities. So we understand the necessity for technology of disaster and we built up the Incheon Disaster Prevention Research Center (IDPRC) in Incheon National University (INU) in 1997.

Accordingly Incheon Disaster Prevention Research Center (IDPRC) in Incheon National University has progressed of research on the prevention of disasters and calamities through the various seminars, conference and lectures. These research could be conducted cooperation with Incheon National University (INU) in various fields structure, soil, hydraulic and environment.

Incheon Disaster Prevention Research Center (IDPRC) will try to be a leader in the disaster of industry through the various research activities and global conference.

Incheon National University

Incheon National University established a global campus by integrating two city council funded colleges into one in March 2010 and became a national university operated by the legal entity. With the start of attracting the branch of Lawrence Berkeley Research Center that is a world famous US national policy research center for education and research, Incheon National University will attract St. Petersburg University in Russia, Polymers University in Britain, and Kent University in Belgium and other foreign universities and research centers. These efforts will make INU spring to a world-wide competitive university.

As a local development leading university, Incheon National University established INU VISON 2020 and improved its competition actively through investment in selection and concentration. In compliance with the geographic feature and areal specialty, it will concentrate on the international trade, goods distribution, applied technology convergence, BNT-based life science, urban science and the sinology- based local humanity fields.

IOP Conf. Series: Materials Science and Engineering **248** (2017) 011001 doi:10.1088/1757-899X/248/1/011001

Lists of Committees

Committee name	Organizer name	Organization
Honorary Chair	J.W. Hu	Incheon National University, South Korea
Conference General Chairs	K. Mosbeh	Incheon National University, South Korea
Honorary Chair	J.W. Hu	Incheon National University, South Korea
Technical Committees	J.K. Ahn	Incheon National University, South Korea
	M. Eldessouki	Technical University of Liberec, Czech Republic
	Mahmoud. M.K. Hedel	Faculty of Engineering, Mansoura University, Egypt

Delegate List

First name	Sure name	Institution	E-mail address
M	**Nabil**	Advanced Technology and New Materials Research Institute	marwamoh2000@yahoo.com
F	**Wang**	Heilongjiang University, Harbin	wang-futong@126.com
Z	**Li**	China Construction Steel Structure Corp	chenmu07040327@sina.com
F	**Yang**	Astronatics Engineering University, Beijing	sarah0824@hotmail.com
Y	**Liu**	Beijing Institute of Technology	liuyang19881225@126.com
P. F.	**Gao**	Anhui University of Science and Technology	568662152@qq.com
Vladislav A.	**Chekanin**	Moscow State University of Technology	vladchekanin@rambler.ru
I M	**Amiruddin**	Malaysia-Japan International Institute of Technology	utm_amir@yahoo.com
R M	**Baitimerov**	South Ural State University	baitimerovrm@susu.ru
Z M	**Jaini**	University Tun Hussein Onn Malaysia	rizuan@uthm.edu.my

IOP Conf. Series: Materials Science and Engineering **248** (2017) 011001 doi:10.1088/1757-899X/248/1/011001

E	Idczak	Poznan University of Technology	eli.idczak@gmail.com
Xiaotao	CHEN	Tsinghua University	chenxiao_qhu@foxmail.com
Liang	Qiao	China Electric Power Research Institute	535215163@qq.com
Jian	Qin	China Electric Power Research Institute	qinjian@epri.sgcc.com.cn
J H	Zhou	China Ship Development and Design Center	jhzhou.2005@163.com
T L	Teng	Hsiuping University of Science and Technology	tlteng@hust.edu.tw
T W	Samadhi	Faculty of Industrial Technology, Institut Teknologi Bandung	twsamadhi@che.itb.ac.id
Moghaddam	Tohid	Toosi University of Technology	tohid.moghaddam@gmail.com
Baoqing	Zhang	Department of Huazhong University of Science and Technology	xmbsk12345@126.com
J	Ismail	University Sultan Zainal Abidin	jailais@petronas.com
Behboodi	Ahad	Biomechanics and Movement Sciences University of Delaware	ahadbeh@udel.edu
X Q	Liu	Tianjin Research Institute for Water Transport Engineering	lxq200@163.com
Z Y	Zhang	Guizhou University	zhangloveall75@foxmail.com
H Z	Zhou	Nanjing University of Aeronautics and Astronautics	2543436974@qq.com
D S	Park	Incheon National University	dspark@inu.ac.kr
M N	Nurhayati	Aerospace Department, Universiti Kuala Lumpur	nmn_3a@yahoo.com
L	Tao	Northwestern Polytechnical University	taoli0401@163.com
Min	Hu	Jiangsu province xuzhou technician institute	humin331947080@qq.com

Peer review statement

All papers published in this volume of *IOP Conference Series: Materials Science and Engineering* have been peer reviewed through processes administered by the proceedings Editors. Reviews were conducted by expert referees to the professional and scientific standards expected of a proceedings journal published by IOP Publishing.

Content from this work may be used under the terms of the Creative Commons Attribution 3.0 licence. Any further distribution of this work must maintain attribution to the author(s) and the title of the work, journal citation and DOI.
Published under licence by IOP Publishing Ltd

IOP Conf. Series: Materials Science and Engineering **248** (2017) 012001 doi:10.1088/1757-899X/248/1/012001

Enhancement of porous silicon photoluminescence using (Ni) treatment

M Nabil[1,*], M Elnouby[1], N Gayeh[2], A H Sakr[2] and H A Motaweh[2]

[1]Advanced Technology and New Materials Research Institute, City for Scientific Research and Technology Applications, New Borg El-Arab City, Alexandria, 21934, Egypt.

[2]Department of Physics, Faculty of Science, Damanhour University, Egypt.

Abstract. A new method has been developed to improve the photoluminescence intensity of porous silicon (PS). Self-organized (PS) can be used in various fields. The deposition of metal nanoparticles (Ni) allows utilizing the obtained nano-composite for numerous applications such as sensor technology, biomedicine, and many more. (Ni/ PS) nano-composite powders are prepared using Ni deposition on the PS powder surface. The (Ni/ PS) powders became hydrophilic by the deposited Ni nanoparticles. At the different percentages of (Ni: PS), at a higher percentage than (3:1) the values of the crystallite size of Ni and PS are nearly equivalent which suggests the saturation case of PS surface by Ni particles. It also depends on the deposition time value. In this study, that is the commercialization of nanostructured materials, it is important that the manufacturing costs are appropriate and inexpensive for large scale production, in addition, the enhancement of the photoluminescence intensity of (PS).

1. Introduction

Crystalline Silicon (c-Si) {as an indirect semiconductor}, which is a non-emitted light. And so it is many alternative materials have been proposed to improve the luminescence from silicon-based materials [1]. Porous silicon (PS) is a versatile material whose fundamental properties is of value to diverse fields, a consequence of specific properties such as visible light emission, tunable surface chemistry, huge surface area, and controllable morphology. Due to its variable pore-size, PS is a suitable host material for loading other nanosized structures, e.g., small molecules, metals, and magnetic nanoparticles, thereby adding properties of the confined spaces to the pores [2]. PS can be considered as a silicon crystal having a network of voids in it. The nanosized voids in the silicon bulk result in a sponge-like structure of pores and channels surrounded by a skeleton of the crystalline Si nanostructure [3]. For improving the electroluminescence and the photoluminescence intensity of PS, several passivation methods have been proposed. [1]

It has been demonstrated that PS with high surface areas may be easily prepared by a simple, safe Chemical technique, which is more energy efficient, cost-effective, and time-saving. Various techniques have been considered to prepare nanocomposite materials, including thermal, plasma spraying and physical and chemical vapor deposition. Among these methods that is widely used, chemical deposition, which offers several advantages when compared with other techniques including precise control, low energy requirements, uniform deposition, low cost in production of large area samples, good reproducibility, versatility, capability to coat complex component geometries, high production rate and reduction of waste. [4]

In one such attempt, attention was focused on nanoparticles-silicon composite materials. Many materials such as silver, gold and the transparent conducting oxides such an indium tin oxide, were

Content from this work may be used under the terms of the Creative Commons Attribution 3.0 licence. Any further distribution of this work must maintain attribution to the author(s) and the title of the work, journal citation and DOI.
Published under licence by IOP Publishing Ltd

tested and proved their ability to improve the PS properties. Also, various metals are deposited onto the PS surface using several techniques as the immersion plating technique or dipping the surface into a solution containing metal ions such as Ag, Cu and Pd [1]. Noticeable, only a few reports are describing the incorporation process of Nickel (Ni) nanoparticles and the surface of a silicon wafer, followed by manipulation of surface chemical bonds. The most useful technique is the deposition from the liquid phase by chemical method, which is actually used in this research as a result of the precipitation process of metal particles "inside and surroundings" of pores. The solution can penetrate the whole porous structure providing the deposition along the full length of the pores. However the deposition process may be complicated by capillary effects and non-uniformity of porous structure.

In the present work, for the first time the (Ni /PS) nanocomposite is prepared in powder form, which facilitate of using in many applications in several fields. In addition, the possibility to enhance PS photoluminescence intensity using chemical techniques for producing the nanocomposite material of PS and (Ni). The properties of the prepared nanocomposite material were studied by various characterization methods such as Raman spectroscopy, FTIR, XRD and Photoluminescence spectra.

2. Material and methods

Firstly: Beginning with commercial Si powder, PS powder was prepared using a stain-etching method, depending on our previous study [5], using alkaline medium. Then, the PS powder was filtered out from the etching solution and dried at 40°C overnight. With this procedure, the porous layer was formed on the Si powder surface.

Secondary: nano-Ni layers were deposited on the PS powder by a chemical deposition technique. The prepared PS powder was immersed in an aqueous $NiCl_2$ solution at 80°C for several (Ni salts: nPS) concentration percentages. Then, nano-Ni was deposited on the PS surface via redox reactions, i.e., Ni ions in the solution were reduced and PS layers were simultaneously oxidized. Finally, the nanocomposite (Ni/ PS) powder was filtered out again from the Ni salt solution and dried at 60° C overnight for two days.

The structure and morphology were characterized using XRD (X-ray 7000 Schimadzu diffractometer). The formation of PS chemical bonds was determined using FTIR (Fourier Transform Infrared Spectrophotometer- Shimadzu FTIR -8400 s, Japan), Raman Spectroscopy (Senteral - Bruker Raman micro-spectroscopy), UV-Vis. Spectrophotometer Double Auto cell (Labomend. INC, USA) and PL (Cary Eclipse Fluorescence Spectrophotometer Photoluminescence- American) analysis of PS, before and after treatment with (Ni).

3. Results & discussion

Figure 1 shows the XRD peaks in the range $2\theta = 20^\circ–100^\circ$ for the (Ni/ PS) samples at different Ni salt concentrations. As can be seen, the (Ni/ PS) samples shows the diffraction peaks of (PS) at 2 (= 28.23°, 47.193° and 56.23° corresponding to Si (111), Si (220) and Si (211), respectively. In addition, some weak diffraction peaks at 2θ = 76.68°, 88.34° and 95.24°, corresponding to the Ni (220), (201) and Ni (311) planes, respectively {JCPDS card 27–1402 and 45-1027, corresponding to Si and Ni, respectively}. Figure 1.a shows the appearance of Ni planes, that in case of the double percentage of the Ni salt concentration with respect to the presence of a PS weight percentage (2 Ni: 1 PS). Noticeable, the decreasing of the Si peak intensities, as a result of the increasing Ni salt concentration at the same value of PS weight percentage (3 Ni: 1 PS), as shown in figure 1.b. Figure 1.c shows the re-increasing in the XRD peak intensities of Si and Ni, which indicates the increasing of Ni layer thickness on the PS surface as increasing of Ni salt concentration till PS surface saturation. In addition, figure 1 shows the directly proportional between the Ni layer and the deposition time (t).

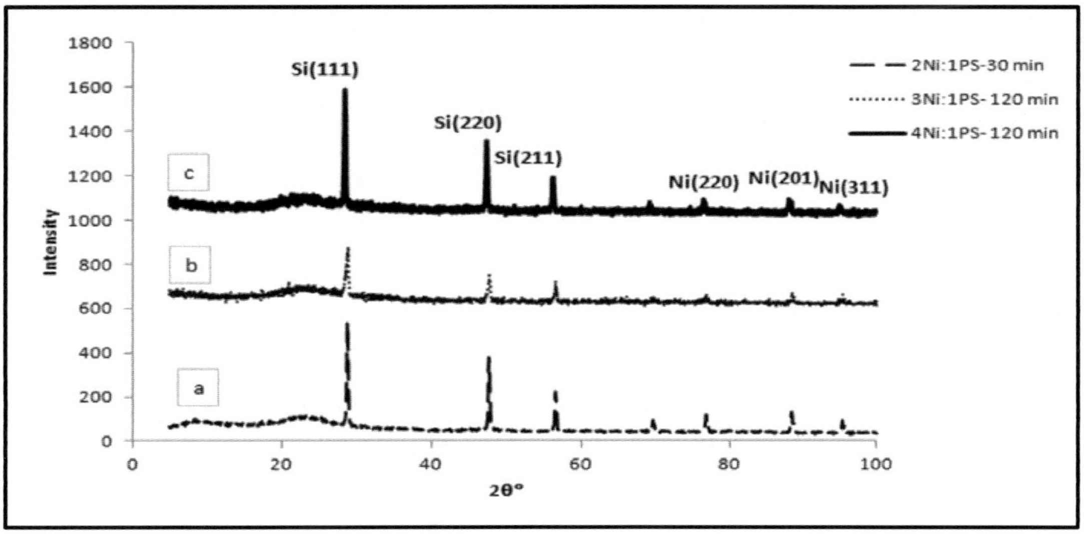

Figure 1. XRD patterns of (Ni/ PS) nano-composite powder chemical deposition process, at (a){(2Ni: 1 PS) at 30 min}, (b) {(3Ni: 1 PS) at 120 min} and (c) {(4Ni: 1 PS) at 120 min}.

Table 1 shows a comparison between the diffraction peaks of Si and Ni. The intensity of XRD signals, which is used for the calculation of the average size of the PS, using Scherrer's equation [3]. It is well known that the crystallize size can be estimated from diffraction pattern analysis by measuring the full width at half maximum (FWHM) measurement and applying the Scherrer's equation:

$$d = \frac{k\ \lambda}{B\ \cos\theta} \tag{1}$$

Where B is the FWHM, K is the Scherrer constant ($1 > K > 0.89$), λ is the wavelength in nanometers, and d the mean crystallite size.

Table 1, also, displays important notes about the smallest value of the crystallite size, which is corresponding to the percentage (3 Ni: 1 PS). In addition, at (4 Ni: 1 PS) percentage, the values of the crystallite size of Ni and PS are nearly equivalent.

Table 1. Typical XRD diffraction peaks of (Ni/ PS) nano-composite powder obtained by electro-less technique.

(Ni: PS)	Crystallite size of PS (nm)			Crystallite size of Ni (nm)		
	Plane (111)	Plane (220)	Plane (211)	Plane (220)	Plane (201)	Plane (311)
2:1	2.48	2.83	2.77	3	3.2	3.3
3:1	1.16	1.58	1.85	1.85	2.39	2.83
4:1	2.17	2.39	2.5	2.83	2.62	2.77

Figure 2 shows the Raman-spectra peak at 520 cm^{-1} for PS powder and a slight shift to 515 and 513 cm^{-1}, Ni deposition within the pores leads to a further shift of the peak to 505 cm^{-1} corresponding to the difference in the preparation percentage {Ni: PS = (1:1), (2:1) and (3:1)}, respectively, for the formation of nano-composite (Ni/ PS). Noticeable, at a percentage (4:1), the value of wavenumber increased again to 510 cm^{-1}, as a result of, the complete coverage (saturation case) on a PS surface with Ni-nanoparticles, which is due to compressive stress and then separation of Ni-nanoparticles as reformation of Ni-salt again.

IOP Conf. Series: Materials Science and Engineering **248** (2017) 012001 doi:10.1088/1757-899X/248/1/012001

The progress of reducing PS absorbance and increasing Ni absorbance is monitored by UV-Vis spectra which displayed in figure 3.a, b, c, d. That is a direct proportionality between Ni-salt concentration and Ni-nanoparticles absorbance peak intensity, which is adsorbed on PS surface, at a constant deposition time. UV spectra explains very important cyclic behavior, confirmation of the absorbance intensity value for concentration percentage (3:1) at deposition time 30 and 90 min, another confirmation for (1:1) at 60 and 120 min. In addition, the absorbance intensity of Ni/ PS nano-composite doesn't affect with varying the concentration percentages at the deposition time more than 120 min, as shown in figure 3.e.

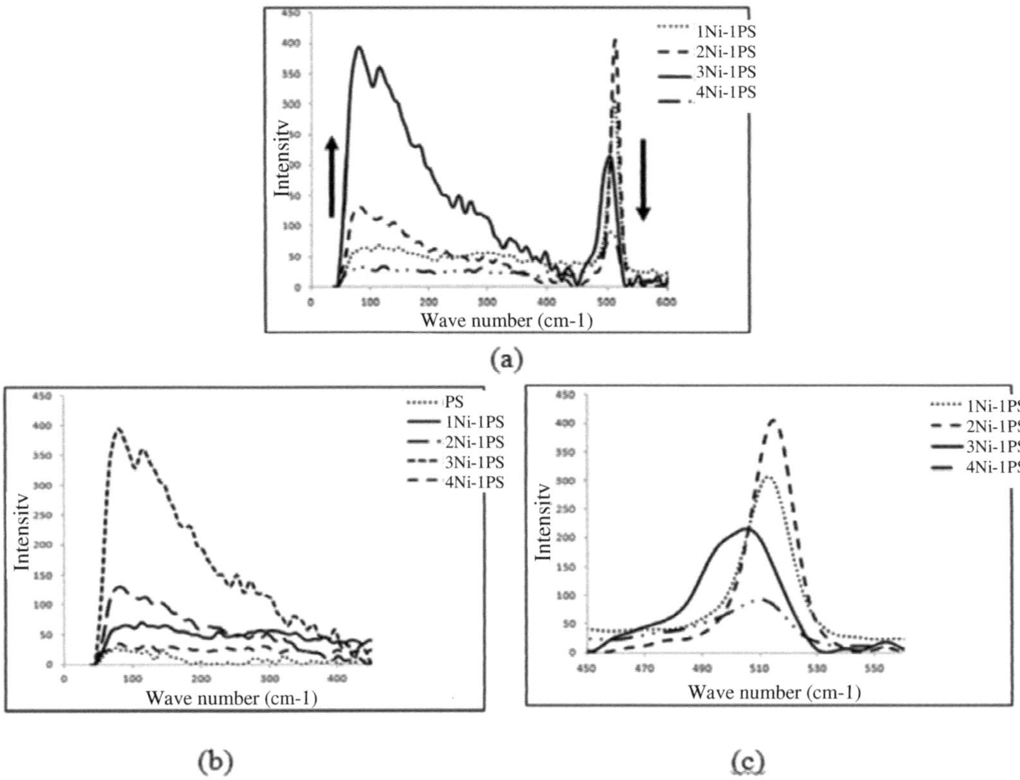

Figure 2. Raman spectrum of (Ni/ PS) nano-composite powder chemical deposition process (a)At different percentage of (Ni: PS), (b) Increasing of NiO spectra at different percentage of (Ni: PS), and (c) Decreasing of PS spectra at different percentage of (Ni: PS).

The PS powder surface {several planes as shown in figure 1} is liable to produce surface Si-H_2 and Si-H_3. These active Si-H_x(x = 1, 2 or 3) species involve in the reduction of Ni(II) while Si-H_x are oxidized into H_2. The redox reactions are depicted in the following chemical reaction mechanism. The surface Si-H_x species donate electrons which accepted by Ni(II) bring in gab out manufacture of Ni on PS surface. That means surface Si-H_x bonds play an important role in the reducing Ni(II). The porous architectures of silicon surface which bring vast surface Si-H_x species play vital roles in producing nano particles of a Ni element on PS surface.

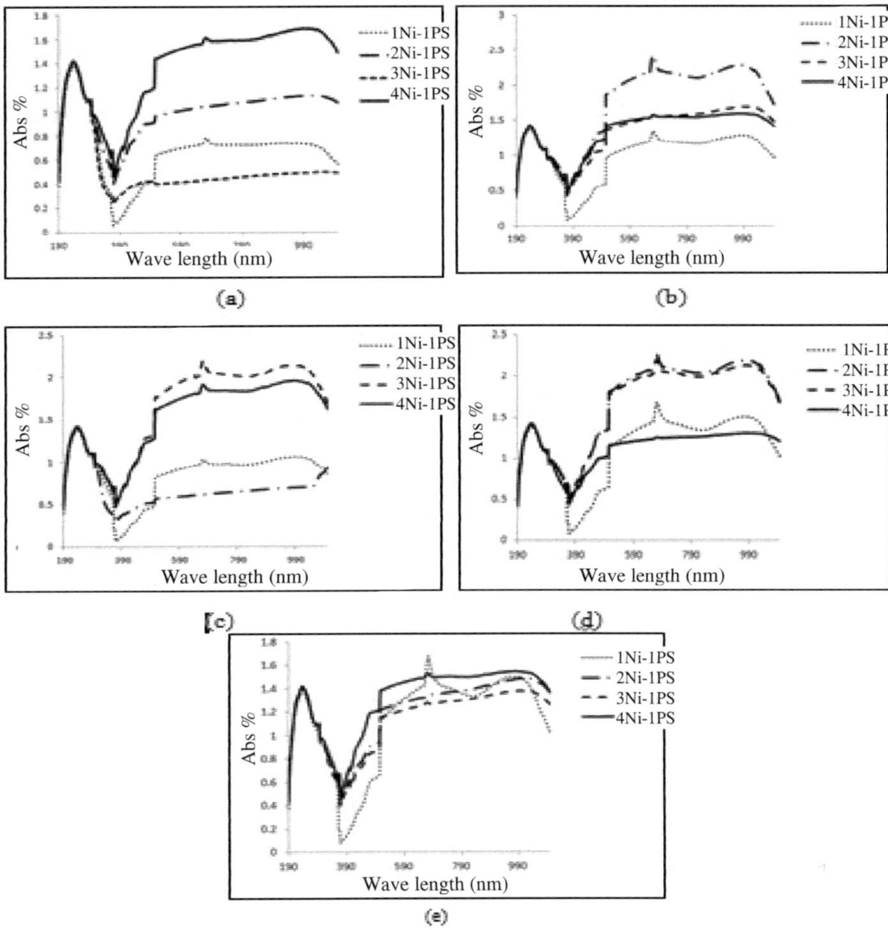

Figure 3. UV spectra of (Ni/ PS) nano-composite powder chemical deposition process at different percentage of (Ni: PS) at different deposition time:
(a)At 30 min, (b) At 60 min, (c)At 90 min,(d) At 120 min, and (e) At over 120 min.

FTIR measure the change in the PS chemical composition after immersion process in Ni-salt solution and the form of Ni which is deposited on the PS surface. As shown in figure 4, the present peaks of PS after immersion process in Ni-solution are the same as those observed in pure PS. The observed FTIR peaks are located at around 600 to 750 cm^{-1} (wagging modes) associated with Si-H_n ($n \geq 1$) bonding, the band at 1,000 to 1,300 cm^{-1} corresponds to the stretching modes of the Si-O-Si bonds in the PS. The obtained results of PS spectra have not given any indication of the presence of Ni on the PS surface at this stage, no peak corresponding to Ni bonds was observed. Identifies the presence of Ni on PS is appearing in three different percentage spectra {(2:1), (3:1) and (4:1)} with different intensities corresponds to the Ni percent, as shown in figure 4 . There appears as a new peak centered at 470 cm^{-1} which is generally attributed to Ni-O bonds [6]. The formation of NiO is an indication that Ni is deposited on PS in its metallic form and not in the form of its oxide (Si-O-Ni). Results of FTIR spectra agree with the above data of different characterization, the (3Ni: 1 PS) is the most suitable percent in the nano-composite preparation technique, which corresponds to the highest value of (Ni-O) peak intensity.

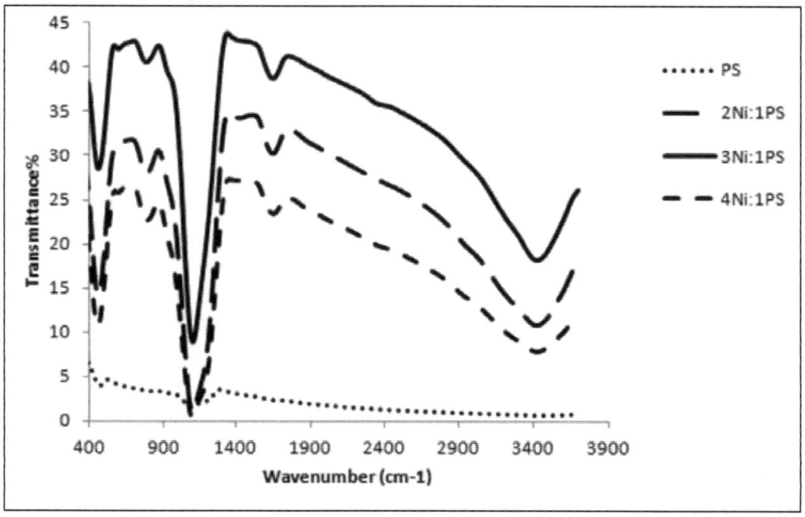

Figure 4. FTIR spectra of nano-composite (Ni/ PS) at different percentages.

The optical properties of the as-formed hybrid nano-structures were evaluated by measuring the PL spectra [7]. Figure 5.a shows the PL spectra of PS powder at 607 nm (2.04eV). Figure 5.b shows the enhancement of PL intensity after the PS chemical treatment at different percentages of Ni. The PL peak is red shifted at 663 nm (1.87eV). This corresponds to a shift towards the longer wavelength. The enhancement of PL intensity after the Ni nanoparticles deposition process may be attributed to the passivation of surface states in PS in a way that reduces the number of dangling bonds. Under this condition, the closely spaced surface states of PS can act as continuous energy levels and thus transfer the energy to the conduction band of Ni, resulting in emission [7].

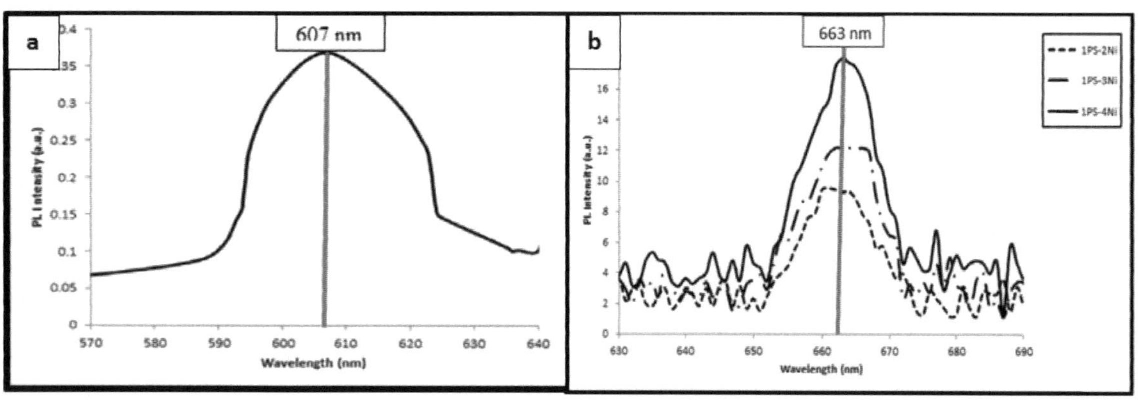

Figure 5. PL spectra of (a) PS, (b) PS powder treated with Ni at different percentages of (Ni).

The shift direction depends essentially on the porosity values, lower porosity would exhibit a red shift. It confirms the completely deposition process of Ni on PS powder surface, which reduces the porosity. The partial oxidation of PS occurring during Ni deposition is believed to induce such shifts, and the degree of oxidation as well as the porosity modification determines the shift direction [8].

Finally, for all nanocomposites studied, a remarkable enhancement in PL intensity is observed. Based on spectral analyses, we propose that exciton transfer can take place from the PS to the Ni nanoparticles, and that this is responsible for the PL enhancement [7].

Then, we have the very simple chemical technique for preparation nanocomposite (Ni/ PS) powder which is used in several fields. That is as a result of the differentiation in PS properties: PS treated with Ni have an enhancement of thermal stability, which is used as a thermal insulator [9]. In addition, it has an enhancement of optical properties as PL spectra which is shown at present work. Nor has there been many differences in PS properties, leading to the use of PS in many applications in several fields.

4. Conclusions

Nano-composite (Ni/ PS) powders were prepared using deposition technique of PS powder in an aqueous $NiCl_2$ solution. The properties of the (Ni/ PS) powders were systematically investigated as functions of the deposition time and Ni-salt concentration. The Raman spectra analysis indicated that the Ni-content deposited on the PS surface {the PS powder has a large surface area with respect to the Si powder} direct proportionality with increasing the deposition time and Ni salt until arriving to the saturation case, which appeared at (4Ni: 1 PS) percentage. After Ni was deposited, the PS powders changed from hydrophobic to hydrophilic because of the oxidation of the PS surface, as shown in FTIR spectra. The synthesis method used here was based on the fully chemical process; (Ni/ PS) is a promise nano-composite for different applications because the simplicity of the preparation method. UV spectra explain very important cyclic behavior, confirmation of the absorbance intensity value. The synthesis of a nano-composite Ni/ PS powder from powdered PS has been reported up to date. In this study suggest a possibility of an efficient passivation of (PS) by means of the incorporation of Ni, which induces a spectacular passivation of the silicon nanoparticles in the (PS) structure.

References

[1] Derbali L and Ezzaouia H 2012 *EPJ. Web. Conf.* **29** 13.
[2] Granitzer P, Rumpf K, Tian Y, Akkaraju G, Coffer J, Poelt P and Reissner M 2013 *Appl. Phys. Lett.* **102** 193110.
[3] Kashyout A H, Soliman H M A, Nabil M and Bishara A A 2013 *Mater. Sci. Appl.* **4** 79.
[4] Ahmad Y H and Mohamed A M A 2014 *Int. J. Electrochem. Sci.* **9** 1942.
[5] Kashyout A H, Soliman H M A, Nabil M and Bishara A A 2013 *Mater. Lett.* **100** 184.
[6] Slama S B, Hajji M and Ezzaouia H 2012 *Nanosci. Res. Lett.* **7** 464.
[7] Harraz F A and Salem A M 2013 *Script. Mater.* **68** 683.
[8] Bao X M, He X, Gao T, Yan Fand Chen H L 1998 *Sol. Stat. Commun.* **109** 169.
[9] Nabil M and Motaweh H A 2016 *Adv. Nanopar.* **5** 199.

IOP Conf. Series: Materials Science and Engineering **248** (2017) 012002 doi:10.1088/1757-899X/248/1/012002

Oxidation mechanism of micron-sized aluminum particles in Al-CO_2 gradually heating system

Y Liu[1], H Ren[1], Q J Jiao[1,]*

[1]State Key Laboratory of Explosive Science and Technology, School of Mechatronical Engi-neering, Beijing Institute of Technology, No.5 Zhongguancun South Street, Haidian District, Bei-jing, 100081, People's Republic of China

Abstract. Micron-sized aluminum powders were heated in carbon dioxide atmosphere through differential scanning calorimetry (DSC) method in this work. Aluminum powders were oxidized into four distinct stages from room temperature to 1500℃. Stage I, amorphous alumina shell turns to γ-Al_2O_3 phase from room temperature to 620℃. Stage II, accompany with the aluminum core melting, alumina shell becomes thicker and fragile at the temperature around 667℃ which is the melting point of aluminum. Stage III, in the temperature range of 690-1150 ℃, alumina shell was broken partially because of the inside pressure. Liquid aluminum spurts out through the weak point which becomes cracks on the surface just like volcano eruption and then oxidized by CO_2 while the temperature increased from 700℃ to 900℃. Stage IV, alumina changes to stable α-Al_2O_3. From what was presented above, a mechanism of micron-sized aluminum particle oxidation in CO_2 under gradually increasing temperature condition was proposed as "eruption model".

1. Introduction

Because of its several practical applications in energetic materials, metal combustion process, especially the aluminum combustion process is a continuing interest of a lot of researchers. Materials that can store large amounts of chemical energy, and are characterized by a high rate of energy release are termed "energetic materials". Micron-sized aluminum powders are often added into various energetic materials, including pyrotechnics, explosives, and rocket propellants. It is because that the performance of energetic materials can be highly improved such as the oxidation enthalpy and the combustion temperature, especially the motor acoustic stability [1-4] In recent several decades, basic understanding of aluminum combustion mechanisms has been improved significantly in experimental and theoretical investigations [5-9]. However, it is difficult to determine the accurate reactive mechanism because the different properties of aluminum powders, for instance, the specific surface area, particle size and distribution, different synthetic methods, environment condition, core@shell structure of aluminum powder [10, 11] will eventually influence the oxidation mechanism of aluminum.

Carbon dioxide is one of the gas products in solid rocket engines, for this reason, it is meaningful to find out the oxidation behavior of aluminum powders in CO_2. Aluminum powder is typical core@shell structure. Although there is a strong thermodynamic driving force for the reaction between aluminum and oxidizers, they can't contact directly because of the alumina oxide shell outside each particle [12]. Several literatures [13, 14] show an idea that the ignition temperature of aluminum particles is in the vicinity of Al_2O_3 melting point (2054℃), however, some others prove that the ignition temperature

Content from this work may be used under the terms of the Creative Commons Attribution 3.0 licence. Any further distribution of this work must maintain attribution to the author(s) and the title of the work, journal citation and DOI.
Published under licence by IOP Publishing Ltd

range of different micron-sized aluminum powders is from 650℃ to 2000℃ [13, 15-18]. Therefore, the oxidation mechanism of micron-sized aluminum particles in CO_2 is not that clear.

Thermal analysis method is mature enough in the study of oxidation kinetics of aluminum powders. Thermogravimetric-Differential Scanning Calorimetry (TG-DSC) was employed to observe the mass and thermal behavior of prepared material during the heating process. TG-DSC can provide useful information on phase transitions through exothermic and endothermic reactions in the studies of aluminum powders [19-23]. Prof. M. A. Trunov [24] presented a characteristic thermo-gravimetric analysis (TGA) curve of oxidizing aluminum powder and sequence of changes of alumina shell. In order to find out the characteristics of micro-sized aluminum in CO_2 under gradually increasing temperature, further studies were done and new mechanism is put forward in this work.

2. Experimental approach

2.1 Materials and apparatus

Aluminum powders in six different sizes numbered from Al#1 to Al#6 which were synthesized through wire electrical explosion (WEE) by Angang Group Aluminum Powder Co., Ltd were investigated in this Al/CO_2 system. Particle size distributions were analyzed by MASTERSIZER2000 Laser particle size analyzer. The experimental setup utilized in this research consists of a combination of Thermo-Gravimetric Analyzer coupled with NETZSCH STA449F3 - DSC200F3 Differential Scanning Calorimeter (TG-DSC) and Hitachi S-4700 field emission gun scanning electron microscopy (FEG-SEM). The corresponding chemical changes are analyzed by Philips PW1710 powder X-ray Diffraction (XRD). The composition of all the samples, including the impurities and their content were analyzed by ELEMENT GD glow discharge mass spectrometer, at the meantime, the specific surface areas of the three samples were calculated by the BET (Brunauer, Emmett and Teller) equation These facilities can closely simulate the residence time, temperature histories a chemical environment of various combustion units allowing for a variety of solid-solid, solid-gas interactions.

2.2 Experiment procedure

All the six samples were heated from room temperature to 1450℃ in the first group. The balance and furnace were purged with argon at 10 and 20mL/min respectively. Heating rate was 20℃/min which was appropriate to ensure high resolution of mass and energy variations as a function of temperature and an ignorable measurement error. The reaction chamber was vacuumed prior to introducing research grade CO_2 with a steady flow rate of 1.2L/h which can ensure aluminum powders completely reacted.

According to the results of TG-DSC curves got in the first round of heating process, experimental conditions will be changed appropriately. (1) Add extra constant high temperature process to special samples. (2) Stop heating process at different temperature and collect the samples. (3) Reheat the samples in a certain condition and compare the different oxidation mechanism. The intermediate and final products will be collected and do further analysis. In order to recover intermediate products without any other impact, samples were cooled down in argon.

3. Results

3.1 Characterization of Al powders

The properties of all the samples are given in table 1. According to the statistical analysis, size distribution of all the samples fully complies with lognormal distribution which is given as equation 1. Figure 1 exhibits the size distribution and fitting curve of Al#3 and Al#6 as examples.

Table 1. Properties of all six aluminum samples used in the experiments.

Sample	Specific surface area m^2/g	Active aluminum(%)	D_{50}(μm)	Impurity content (%)			
				Fe	Si	Cu	H_2O
Al#1	1.35	98.21	1.89	0.080	0.067	0.002	0.025
Al#2	1.01	98.43	2.51	0.081	0.062	0.002	0.023
Al#3	0.45	99.01	5.24	0.082	0.061	0.002	0.022
Al#4	0.17	99.32	13.35	0.079	0.065	0.002	0.020
Al#5	0.10	99.37	24.02	0.082	0.062	0.002	0.021
Al#6	0.06	99.35	40.05	0.080	0.063	0.002	0.019

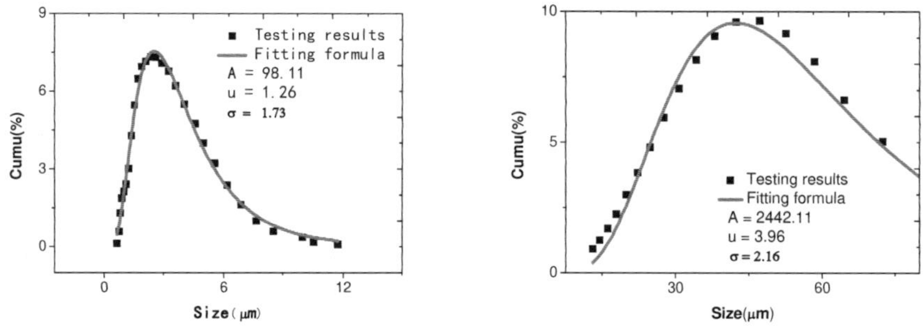

(a) (b)

Figure 1. Size distribution and fitting curves of two samples. (a) Al#3, (b)Al#6.

$$f(x) = \frac{A}{\sqrt{2\pi}\sigma x} \cdot \exp\left(\frac{-(\ln x - \mu)^2}{2\sigma^2}\right) \qquad x \in [a, b] \tag{1}$$

Figure 2 is the SEM pictures of four different sized particles. The surface of the particles is not that smooth (see the inset of figure 2 (b)), some cracks can be seen clearly.

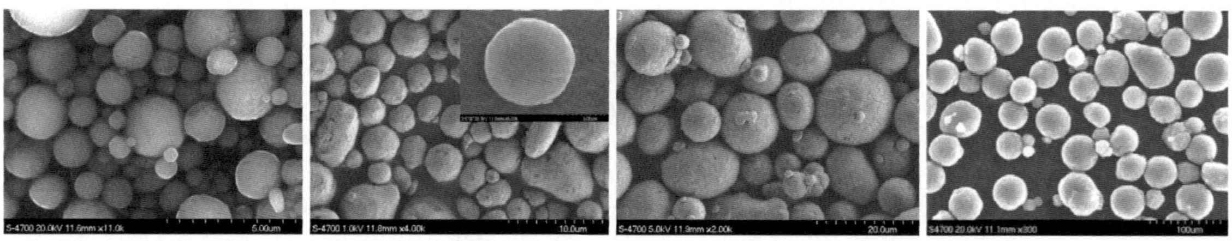

(a) (b) (c) (d)

Figure 2. SEM characterization of aluminum powders (1) Al#2 is roughly equiaxial. (2) Al#3 particle size distribution is in the range of 2-8μm. (3) Al#4 slightly agglomerate together. (4) Al#6 contains a wider range of size distribution from 4μm to 50μm.

3.2 Thermal behavior of aluminum powders

Assuming that alumina was the only solid product of the reaction, the mass increasing can be calculated easily according to the chemical reaction equation. Under ideal condition, mass increase should be 88.89%, however, all the aluminum powders utilized in the experiment consist several kinds of impurities (table 1). Therefore, even though the active aluminum was oxidized completely, the mass increase is less than 87.30%. In order to keep it clear enough to read, TG curves of Al#2, Al#3, Al#4 and Al#6 were demonstrated in figure 3, the changing tendency of all six samples is highly related to the particle sizes. In figure 3 (a), the DSC records in the range from 300℃ to 1450℃ are presented (there are no mass changes before 300℃). The mass increase of the aluminum samples doesn't stop except the ones which are smaller than 2μm. Then the second round of DSC experiments is designed as follows, heating process starts from room temperature to 1500℃ and kept in 1500℃ for additional

15 minutes. Under this condition, figure 3 (b) presents clearly that the particles smaller than 5μm could be oxidized completely and kept in a constant mass eventually, particles larger than that still stayed in a relatively low reaction rate. Therefore, sample Al#3 are chosen as a representative in the following discussion.

The entire oxidation process can be divided into four stages. Oxidizing reaction of micron-sized aluminum particles observed in this work could be divided into 4 stage. Stage I, from room temperature to 620℃, the weight of the samples changed little, but DSC curve had a small exothermic peak at the end of this stage, amorphous alumina changed its phase to γ-Al_2O_3. Stage II, TG curve increased a little and an obvious endothermic process was observed at the same time. This stage happened from 620℃ to 690℃, aluminum reach its melting point (667℃) during this stage. Stage III, vigorous oxidation reaction was happened from 690℃ to 1150℃, peak C corresponded to the increasing of weight. Stage IV, from the beginning of 1150℃, oxidation processes went on and all the alumina turned to α- Al_2O_3.

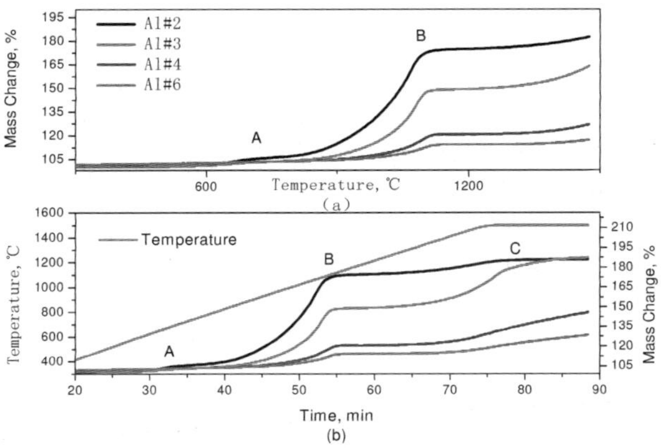

Figure 3. TG curves of four different sizes aluminum samples. (a) Samples heated to 1450℃, (b) Samples heated to 1500℃ and kept in 1500℃ for another 15 minutes.

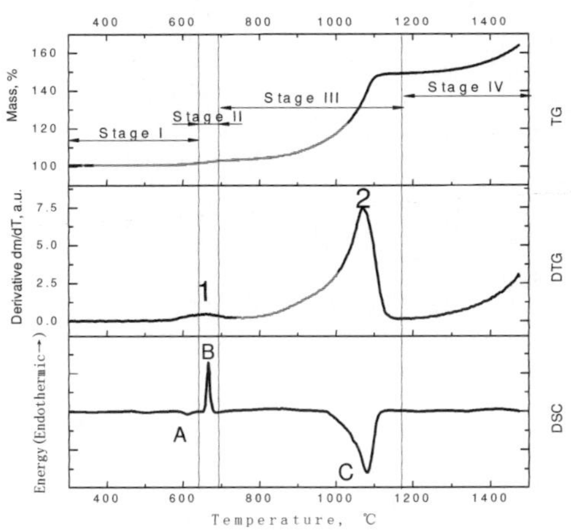

Figure 4. TG, DTG and DSC curves of Al#3 from 300℃ to 1450℃. Stage I, from room temperature to 620℃. Stage II, 620-690℃. Stage III, 690-1150℃.

In the whole heating process according to figure 4, there are two steps of mass increasing which is presented by DTG curve. The first step (peak 1) begins around 660℃ and the accurate value of mass increasing which is considered as a thickness increasing of the alumina shell is measured for 4%. The second step (peak 2) emerges in the vicinity of 800-930℃, the mass changes quite obviously and reaches 150% of the original weight. DSC curve presents three peaks which are marked by A, B and C. Peak A and C are exothermic process which are known to be respectively associated with a phase transformation alumina and with ensuing thermally activated oxidation, peak B is an endothermic process which is considered to be the melting of aluminum core.

3.3 Chacterization of intermediate products

How the samples changed in the previous system could not be determined just by DSC experiments, therefore XRD and SEM are required as support. Figure 5 shows the X-ray patterns of samples. Strong peaks of metallic aluminum of the raw particles are observed.

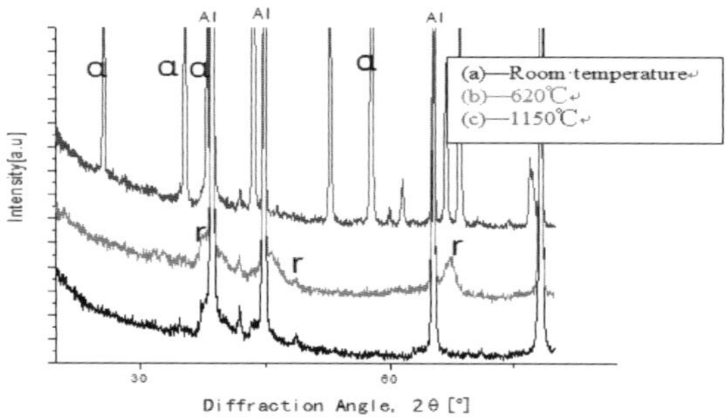

Figure 5. XRD patterns of Al#3 sample. The observed XRD peaks are identified for different polymorphs of Al_2O_3. (a) as-prepared; (b) after calorimetric measurements in the 20–620 ℃ temperature range; (c) after Calorimetric measurement in the 20–1150 ℃ temperature range.

SEM pictures of Al#3 after calorimetric measurement in different temperature are presented in figure 6, in general, the stable oxide exterior shells still kept the initial sphere shape. Empty shell structure could be seen clearly and the size of the particle is as big as the initial aluminum particles. The thickness of the shell was measured in SEM pictures which increased to almost 0.1μm. Figure 7 describes the monograph of particles that heated from room temperature to 1450℃. Small particles could be oxidized completely; the reaction products consist of fragment of alumina shell and empty shells of the aluminum particles. When it comes to the larger sized powders (>10μm), the sphere particles with a few cracks on the surface could be seen instead of empty shell structure. What was more, compared with the initial particles, the size did not change as well.

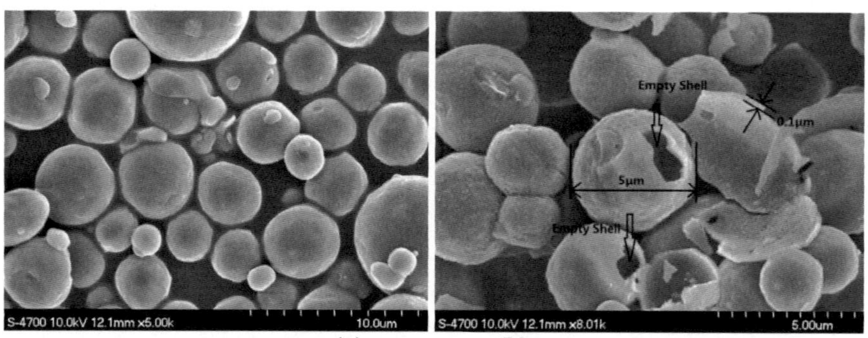

(a) **(b)**

Figure 6. Monograph of Al#3 heated to different temperature. (a) 620℃. (b) 1150℃.

(a) **(b)**

(c) **(d)**

Figure 7. Monograph of aluminum particles heated from room temperature to 1450℃ (a) Al#1 (b) Al#3 (c) Al#5 (d) Al#6.

4. Discussions

4.1 Oxidation process of aluminum powders

For the recovered samples that only heated to 620℃, weak peaks of γ-Al_2O_3 are observed. For the XRD patterns of 1150℃ heated samples, the peaks of α-Al_2O_3 become comparatively strong and the peaks of γ-Al_2O_3 disappeared. According to the XRD patterns, during the entire gradually heating process, the alumina shell changed its phase stages by stages. The whole process includes a history of amorphous $\rightarrow \gamma \rightarrow \alpha$-$Al_2O_3$. γ-phase alumina can be synthesized industrially with the temperature of 500-750℃, 800-1050℃, 900-1100℃ respectively, which is accordance with the phenomenon in the experiment. α-Al_2O_3 is the most stable and hard kind of alumina phase. In the present results of XRD pattern, γ- and α-Al_2O_3 could be seen clearly at the end of the first stage and the fourth stage respectively.

SEM observations after DSC scans show that the granular structure of the powder was modified by the thermal treatment. It is noted that if the molten aluminum is pushed into these cracks by the high internal pressure characteristic of the stress field, its surface, where reaction occurs, could be fractal in nature and affect the fundamental nature of the burning law.

4.2 Comparation of oxidation behavior in different heating process

Firstly, all the samples were heated in CO_2 atmosphere from room temperature to 620℃ which is the point of the second stage, then cooled down to room temperature in argon. After the first preheated process, the samples experienced another whole DSC test from room temperature to 1450℃ again. Mass increase of 53% is observed on one-time heated process from figure 8, whereas only a value of 0.37% is recorded for the reheated Al#3 sample. In figure 8, there is only one endothermic peak at the melting point of aluminum. Because of consumption of aluminum in the first heated process, the endothermic peak is smaller than the one-time heated samples. It is shown that, whatever is the size of particles, the thermal expansion of the metallic core is not affected by an alumina shell with a 3 ± 2 nm thickness which is the normal value. In contrast, when the thickness of the crystalline alumina shell exceeds 20 nm, strong strain effects are evidenced. This different results between the two round of DSC test demonstrate that the thickness increase of alumina shells do causes apparent restriction of the oxidation of aluminum particles.

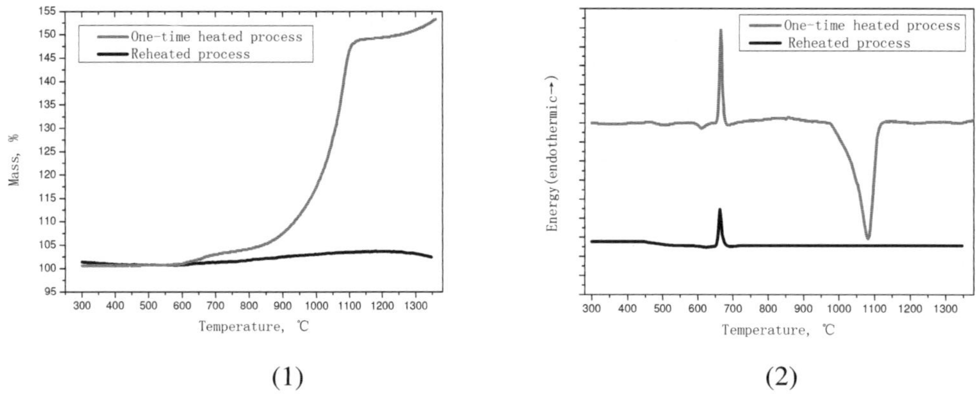

Figure 8. TG-DSC scanning curves of Al#3 samples in one-time heated and reheated process respectivly. Comparation of TG (1) and DSC (2) between two different heating processes.

4.3 Oxidation mechanism of aluminum particles

In the first stage, from room temperature to 620℃, XRD patterns showed that alumina shell changed from amorphous phase to γ- Al_2O_3, this process was corresponded to the small exothermic peak A in DSC curve. The microscopic morphology of aluminum powders changed little; therefore, the main difference between the aluminum heated to 620℃ and the initial ones was that the γ- Al_2O_3 replaced the amorphous alumina.

The thickness of metal-substrate crystalline-oxide film will increase with the growing of temperature, a significant increase in the metal's reactivity leads to the fracture of the protective oxide film. The thermal expansion coefficient of aluminum is approximately ten times larger than that of γ-Al_2O_3. Then the pressure in the oxide shell increased and the overwhelming force formed between aluminum core and alumina film was becoming larger and larger. Alumina shell was in the stretch state, and its counterpart aluminum core, was in the compression state. This kind of internal stress was increasing with the elevating temperature. The influence of mechanical stresses is quantitatively described, arising due to phase and polymorphic transformations of the metal, and also of the difference in the thermal-expansion coefficients and densities of the metal and its oxide. The smaller the particle sizes were, the higher radius of curvature of particle surface had. With the existing of inner stress, alumina shell was kinetically unstable and the tendency of forming cracks on the alumina shell was large enough. In addition, γ- Al_2O_3 was not the same dense and stable as the amorphous alumina as well. The cracks were a kind of passageway for gas; oxidizing gas could go through the passageway and reacted with the aluminum in the shell. So the early oxidation reaction is happened around the melting point of aluminum. When the particle is immersed in a hot atmosphere containing oxygen atom, dissociative adsorption occurs and then the O atoms at the surface diffuse inwards towards the aluminum surface which can be expressed as equation 2.

$$2Al + 3O = Al_2O_3 \tag{2}$$

Stage III, from 690℃ to 1150℃, when the samples were heated up to the melting point of aluminum, the solid aluminum core melted and its volume was becoming larger. Because of the cracks on the shells as well as the impurity content of the samples, defects existed on the surface of the particle inevitably. When the inside pressure became large enough to break the defect, liquid phased aluminum came out. When aluminum powders was larger than 10μm, the alumina shell was thicker than the small particles, the melting aluminum could not erupt easily, the liquid aluminum flowed from the cracks and reacted with the CO_2. With the reaction of the aluminum, the alumina deposit formed on the particle surface. At last, the volume of aluminum inside the shell was too weak to flow out. Therefore, larger aluminum powders did not react completely. During heating, the interference, caused by the difference of thermal expansion coefficients between aluminum core and its oxide shell, can be calculated. Even though the particle sizes are different, the increment caused by increasing of the temperature in the radial direction is unchanged. Therefore, from the microcosmic point of view, the oxide films of smaller particles that owe larger radius of curvature are more likely to break. So the smaller the particles are, the more completely of the oxidation processes happen. And the big peak C of DSC curve appeared at the same period that stands for the degree of the reaction.

Stage IV, from 1150℃ to a higher degree, the specific processes responsible for the individual oxidation stages of aluminum powders must be understood in order to accurately model aluminum ignition. XRD patterns show that α-Al_2O_3 is the main product and the structure of different sized powders changed differently. The oxidation stages observed in this work are generally consistent with previous reports on aluminum powder oxidation.

The mechanism influenced by particle sizes were firstly take into consideration, figure 9 shows the whole history of the oxidize process of aluminum particles.

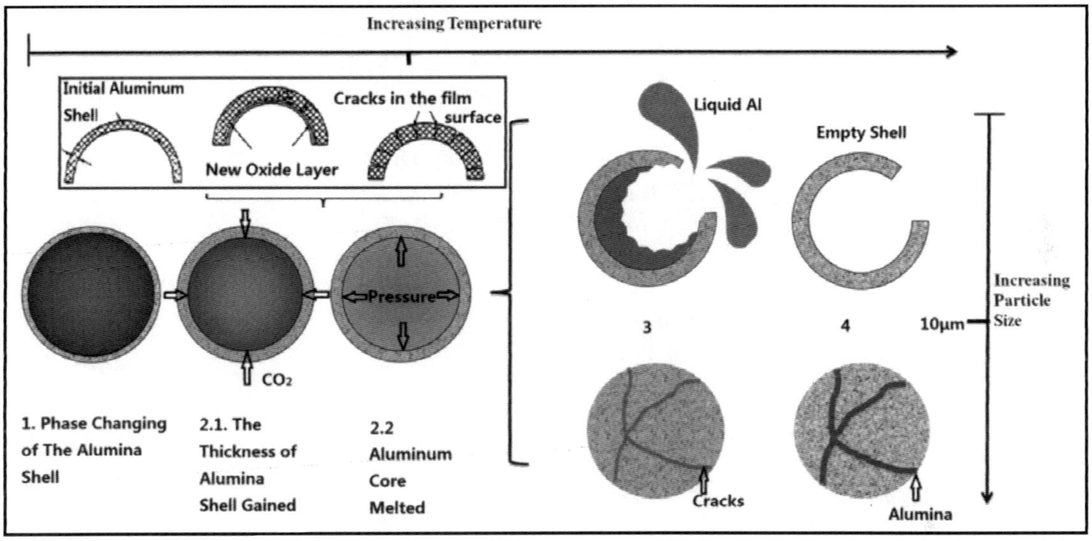

Figure 9. The whole history of oxidation processes about micro-sized Aluminum particles.

5. Conclusion

By integrating sophisticated experimental and testing methods, history of slow heating reaction and changing of single aluminum particle were characterized. An "eruption model" which is accounted for multiple processes occurring in aluminum combustion and is not expected to be limited to quasi-steady description of the single micron-sized aluminum particle oxidizing process in gradually heated oxidizing atmosphere was developed. Also, the present conclusion makes an emphasis on the relationship between different steps of oxidation processes and the change of the particle sizes. During the heterogeneous oxidation of metallic particles in a gaseous oxidant, an oxide film usually separates the metal and oxidant, so the rate of reaction is governed by the protective properties of that film. Reaction history of the micron-sized aluminum particles on the gradually heating environment includes 4 stages. The whole process can be summarized to be an eruption model; also, the particle sizes are taken into consideration. For the aluminum powders that the impurities are less than 2% of the total weight, this conclusion can be referred in practical application.

6. References

[1] Matysik P, Jozwiak S and Czujko T 2012 *J. Alloy. Comp.* **549** 92-9.
[2] Palaszewski B, Ianovski L S and Carrick P 1998 *J. Prop. Pow.* **14(5)** 641-8.
[3] Zhang F, Murray S and Gerrard K 2006 *Shock Wav.* **15(5)** 313-24.
[4] Puri P. 2008. The Pennsylvania State University: Dissertations & Theses - Gradworks, 2008.
[5] Belyaev A, Frolov Y V and Korotkov A 1968 *Combust. Explos. Shock Wav.* **4(3)** 182-5.
[6] Boreisho A, Ivashchenko A and Shelukhin G 1975 *Combust. Explos. Shock Wav.* **11(4)** 561-2.
[7] Mueller M et al 1999 *Int. J. Chem. Kin.* **31(2)** 113-25.

[8] Yetter R, Dryer F and Rabitz H 1991 *Combust. Sci. Tech.* **79(1-3)** 97-128.
[9] Zhu X, Schoenitz M and Dreizin E L 2009 *J. Phys. Chem. C* **113(16)** 6768-73.
[10] Zhang S, Schoenitz M and Dreizin E L 2013 *Combust. Sci. Tech.* **185(9)** 1360-77.
[11] Ilyukhina A V, Kravchenko O V and Bulychev B M 2017 *J. Alloy. Comp.* **690** 321-9.
[12] Puri P and Yang V 2008 *Thermo-mechanical behavior of nano aluminum particles with oxide layers in 46th AIAA Aerospace Sciences Meeting and Exhibit Reno NV.*
[13] Rozenband V and Vaganova N 1992 *Combust. Flame* **88(1)** 113-8.
[14] Merzhanov A, Grigorjev Y M and Gal'chenko Y A 1977 *Combust. Flame* **29** 1-14.
[15] Friedman R and Maček A 1962 *Combust. Flame* **6** 9-19.
[16] Johnson C et al 2000 *Combustion and oxidation of metal nanoparticles and composite particles.*
[17] Rozenband V, Afanas'eva L and Lebedeva V 1990 *Fizika Goreniya i Vzryva* **26(5)** 13-5.
[18] Ermakov V et al 1982 *Combust. Explos. Shock Wav.* **18(2)** 256-7.
[19] Aly Y, Schoenitz M and Dreizin E L 2011 *Combust. Sci. Tech.* **183(10)** 1107-32.
[20] Starink M and Li X 2003 *Metallurg. Mater. Trans.* A **34(4)** 899-911.
[21] Li X and Starink M J 2000 *Analysis of precipitation and dissolution in overaged 7xxx aluminium alloys using DSC.*
[22] Ramesh A and Koziński J 2001 *Combust. Flame* **125(1)** 920-30.
[23] Mench M et al 1998 *Combust. Sci. Tech.* **135(1-6)** 269-92.
[24] Trunov M A et al 2005 *Combust. Flame* **140(4)** 310-8.

Acknowledgements
It's very appreciate to State Key Laboratory of Explosive Science and Technology of Beijing Institute of Technology.

IOP Conf. Series: Materials Science and Engineering **248** (2017) 012003 doi:10.1088/1757-899X/248/1/012003

Design and development of the Macpherson Proton Preve Magneto rheological damper with PID controller

I M Amiruddin[1], M Pauziah[1], A Aminudin[1] and M H Unuh[1]

[1]Intelligent Dynamic and System (IDS) i-Kohza, Malaysia-Japan International Institute of Technology (MJIIT), Universiti Teknologi Malaysia, Jalan Sultan Yahya Petra, 54100 Kuala Lumpur, Malaysia

Abstract. Since the creation of the first petrol-fuelled vehicle by Karl Benz in the late nineteenth century, car industry has grown considerably to meet the industrial demands. Luxurious looks and agreeable rides are the primary needs of drivers. The Magneto-rheological damper balanced their damping trademark progressively by applying the damping coefficient depending on the control system. In this research, the control calculations are assessed by utilizing the MR damper. The capacity and reliably of the target force for the damper speed is investigated from control algorithm. This is imperative to defeat the damper limitation. In this study, the simulation results of the semi-dynamic MR damper with the PID controller shows better performance in sprung mass acceleration, unsprung mass acceleration and suspension dislodging with permitting over the top tyre acceleration. The altered model of the MR damper is specially designed for Proton Preve specifications and semi-active PID control. The procedure for the advancement incorporates the numerical model to graphically recreate and break down the dynamic framework by utilizing Matlab.

1. Introduction

The suspension of a ground vehicle includes parts that certify a versatile association amongst tyres and vehicle body. At the point when vibration is induced by the abnormalities on the road profile, this flexibility secures the auto body and the passenger [1]. One crucial part in a vehicle suspension that controls oscillations is the damper. Its primary function is to hold the abundance imperativeness from springs and tyres. Also, it removes the skipping impact of each wheel. Vibrations in the passenger compartment will be automatically minimized. The essential goals of the suspension are to give comfort to passengers and driver [2]. A semi-dynamic suspension is lighter, more passive and less complex than their dynamic accomplices [3]. In addition, semi-dynamic dampers are more ideal than current passive dampers in various applications. Semi-dynamic dampers are the cutting-edge car applications and they are observed to be marketable [4]. The diverse method for measuring information from a quarter car suspension framework [3] includes utilizing acceleration and displacement information. In this research, the damper is assessed in frequency and time domain. The capacity and consistency given by the MR damper are explored with control algorithm. This is essential to give a similar course of damper velocity and target force. To overcome the damper constraint, the control algorithm is adjusted and enhanced for its power in the quarter car structure. The segment of the PID controller utilized as a part of this research is for magneto rheological damper. The MR damper is specially designed for Proton Preve specifications with semi-active PID control which is new for Malaysian vehicles technology.

Content from this work may be used under the terms of the Creative Commons Attribution 3.0 licence. Any further distribution of this work must maintain attribution to the author(s) and the title of the work, journal citation and DOI.
Published under licence by IOP Publishing Ltd

IOP Conf. Series: Materials Science and Engineering **248** (2017) 012003 doi:10.1088/1757-899X/248/1/012003

2. Damper design and modelling

The prototype design is developed using a three-dimensional computer-aided design (CAD) package in SolidWorks software. The modelling provides an overview of the prototype design equipped with precise and accurate dimensions of the original equipment manufacturer (OEM) for fabrication process and the section view of parts and assembly of the prototype design for simulation study to be carried out using ANSYS Mechanical APDL 15.0 software. The CAD modelling is shown in figure 1. The finite element analysis (FEA) tool in the software is employed to model the behaviour of the magneto rheological (MR) fluid that was exploited in the prototype design during induction of electromagnetic force and simulating the wave of electromagnetic field. The selected unit for the dimensions of the prototype design is metre (m). The conceptual design shown in figure 1 was developed using the Solidworks CAD software. The objective is to establish a drawing with full specification before fabrication. This approach is vital to prevent failure and argument of subsequent approaches. The design development with precise and accurate dimensions is important to be initially achieved before the next stage. The dimensions are as follows:

Table 1. Dimensions of the proposed design.

Part/ Assembly	Dimension (mm)
External housing outer diameter, $\emptyset_1$	39.00
External housing inner diameter, $\emptyset_2$	36.63
External housing height, $\boldsymbol{h_1}$	342.29
Internal housing height, $\boldsymbol{h_2}$	288.01
Piston head height, $\boldsymbol{h_3}$	24.96
Stroke distance, $\boldsymbol{h_4}$	299.52

2.1 Design and prototype

The main concern for the fabrication process is the winding of electrical copper coil for the induction of magnet. The winding is anticlockwise which starts from the beginning of the wire at the upper ring to the bottom ring until 450 turns are achieved.

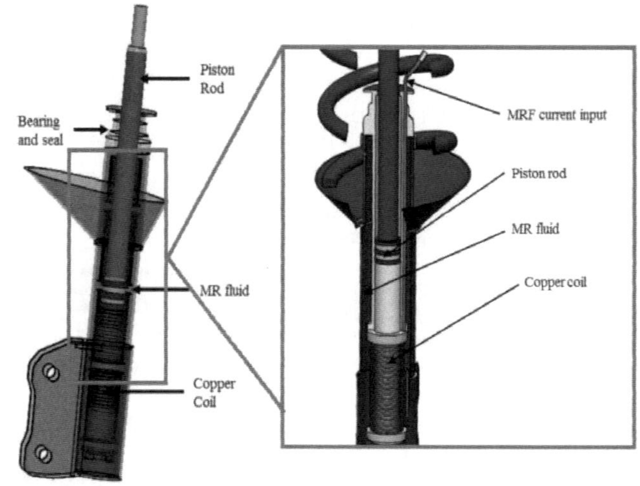

Figure 1. Sectional view of the Prototype of twin tube MR damper (Proton Preve).

2.2 Magnetic flux line

The analysis is simplified to a single iteration analysis. The outside of the model parameter is presumed to be free of any flux leakage, and saturation of the material does not ensue. The ramification of flux leakage is demonstrated by creating a layer of air bordering the iron. The maximum radius of the iron shall be equal to or smaller than the air layer. The flux will work in parallel to the surface of the model's parameter to satisfy the assumption defined in the previous section. The parallel flux boundary condition will enforce the assumption to be realized. Zooming to the flux path where the MR fluid is stored is shown in figure 2, illustrating that the flux lines concentration increase as the distance between adjacent flux lines becomes smaller. For this reason, it demonstrates the influence of the rheology properties of the MR fluid on the electrical coil in general and to the generated magnetic field for a more focused subject.

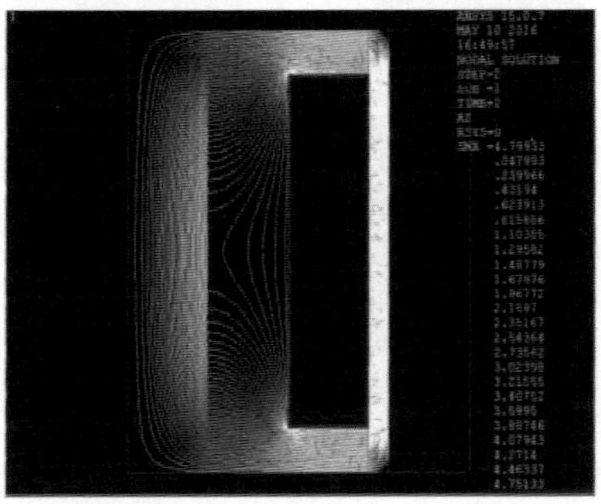

Figure 2. Magnetic flux lines.

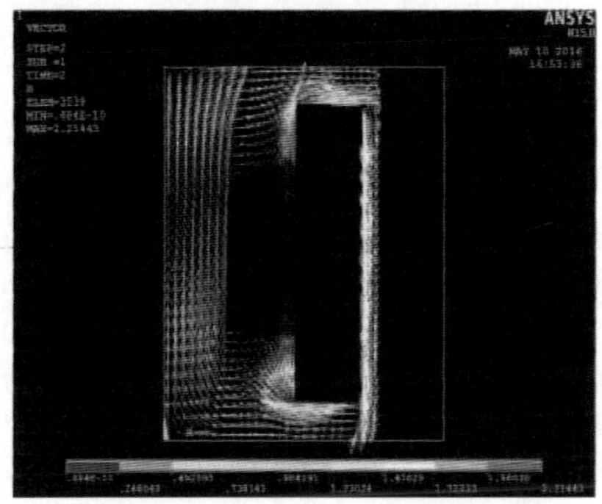

Figure 3. Magnetic flux lines in vector form.

In addition, figure 3 illustrates the direction change of magnetic flux lines in vector form upon entering the area that stores the MR fluid. Thus, the highest value of magnetic flux density is recorded along the area where the MR fluid is stored. At the moment the MR fluid becomes saturated, the concentration of the magnetic flux lines concentration decreases.

3. Modelling quarter car suspension system

Active suspension system for PROTON PREVE includes force actuators to generate the wavering force. The actuator is designed to fit the conventional PROTON springs and MR damper in the vertical connection for the two degree of freedom system. To improve the PROTON PREVE suspension system, the Arduino microcontroller is used. By tuning the PID controller designed in the system, the ride control and comfort of the vehicle will be better. A quarter car models is used in designing the suspension system as shown in figure 4, to simplify the one-dimensional spring-damper system.

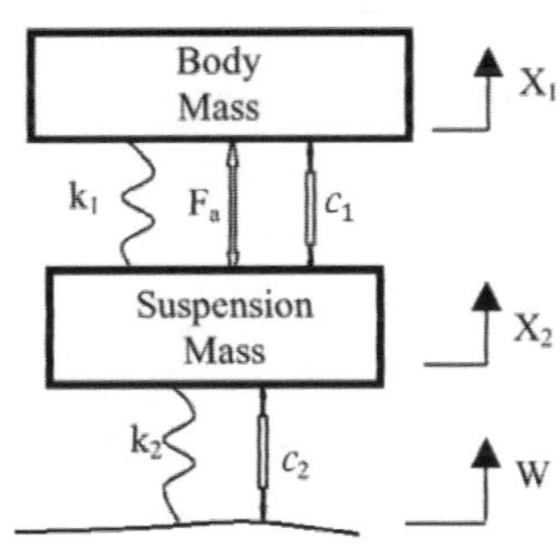

Figure 4. Quarter car model of Proton Preve

The passive suspension system, identified as the conversional suspension system is normally unable to dissipate large oscillations quickly. As tyre deformation is rather difficult to be measured in real condition, the simulation will help to generate and analyse the data. Road input which is also known as road disturbance W is represented as step input and speed bump. These inputs comply with the Malaysian standard of a road bump height i.e. 10 cm.

3.1. Transfer function modelling

The assessment of the system behaviour is crucial in estimating the feedback fault between simulation and experimental values of a designed controller. The stationary and transitional work regime systems can be defined with this approach. The mathematical model of the dynamics system obtained from the dynamic equation, which represents the Newton's Law as shown equation 1 and 2:

$$\mathrm{m_1\ddot{x}_1 = -k_1(x_1 - x_2) - c_1(\dot{x}_1 - \dot{x}_2) + F_a} \tag{1}$$

$$\mathrm{m_2\ddot{x}_2 = -c_1(\dot{x}_1 - \dot{x}_2) + k_2\ (W - x_2) - k_1(x_1 - x_2) + c_2(\ddot{W} - \dot{x}_2) - F_a} \tag{2}$$

The application of Laplace transforms can be simplified by solving these equations. The linear differential equation with a constant coefficient solution can be obtained from the given initial conditions. The forms of an appropriate block diagram are represented by the transforms provided. The main procedure in the evaluation of an automatic control system for the selection of Laplace transforms is less suitable for the algorithm that looks quite complex. In this way do not use direct non-Transform scheme and equation involved is as below, the dynamic equation above can be expressed in a form of transfer function. The derivation from above equations of the two inputs, F_a and W and Transfer functions $G_1(s)$ and $G_2(s)$ of output, $x_1 - x_2$, are as follows:

$$\begin{bmatrix}(m_1s^2 + c_1s + k_1) & -(c_1s + k_1) \\ -(c_1s + k_1) & m_2s^2 + (c_1 + c_2)s + (k_1 + k_2)\end{bmatrix} * \begin{bmatrix}X_1(s) \\ X_2(s)\end{bmatrix} = \begin{bmatrix}F_a\,(s) \\ (c_2s + k_2)W(s) - F_a(s)\end{bmatrix} \tag{3}$$

$$\Delta = \det[M] = (m_1s^2 + c_1s + k_1) * (m_2s^2 + (c_1 + c_2)s + (k_1 + k_2)) - (c_1s + k_1)^2 \tag{4}$$

Next, the input $F_a(s)$ and $W(s)$ is multiplied with the right hand side inverse of the matrix A as follows:

$$\begin{bmatrix}X_1(s) \\ X_2(s)\end{bmatrix} = \frac{1}{\Delta} * \begin{bmatrix}m_2s^2 + b_2s + k_2 & c_1c_2s^2 + (c_1k_2 + c_1k_1)s + k_1k_2 \\ -m_1s^2 & m_1c_2s^3 + (m_1k_2 + c_1c_2)s^2 + (c_1k_2 + c_1k_1)s + k_1k_2\end{bmatrix} * \begin{vmatrix}F_a(s) \\ W(s)\end{vmatrix} \tag{5}$$

The input considers as $F_a(s)$ only, and $W(s) = 0$. As the result the transfer functions $G_1(s)$ as the following:

$$G_1(s) = \frac{X_1(s) - X_2(s)}{F_a(s)} = \frac{(m_1 + m_2)s^2 + c_2s + k_2}{\Delta} \tag{6}$$

If the input is considered as $W(s)$ only, and $F_a(s) = 0$. Thus, the transfer functions $G_2(s)$ as the following equation:

$$G_2(s) = \frac{X_1(s) - X_2(s)}{W(s)} = \frac{-m_1c_2s^3 - m_1k_2}{\Delta} \tag{7}$$

The input of Transfer Function equation (6) and (7) into Matlab and the standard transfer function $G_1(s)$ and $G_2(s)$ is as follows:

$$\mathrm{G_1(s) = \frac{nump}{denp} \qquad G_2(s) = \frac{nums}{dens}} \tag{8}$$

The second reason is that the PID controller has less complexity thus fulfilling the design controller requirement for the Malaysian quarter car model. The schematic diagram of the close-loop controller system is shown in figure 5:

IOP Conf. Series: Materials Science and Engineering **248** (2017) 012003 doi:10.1088/1757-899X/248/1/012003

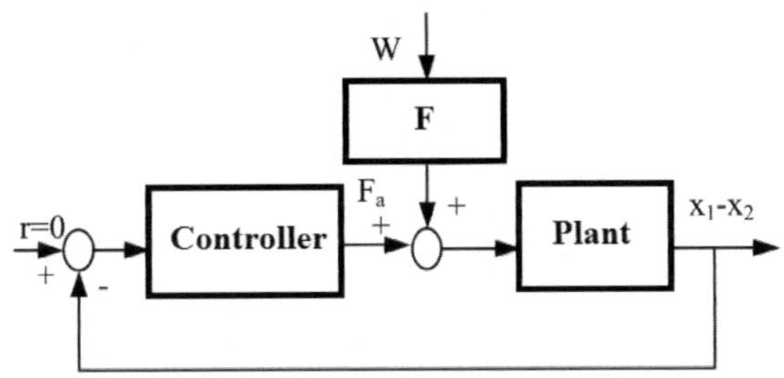

Figure 5. Diagram of the Closed-loop system.

4. Basic method of the semi-active suspension control

Figure 6 shows the complete system incorporating the application of the PID controller to the passive quarter car Proton Preve suspension system. The mathematical model approach for the quarter car suspension system is developed using the Matlab *Simulink* application.

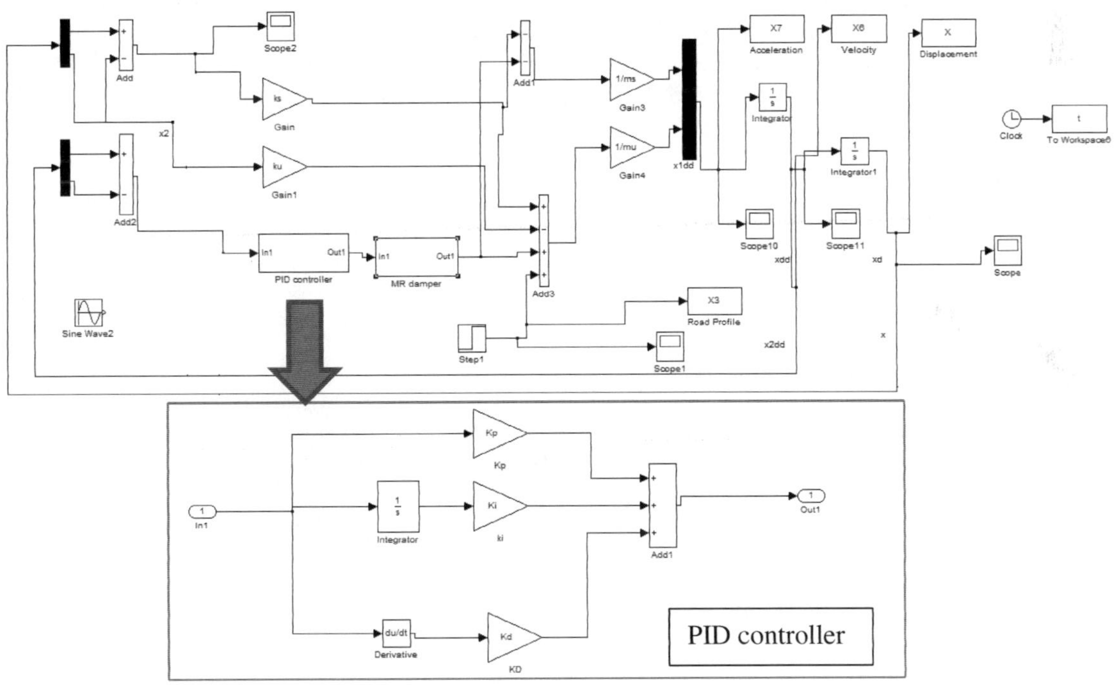

Figure 6. Two degree of freedom quarter car model with PID controller designed in the Matlab and Simulink.

5. Validating of the MR damper model with PID control

Simulation is performed in Matlab-Simulink to acquire the legitimacy of non-parametric information that is used on the MR damper. A road profile (input signal) is tested in this research. Only step profiles are tested for the simulation work as shown in figure 7. This step profile has 0.03m height which is similar for the experiment test. Each simulation produced different types of interesting results depending on K_P, K_I and K_D values. The basic parameters K_P, K_I and K_D are chosen by the trial and error method. Although tuning the PID controller is challenging, the performance of the quarter car system slightly increases.

IOP Conf. Series: Materials Science and Engineering **248** (2017) 012003 doi:10.1088/1757-899X/248/1/012003

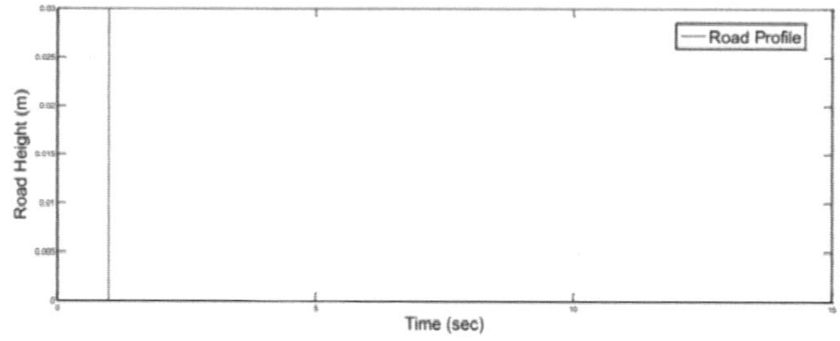

Figure 7. Road profile 3cm step.

The effectiveness of MR damper using PID controller in reducing time taken to isolate the resultant motion into an equilibrium position are shown in Figure 8 until Figure 13. The result of the conventional system is far from pleased in isolating vibration. It exceeds 15 seconds of MR damper integrated with the PID controller time taken to completely absorb. It follows that the PID model can reduce the oscillation as early as 3 seconds up to 10 seconds. However, this depends on the parameters i.e K_P, K_I and K_D. It is necessary to establish the optimum parameter of PID before manipulating electrical current level to fully verify the effectiveness of the system. Referring to Figure 8 to Figure 13, value of K_P varies from 0.1 to 0.7 with an increment of 0.1. As seen from figures, the oscillation is reduced to 0.015 m and 0.01 m from 0.025 m.

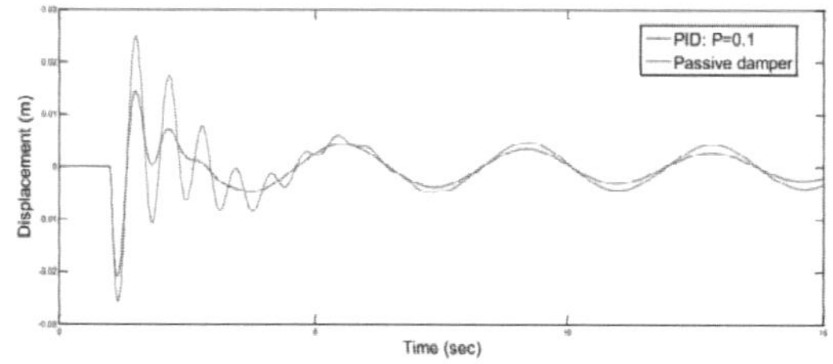

Figure 8. Sprung mass displacement vs time, (PID: P=0.1and Passive damper).

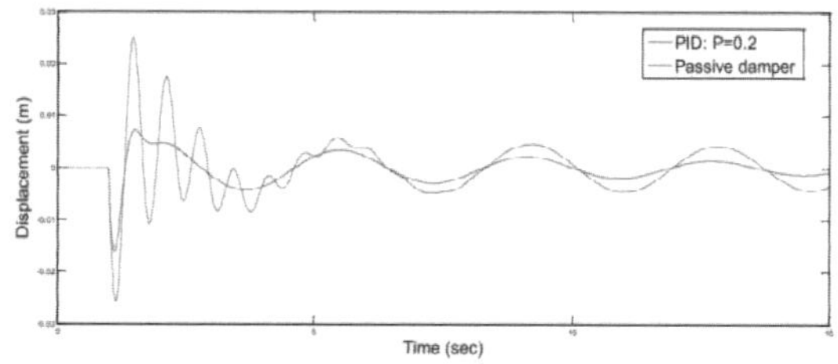

Figure 9. Sprung mass displacement vs time, (PID: P=0.2 and Passive damper).

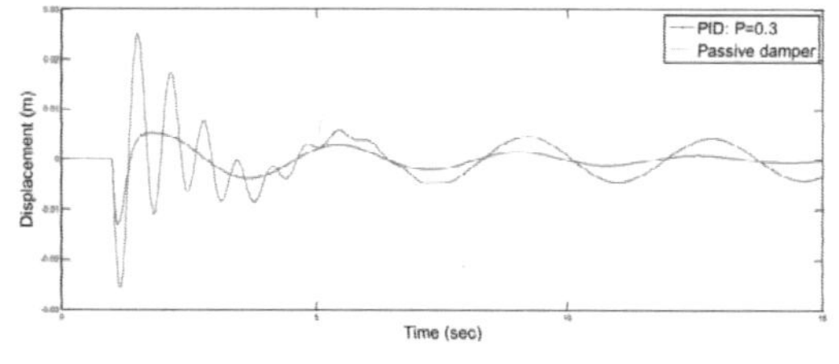

Figure 10. Sprung mass displacement vs time, (PID: P=0.3 and Passive damper).

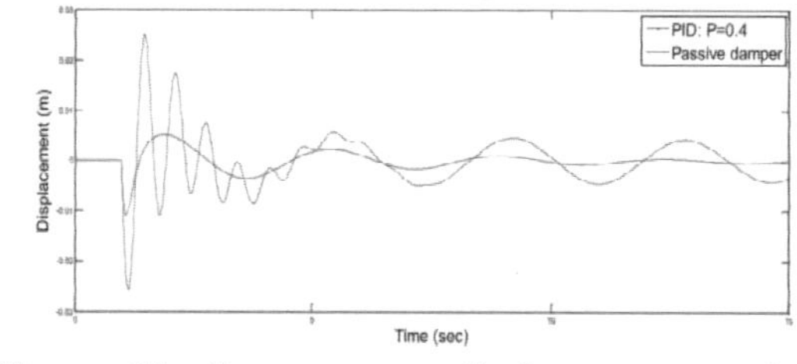

Figure 11. Sprung mass displacement vs time, (PID: P=0.4 and Passive damper).

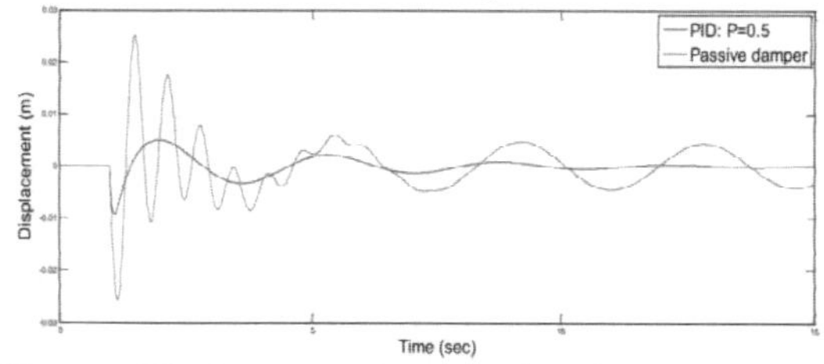

Figure 12. Sprung mass displacement vs time, (PID: P=0.5 and Passive damper).

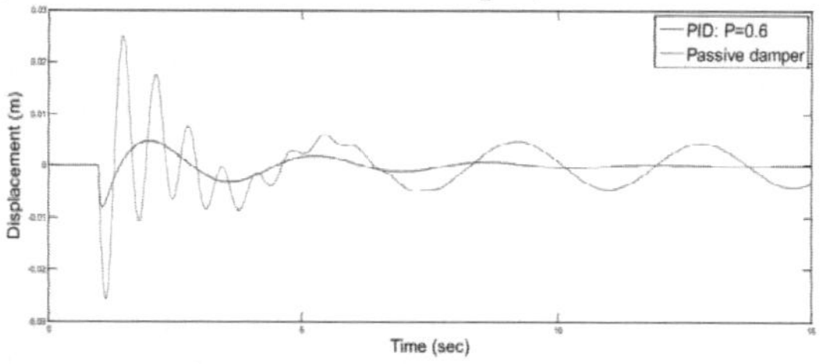

Figure 13. Sprung mass displacement vs time, (PID: P=0.6 and Passive damper).

Although the oscillation becomes lower as the value of K_P increases and the percent of the overshoot shown by these figures is still considerably high enough to give unsatisfactory time settling. These figures illustrate the incapability of the system to give the best performance if only the value of K_P alone is manipulated.

6. Conclusion

Research has been carried out, starting with the designing of the algorithm and physical model based on the quarter car model. The MR damper system is built with the PID controller in Matlab and Simulink. The prototype of MR suspension system is designed following the specifications of PROTON Preve. At the end, this system managed to provide good feedback to the suspension system by tuning in K_P, K_I and K_D. The correlation demonstrates the ability of the MR damper that can offer comfort of more than 60% for the driver and passenger. By looking at these MR and OEM dampers, it can be inferred that the MR model with the PID controller gives the best outcome and it is reasonable for the Malaysian car suspension system. In further research, the dynamic control framework will be be adapted into the suspension system.

References

[1] Zhang Y, Zhan M, Zhao F, Liuc Z W 2015 *J. Franklin Inst.* **352(2)** 485-99.

[2] Yeh F K, Huang C W and Huang C W 2010 *Adaptive-Sliding-Mode Semi-Active Bicycle Suspension Fork*, International Conference on Instrumentation Publication, Taipei, Taiwan.

[3] Patil S A, Joshi S G 2014 *Syst. Contr. Eng: Open Access J.* **2(1)** 621-31.

[4] Gao B Z, Lei Y L, Ge A L, Chen H and Sanada K 2011 *Vehicle Syst. Dyn.: Int. J. Vehicle Mech. Mobil.* **49(5)** 685-701.

[5] Occhiuzzi A, Spizzuocco M and Serino G 2003 *Smart Mater. Struct.* **12(5)** 703-11.

[6] Poyner J C 2001 *Innovative Designs for Magneto-Rheological Dampers,* Advanced Vehicle Dynamics Lab, Virginia Polytechnic Institute and State, University, Blacksburg, VA, MS Thesis, August 7, 2001.

[7] Wu L, Cao Y L and Chen H L 2008 *Hierarchical Model Control of A Motorcycle Semi-Active Suspension with Six Degree-Freedoms.* Proceeding of the 2008 IEEE/ASME International Conference on Advanced Intelligent Mechatronics. July 2-5, 2008, Xi'an China.

IOP Conf. Series: Materials Science and Engineering **248** (2017) 012004 doi:10.1088/1757-899X/248/1/012004

The manufacturing of TiAl6V4 implants using selective laser melting technology

P A Lykov[1], R M Baitimerov[1], A V Panfilov[1], A O Guz[2]

[1] South Ural State University, Lenin avenue 74, Chelyabinsk, Russian Federation
[2] GBUZ Chelyabinsk Regional Clinical Oncology Center, Bluhera Street 42, Chelyabinsk, Russian Federation

Abstract. In this article we study the technique for creating medical implants using additive technologies. A plastic skull model was made. The affected part of the skull was identified and removed. An implant was made of titanium alloy. The implant was installed in the model skull.

1. Introduction

Additive manufacturing is often used in high-tech areas such as aircraft engineering [1, 2] and the space industry [3, 4] for the manufacture of geometrically-complex products. At the same time, they are becoming more widespread in medicine [5, 6]. The main reason for the use of additive manufacturing in medicine is the possibility of producing a personalized implant within a short time. As a rule, selective laser melting (SLM) and electron beam melting (EBM) are used for the manufacture of implants [7]. Depending on the area of application, CoCr [8, 9] or titanium alloys [10-12] are more widely used as a construction material. These materials have high biocompatibility. There are studies on the use of additive manufacturing in craniofacial surgery [13, 14]. The aim of this work is the fabrication of a titanium implant to replace the affected area of a skull cancer patient.

2. Main part

In the first stage of work (figure 1), a model of the fragment of the skull with the affected part was made. Malyan Desktop 3D printer was used. The material is plastic ABS. The diameter of the plastic wire is 1.75 mm. The plastic feed rate is 50 mm/sec. The feed speed of the table is 90 mm/s. The temperature of heating the plastic is 230 °C. The temperature of the heating base plate is 85 °C. The fill factor of the volume of the walls of the sample is 30%.

The second stage of the work is determining the part of the skull, which we planned to remove (figure 2).

A metal implant was made to replace the affected part of the skull. The implant was manufactured using selective laser melting. A powder of titanium alloy TiAl6V4 was used as the feedstock.

Selective laser melting (SLM) is based on layer-by-layer laser fused metal powder. This technology allows us to produce parts with complex shapes in a short time. A schematic of the SLM process is shown in figure 3. The part is divided into elementary layers (with a thickness from 20 to 100 microns). A two-dimensional sketch is formed for each layer. The plate is lowered to the thickness of the layer and the layer is formed by the leveller. The manufactured part has a near-net shape.

Content from this work may be used under the terms of the Creative Commons Attribution 3.0 licence. Any further distribution of this work must maintain attribution to the author(s) and the title of the work, journal citation and DOI.
Published under licence by IOP Publishing Ltd

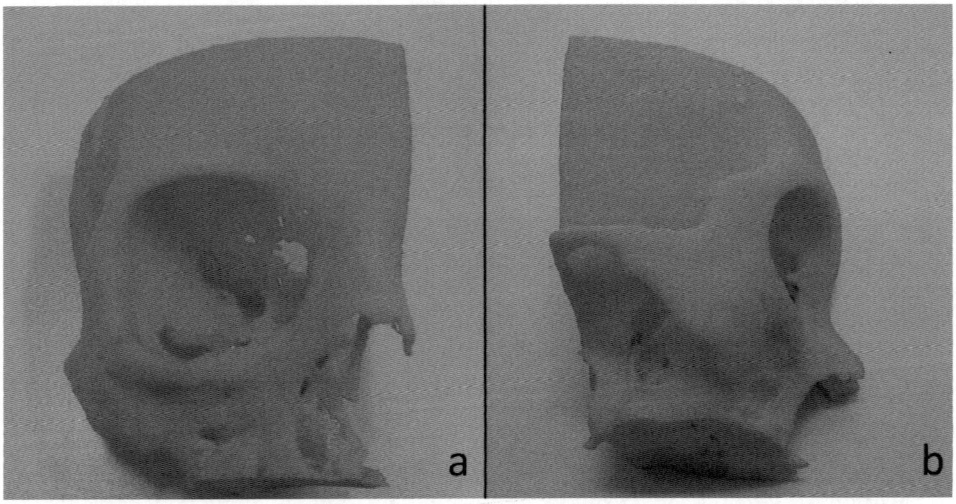

Figure 1. The skull fragment model.

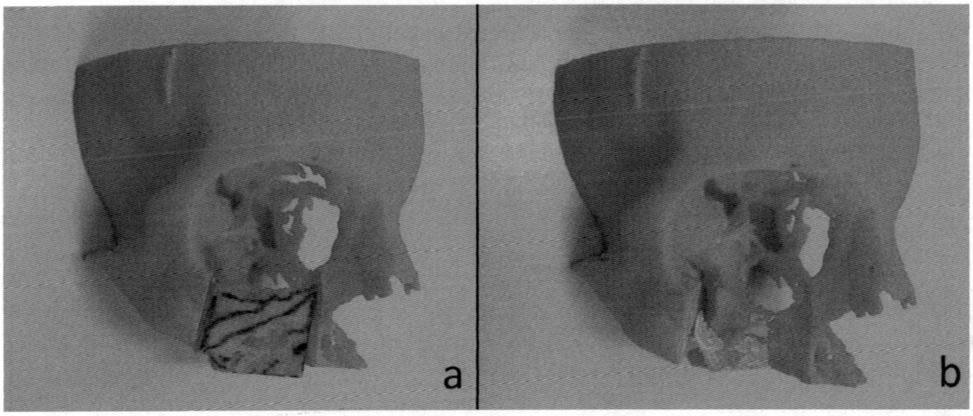

Figure 2. The affected part of the skull (model): a - before removal, b - after removal.

The metal implant was made for replacement of the affected part of the skull. The implant was manufactured using selective laser melting. The powder of titanium alloy TiAl6V4 was used as feedstock.

The process of selective laser melting of metal is determined by a large number of parameters. Each parameter influences the properties of products (roughness, strength, density, hardness, wear resistance, etc.). The most important parameters are:

- Power Output (P) – laser output, W;
- Speed (V) – scanning speed (when using continuous melting), mm/s;
- Layer Thickness (h) – the thickness of powder layer, μm;
- Point Distance (PD) – the point spacing of the dotted laser path (if the machine allows dotted scanning of the exposure vectors, the spacing of points may be specified with this function), μm.
- Exposure time (ET) – exposure time on one point of the dotted laser path, μs.
- Hatch Space (HS) – the spacing between individual hatch lines, μs.

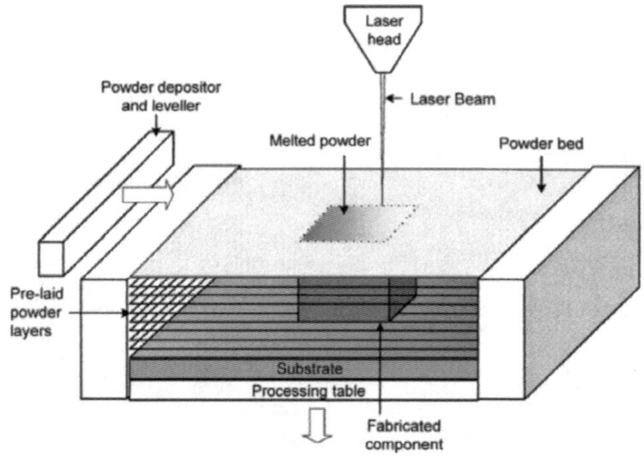

Figure 3. Selective Laser Melting process [15].

There are 2 groups of SLM machines:

- with continuous melting mode: the laser operates continuously; the laser beam runs over the surface of the powder with a constant speed;
- with "dotted melting" mode: the laser pulses; the laser beam moves from point to point.

The first type of machine requires the specification of one parameter – Speed (V). The second type of installation requires the specification of two parameters: Point Distance (PD) and Exposure time (ET). All other parameters for both types of machines are identical.

The machine SINTERSRATION ® Pro DM125 SLM [16] is one of the second type of machines. This machine is equipped with a CO2 laser redPower SP-200C-04 with output power up to 200 W. The scanning speed can vary, up to 1000 mm/s. The laser beam diameter at powder surface is 35 μm. Layer thickness can be vary from 20 to 100 μm.

Alloy TiAl6V4 is characterized by low density, satisfactory ductility, and high biocompatibility. At the same time, the alloy shows a tendency to residual stresses, cracks, and microcracks [17-23]. Residual stress and cracks may appear in the process with high cooling rates. At the stage of implant manufacturing, the aim was to select parameters of selective laser melting which provide for the density of the product structure.

The large number of SLM process parameters makes it complicated to analyze the influence of mode selective laser melting on the characteristics of the resulting products. Often, researchers use , density of absorbed laser energy to describe the mode of SLM. In this paper, two types of energy densities are used - surface and volume - to describe SLM modes.

$$E_s = \frac{P \cdot ET}{PD \cdot HS}, \tag{1}$$

$$E_v = \frac{P \cdot ET}{PD \cdot HS \cdot h}. \tag{2}$$

Parameters used for implant manufacturing are shown in table 1.

Table 1. Parameters of SLM.

P, (W)	h, (μm)	PD, (μm)	ET, (μs)	HS, (μm)	E_s, (J mm^{-2})	E_v, (J mm^{-3})
100	50	50	100	75	2,7	53

Apart from SLM modes, scanning strategy also affects the properties of the produced material. In this paper the "alternating" scanning strategy was used (figure 4). Hatch direction for the new layer was changed to 90 degrees.

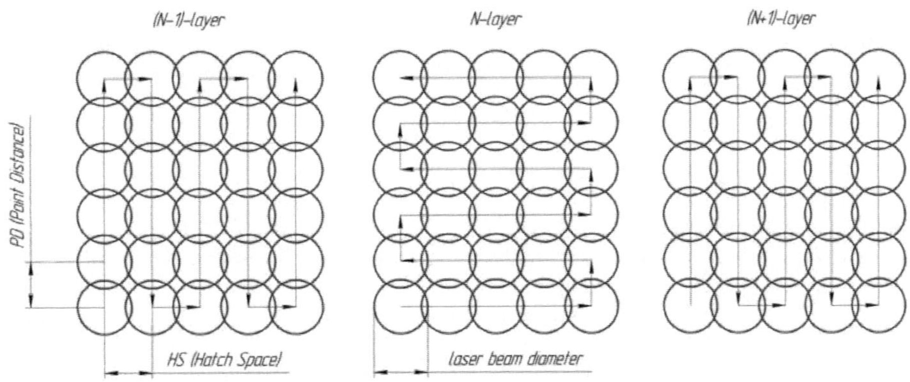

Figure 4. Alternating scanning strategy.

The metal implant (figure 5) was installed in the skull model in a remote area (figure 6). The analysis showed a satisfactory pair of surfaces. The total duration of operations for 3d-modeling and production of the model and the titanium implant was 4 days. Density of obtained implant is 93.2% of the theoretical. Density was measured using Archimedes technique.

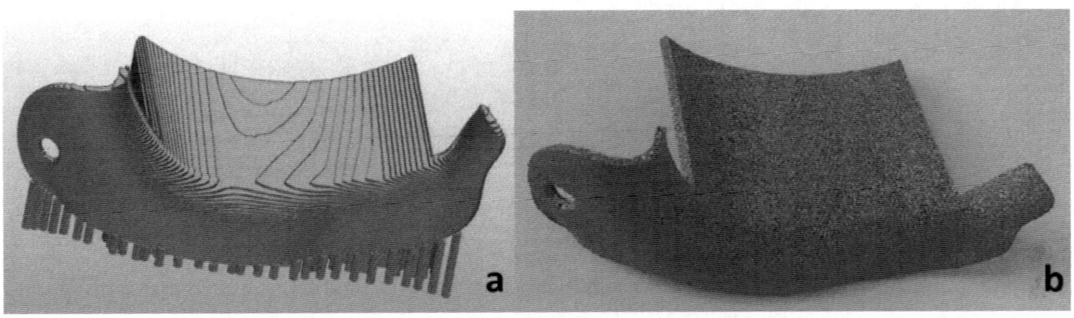

Figure 5. Medical implant: a - CAD-model, b – metal product.

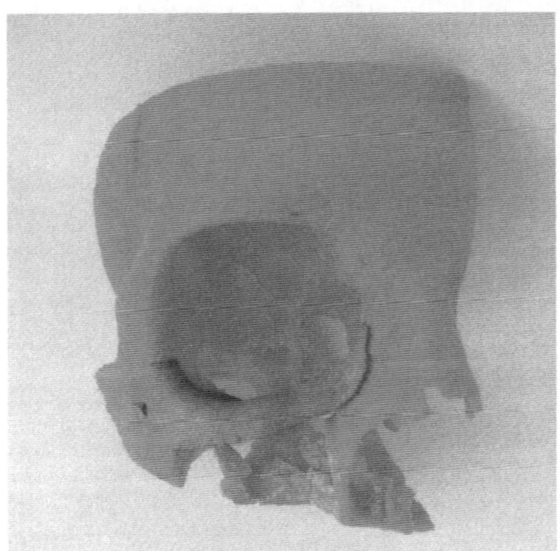

Figure 6. The installation of the implant in the skull model.

3. Summary

We developed the technique to produce titanium implants using additive technologies. The analysis showed a satisfactory pair of surfaces between the skull model and implant. The production time of 4 days.

Acknowledgements
This research was financially supported by a grant of the President of the Russian Federation (No. MK-1881.2017.8).

References

[1] Wagner S M, Walton R O 2016 *Prod. Plan. Control.* **27** 1124.
[2] Uhlmann E. Kersting R, Klein T B, Cruz M F, Borille A V 2015 *Procedia CIRP* **35** 55.
[3] Haynes J 2014 *Proc. of the International Astronautical Congress, IAC* vol 9 (International Astronautical Federation, IAF) p 6120.
[4] Liang X, Chen J, Ding P, Dong P, He J 2013 *Proc. of the International Astronautical Congress, IAC* vol 8 (International Astronautical Federation, IAF) p 6287.
[5] Wauthle R, van der Stok J, Amin Yavari S, Van Humbeeck J, Kruth J P, Zadpoor A A, Weinans H, Mulier M, Schrooten J 2015 *Acta Biomater.* **14** 217.
[6] You J, Fang L, Zhang Q, Gao Y, Peng W 2015 *Chinese J. Biomed. Eng.* **34** 315.
[7] Wang H, Zhao B, Liu C, Wang C, Tan X, Hu M 2016 *PLoS One* **11** e0158513.
[8] Lu Y, Wu S, Gan Y, Li J, Zhao C, Zhuo D, Lin J 2015 *Mater. Sci. Eng. C.* **49** 517.
[9] Li J, Chi S, Xu J, Wang Y, Zhan D 2014 *West China J. Stomatol.* **32** 115.
[10] Popovich A A, Sufiiarov V S, Polozov I A, Borisov E V, Masaylo D V, Vopilovskiy P N, Sharonov A A, Tikhilov R M, Tsybin A V, Kovalenko A N, Bilyk S S 2016 *Biomed. Eng. (NY).* **50** 202.
[11] Armencea G, Berce C, Rotaru H, Bran S, Leordean D, Coada C, Todea M, Jula C A, Gheban D, Baciut G, Baciut M, Campian R S 2015 *Clujul Med.* **88** 408.
[12] Rana M, Gellrich M M, Gellrich N C 2015 *Br. J. Oral Maxillofac. Surg.* **53** 208.
[13] de Moraes P H, Olate S, Cantin M, Assis A F, Santos E, de Oliveira Silva F, de Oliveira Silva L 2016 *Int. J. Morphol.* **33** 826.
[14] Choi J W, Kim N 2015 *Arch. Plast. Surg.* **42** 267.
[15] Mumtaz K, Hopkinson N 2014 J. *Mater. Process. Technol.* **210** 279.
[16] Information on http://www.3dsystems.com/products/datafiles/sinterstation_pro_slm/SinterstationPro_DM125_DM250_SLMSystem.pdf.
[17] Yadroitsev I, Yadroitsava I 2015 *Virtual Phys. Prototyp.* **10** 67.
[18] Wycisk E, Emmelmann C, Siddique S, Walther F, 2013 *Adv. Mater. Res.* **816-817** 134.
[19] Vilaro T, Colin C, Bartout J D 2011 *Metall. Mater. Trans. A Phys. Metall. Mater. Sci.* **42** 3190.
[20] Van Hooreweder B, Moens D, Boonen R, Kruth J P, Sas P 2012 *Adv. Eng. Mater.* **14** 92
[21] Song B, Dong S, Liao H, Coddet C 2011 *Int. J. Adv. Manuf. Technol.* **61** 967.
[22] Gong H, Rafi K, Gu H, Starr T, Stucker B 2014 *Addit. Manuf.* **1** 87.
[23] Cain V, Thijs L, Van Humbeeck J, Van Hooreweder B, Knutsen R 2014 *Addit. Manuf.* **5** 68.

IOP Conf. Series: Materials Science and Engineering **248** (2017) 012005 doi:10.1088/1757-899X/248/1/012005

Strength and fracture energy of foamed concrete incorporating rice husk ash and polypropylene mega-mesh 55

Z M Jaini[1,2], R H M Rum[3] and K H Boon[1]

[1]Jamilus Research Center, Universiti Tun Hussein Onn Malaysia, 86400 Parit Raja, Johor, Malaysia
[2]Civil and Structural Engineering Branch, Public Works Department, 50582 Kuala Lumpur, Malaysia
[3]Faculty of Civil and Environmental Engineering, Universiti Tun Hussein Onn Malaysia, 86400 Parit Raja, Johor, Malaysia

Abstract. This paper presents the utilization of rice husk ash (RHA) as sand replacement and polypropylene mega-mesh 55 (PMM) as fiber reinforcement in foamed concrete. High pozzolanic reaction and the ability to become filler make RHA as a strategic material to enhance the strength and durability of foamed concrete. Furthermore, the presence of PMM optimizes the toughness of foamed concrete in resisting shrinkage and cracking. In this experimental study, cube and cylinder specimens were prepared for the compression and splitting-tensile tests. Meanwhile, notched beam specimens were cast for the three-point bending test. It was found that 40% RHA and 9kg/m^3 PMM contribute to the highest strength and fracture energy. The compressive, tensile and flexural strengths are 32MPa, 2.88MPa and 6.68MPa respectively, while the fracture energy achieves 42.19N/m. The results indicate high potential of RHA and PMM in enhancing the mechanical properties of foamed concrete.

1. Introduction

Nowadays, foamed concrete has received high demand in construction industry due to primary characteristics of versatility and lightness. As a lightweight material, foamed concrete has density in around 400kg/m^3 to 1800kg/m^3. However, foamed concrete possesses low compressive strength in the range of 10MPa to 15MPa [1, 2]. Meanwhile, the fracture energy can only reach up to 35N/m [3]. These drawbacks have limited the application of foamed concrete as bearing structural components. The minimum compressive strength that permitted by Eurocode 2 [4] for reinforced concrete structures is 22MPa. If foamed concrete is employed as composite structures, Rackham et al. [5] insisted that the compressive strength should achieve at least 28MPa. Therefore, there are huge attempts to improve the compressive strength of foamed concrete.

Various types of fibre such as polypropylene, steel, glass and polyvinyl [6-8] were used to improve the mechanical properties of foamed concrete. However, the modification of cementitious using fly ash (FA), palm oil fuel ash (POFA) and rice husk ash (RHA) showed the most favourable outcomes [9, 10]. Therefore, this study intends to utilize RHA and polypropylene mega-mesh 55 (PMM) to produce high-performance foamed concrete that can comply with the code of practise. More focus is given to investigate the effect of RHA and PMM on the strength and fracture energy of foamed concrete. The fascination toward green and sustainable construction material has prompted the use of RHA and PMM. The inspiration is to reduce environmental pollution, improve the quality of foamed concrete and reduce the cost of production, as well as become a novel solution of problems in agriculture-waste.

Content from this work may be used under the terms of the Creative Commons Attribution 3.0 licence. Any further distribution of this work must maintain attribution to the author(s) and the title of the work, journal citation and DOI.
Published under licence by IOP Publishing Ltd

2. Experimental study

2.1. Material preparation

Material preparation involved the provision of mix design and required quantity of cement, sand, RHA, PMM, foam agent, water and admixture. Figure 1 shows the RHA and PMM as main materials in foamed concrete. The ratios of cement-sand and water-cement are 0.50 and 0.55 respectively. In Stage 1, RHA was used as partially replacement of sand at the range of 0% to 50%. In Stage 2, the optimum contain of RHA and PMM with dosage up to 12kg/m^3 were utilized in foamed concrete. RHA was obtained from the combustion process at 700°C for 6 hours. The chemical composition of RHA consists of 89.90% SiO_2 which is highly amorphous silica. The pre-forming method was used in the production of foamed concrete. Foam agent of synthetic-based Sika AER 50/50 was diluted with water using concentration ratio of 0.05.

a)

b)

Figure 1. Main materials in foamed concrete: a) rice husk ash (RHA), and b) polypropylene mega-mesh 55 (PMM).

2.2. Specimen preparation

In Stage 1, a total 45 cube specimens were moulded using size of 100mm length, 100mm width and 100mm depth. Meanwhile, Stage 2 requires 45 cube specimens, 45 cylinder specimens and 10 notched beam specimens. The cylinder specimens were prepared based on size of 350mm depth and 150mm diameter. The notched beam specimens were fabricated with size of 500mm length, 100mm width, 100mm depth and 30mm V-notch depth. The cube and cylinder specimens were underwent air curing for 7, 14 and 28 days. On the other hand, the notched beam specimens were placed at the ambient condition for 28 days.

2.3. Test programme

The compression test on cube specimens was conducted according to BS EN12390-3:2009. On the other hand, BS EN12390-6:2001 was referred for the splitting-tensile test to obtain the tensile strength of foamed concrete. Both compression and splitting-tensile tests were performed using Ele Compact Machine 1500 as can be seen in figure 2. Before the tests, the cube and cylinder specimens were visually inspected, labelled, sized and weighted. The three-point bending test was conducted on the notched beam specimens to determine the flexural strength and fracture energy. The three-point bending test, as shown in figure 3, was equipped with load cell, LVDT and CMOD gauge to capture the load, deflection and crack width respectively.

3. Results and discussion

3.1. Compressive strength

Figure 4 shows the compressive strength of foamed concrete containing RHA. Plain foamed concrete accomplishes the compressive strength of 15.1MPa. The value complies with that determined by Jaini et al. [1] and Abd Rahman et al. [3], where plain foamed concrete with similar mix design has compressive strength around 12MPa to 18MPa. The utilization of RHA as sand replacement indicates a statistically significant improvement. At the optimum contain of 40% RHA, the compressive strength increased by up 48% compared to plain foamed concrete. However, the presence of large portion RHA was found to delay the pozzolanic reaction, thereby contributes to the adverse effect.

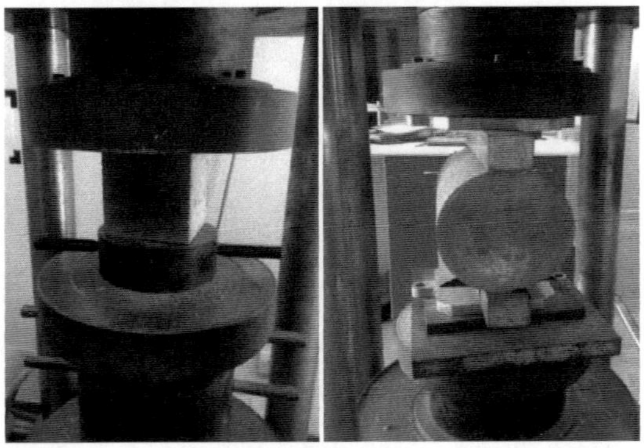

Figure 2. Compression and splitting-tensile tests on cube and cylinder specimens.

Figure 3. Three-point bending test on notched beam specimens.

On the other hand, figure 5 shows the compressive strength of foamed concrete containing 40% RHA and PMM. There is an apparent effect on the compressive strength, where incorporating 9kg/m^3 PMM lead to the increment of 30%. This improvement occurs due to the confinement provided by PMM which increases the bonding of foamed concrete. However, higher dosage of PMM decreases the compressive strength due to interfering in cohesiveness of foamed concrete. It was observed that the good binder composition gave high influence to the compressive strength. In addition, PMM also acts as filler that reduce the volume fraction of air-void and subsequently provides an intact condition.

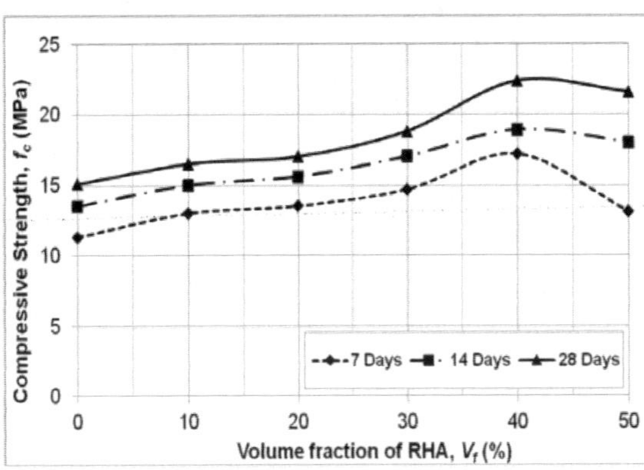

Figure 4. Compressive strength of foamed concrete incorporating RHA.

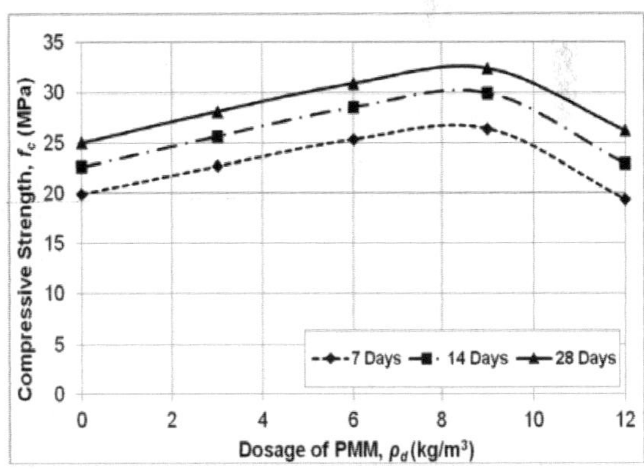

Figure 5. Compressive strength of foamed concrete with 40% RHA and PMM.

3.2. Tensile strength

The tensile strength obtained from splitting-tensile test is shown in figure 6. It can be confirmed that the tensile strength increases due to the dosage of PMM. The optimum dosage is 9kg/m^3 that led to the tensile strength of 2.87MPa. It is a very significant increment around 44% compared to foamed concrete containing RHA only. On the other hand, the relationship between compressive strength (f_c) with tensile strength (f_t) is shown in figure 7, which can also be expressed as:

$$f_t = 0.034 f_c^{\,1.27} \tag{1}$$

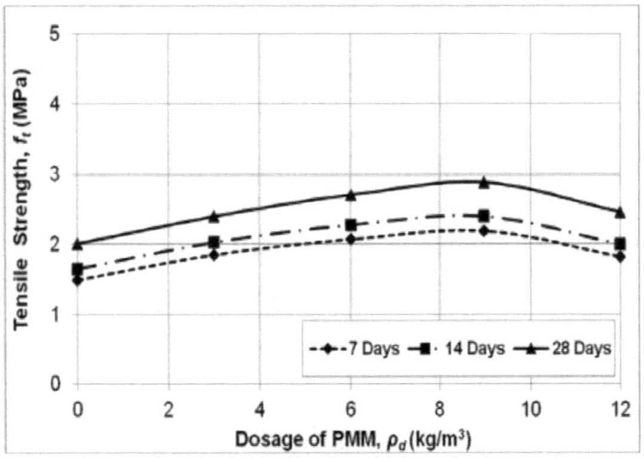

Figure 6. Tensile strength of foamed concrete with 40% RHA and PMM.

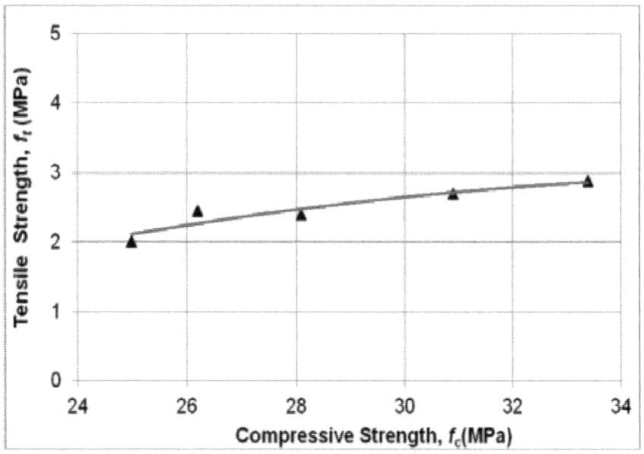

Figure 7. Relationship between compressive strength and tensile strength.

3.3. Flexural strength

The flexural strength was determined using an empirical formulation as depicted in equation (2):

$$f_f = \frac{3P_{\max} S}{2B(W - a_o)^2} \tag{2}$$

where P_{max} is the maximum load while S, B and W are the span, width and depth of specimen respectively and a_o is the notch depth. Table 1 shows the flexural strength of foamed concrete containing RHA and PMM. The presence of PMM resembles significant resistance against the propagation of cracks. In addition, PMM reduces stress concentration accumulated around the cracks. The flexural strength is effectively increased by incorporating 9kg/m^3 PMM, where it conceded 6.68MPa that correspond to the increment of 40%. However, the flexural strength decreases at dosage of 12kg/m^3, a similar condition that was observed on the compressive and tensile strengths.

Table 1. Flexural strength of foamed concrete incorporating 40% RHA and PMM.

Dosage of PMM (kg/m^3)	P_{max} (kN)	f_f (MPa)
0	5.08	4.76
3	6.24	5.85
6	6.52	6.12
9	7.12	6.68
12	6.80	6.38

3.4. Fracture energy

The Hillerborg model as formulated in equation (3) was used to calculate the fracture energy of foamed concrete containing RHA and PMM.

$$G_f = \frac{U_o + m_g d_o}{B(W - a_o)} \tag{3}$$

where U_o is the area of softening phase from load-deflection curve, m_g is mass of specimen, d_o is the fracture deflection, a_o is the notch depth, B is span of specimen and W is thickness of specimen. Table 1 shows the fracture energy of foamed concrete incorporating RHA and PMM. A comparison with Bazant model shows favourable agreement. The presence of RHA and PMM have tendency to increase fracture energy up to 42.19N/m, where the dosage of 9kg/m^3 at 40% RHA contributes to the highest value. It is improved approximately 35% above foamed concrete with 40% RHA, but higher 65% than plain foamed concrete.

Table 2. Fracture energy of foamed concrete incorporating RHA and PMM.

Dosage of PMM (kg/m^3)	U_o (Nm)	d_o (m)	G_f (N/m)	
			Hillerborg	Bazant
0	4.117	2.171	31.33	35.32
3	16.505	1.686	39.96	37.01
6	26.831	0.738	41.01	38.44
9	18.033	1.548	42.19	39.18
12	4.167	2.220	35.90	35.99

4. Conclusions

The investigation on the compressive, tensile and flexural strengths as well as the fracture energy of foamed concrete containing RHA and PPM were conducted by using the compression, splitting-tensile and three-point bending tests. The results indicate that RHA and PPM greatly improve the mechanical properties of foamed concrete. The addition of 40% RHA in foamed concrete contributes to the enhancement of compressive strength around 48%. Moreover, the utilization of 40% RHA and 9kg/m^3 PPM increase the compressive, tensile and flexural strength around 30% to 40% while the fracture energy improve around 35%. However, the presence of large volume fraction of RHA and dosage of PPM dramatically reduce the strength and fracture energy of foamed concrete.

5. References

[1] Jaini Z M, Mokhatar S N, Mohd Yusof A S, Zulkiply S and Abd Rahman M H 2016 *MATEC Web of Conf.* **47** 01013.

[2] Mugahed Amran Y H, Farzadnia N and Abang Ali A A 2015 *Constr. Build. Mater.* **101** 990-1005.

[3] Abd Rahman N, Jaini Z M and Mohd Zahir N N 2015 *ARPN J. Eng. Appl. Sci.* **10** 6562-70.

[4] EN 1992-1-1:2004 *Design of Concrete Structure Part 1-1: General Rules and Rules for Buildings* (London: British Standards Institution).

[5] Rackham J W, Couchman G H and Hicks S J 2009 *Composite Slabs and Beams Using Steel Decking: Best Practise for Design and Construction* (Berkshire: The Steel Construction Institute).

[6] Daneti S B, Wee T H and Thangayah T 2011 *Mag. Concr. Res.* **63** 871-81.

[7] Bing C, Zhen W and Ning L 2011 *J. Mater. Civ. Eng.* **24** 113-18.

[8] Hamad E T and Dawood A J 2015 *Struct. Concr.* **16** 496-507.

[9] Sathwanea S H, Vairagadeb V S and Kenec K S 2013 *Proc. Eng.* **51** 35-44.

[10] Lim S K, Tan C S, Lim O Y and Lee Y L 2013 *Constr. Build. Mater.* **46** 39-47.

Acknowledgments

Authors would like to thanks Universiti Tun Hussein Onn Malaysia (UTHM) and Ministry of Education Malaysia for the continuous supports in term of facilities and financial. This experimental study was conducted under Research Acculturation Grant Scheme (RAGS), Vot. No. R068.

IOP Conf. Series: Materials Science and Engineering **248** (2017) 012006 doi:10.1088/1757-899X/248/1/012006

Minimization of Poisson's ratio in anti-tetra-chiral two-phase structure

E Idczak[1], **T Strek**[1]

[1] Institute of Applied Mechanics, Poznan University of Technology
ul. Jana Pawla II 24, 61-139 Poznan, Poland

Abstract. One of the most important goal of modern material science is designing structures which exhibit appropriate properties. These properties can be obtained by optimization methods which often use numerical calculations e.g. finite element method (FEM). This paper shows the results of topological optimization which is used to obtain the greatest possible negative Poisson's ratio of the two-phase composite. The shape is anti-tetra-chiral two-dimensional unit cell of the whole lattice structure which has negative Poisson's ratio when it is built of one solid material. Two phase used in optimization are two solid materials with positive Poisson's ratio and Young's modulus. Distribution of reinforcement hard material inside soft matrix material in anti-tetra-chiral domain influenced mechanical properties of structure. The calculations shows that the resultant structure has negative Poisson's ratio even eight times smaller than homogenous anti-tetra chiral structure made of classic one material. In the analysis FEM is connected with algorithm Method of Moving Asymptote (MMA). The results of materials' properties parameters are described and calculated by means of shape interpolation scheme – Solid Isotropic Material with Penalization (SIMP) method.

1. Introduction

One of the most popular structures which exhibit the negative Poisson's ratio (auxetics) are periodic lattice structures. These structures are built of single unit cells. The main class of auxetic unit cells are: 2-dimensional and 3-dimensional re-entrant honeycombs – first suggested by Gibson [1], rotating rigid and semi-rigid units (triangles or rectangles) – proposed by Grima et al. [2], chiral structures – first introduced by Prall and Lakes [3]. During the last three decades the researchers developed more sophisticated and complicated geometries of auxetic unit cell. Furthermore, there are also known structures which are the combination of two or more different unit cells. This work will focus on anti-tetra-chiral structures.

The adjective "chiral" means physical property to spin. Basic unit of this kind of structures consists of straight ligaments connecting to central nodes (see figure 1). The node may be of any geometrical form, usually circles or rectangles. The mechanism of "auxeticity" is obtained by the wrapping and unwrapping of the ligaments around the nodes (circles or rectangles). The whole chiral network is built of many single unit cells joined together (see figure 1a). In the papers devoted to chiral structures the value of Poisson's ratio is estimated close to -1. The geometry of unit cell is named by the Greek numbers: tri-, tetra-, penta- or hexa-chiral – the number depends on amount of ligaments joined to one node. Apart from chiral lattices it is possible to develop anti-chiral structures. Anti-chiral networks are generated if any two adjacent nodes share the same side of the common tangential ligament. Chiral and anti-chiral geometries can be clearly defined by ratio of the parameter of node (e.g diameter for the circle) and the length of ligament. The behaviour, deformation process and mechanical properties of chiral geometries are investigated in many works of researchers.

Content from this work may be used under the terms of the Creative Commons Attribution 3.0 licence. Any further distribution of this work must maintain attribution to the author(s) and the title of the work, journal citation and DOI.
Published under licence by IOP Publishing Ltd

Grima et al [4] described in details the mechanism of deformation responsible for auxetic functionality. The discussion of results was made for the whole family of the chiral structures and their auxeticity. The Poisson's ratio and other in-plane elastic constants were various for the different amount of ligaments connecting to nodes and different shape of node (circle, rectangle and other). Alderson et al. [5] studied the in-plane linear elastic response and process of out-of-plane bending of tri- and anti-tri-chiral honeycombs and their re-entrant counterparts using FE analysis and experiments.

The out-of-plane linear elastic mechanical properties of tri-, tetra- and hexa-chiral honeycomb configurations was described by Loreto et al. [6]. Analytical models were developed to compute the transverse Young's modulus and the Voigt and Reuss bounds for the transverse shear stiffness. Besides the analytical solutions numerical models was created for validation the analytical results, and to find the correlation between the transverse shear stiffness and the gauge thickness of the honeycombs. Miller et al. [7] investigated the flatwise compressive behaviour of tetra- and hexa-chiral honeycombs using FEM.

The researchers investigated also chiral structures for their use in the industrial applications. The best way to obtain this goal is to optimize specified mechanical properties using numerical methods. Kolla et al. [8] investigated auxetic structures for its use in shear layer of non-pneumatic wheel. The objective of the study was to find an ideal geometry for the shear layer while meeting its requirements of shear properties with polycarbonate as base material. Joshi [9] studied the mechanical, acoustic and vibration properties of many chiral and anti-chiral structures with various geometrical parameters, so the dynamic response of shapes in chiral fulfilments was investigated.

Chiral lattices were optimized also to combine the best mechanical properties with the best technology of production. Airoldi et al. [10] presents the advancements in the technological processes developed to produce chiral honeycombs made of thin composite laminates. The objective of such approach is the production of thin-walled chiral composite structures with enhanced strength properties by using a more feasible technology.

Mousanezhad et al [11] used FEM and optimization method to optimize several chiral, anti-chiral, and hierarchical honeycombs with hexagon and square based networks. Simple closed-form expressions were derived for the optimization of elastic moduli.

Undoubtedly, topology optimization is big a challenge for researchers and many works have already been devoted to this subject. Important aspect of optimization is spatial distribution of material with the gaol to obtain optimal structure which fulfils assumed criteria [12, 13]. This definition of properties' optimization is made by means of material density-like function. Single material and multi-material topological design in elasticity were solved using continuous control variable function [14]. The distribution of two materials with non-vanishing stiffness and voids in composite-like structure was investigated by Thomson [15]. Sigmund [16] also optimized the two-phase composite structure to obtain the structure with extreme behaviour parameters. The multi-phase composite structure can also be characterized by the negative Poisson's ratio. This possibility has been investigated by Evans [17] and Milton [18]. Milton in his work presented that negative Poisson's ratio materials can be built of symmetrized laminates as well as in two as in three dimensions. Evans et al. described that a negative Poisson ratio can be obtained by insertion in a network-embedded composite using materials that have the large ratio of the reinforcement to matrix modulus.

Lim [19] has been presented mechanical properties of auxetic sandwich panels made by different materials. It has been built also topologies which have effective properties for vibration damping inserts in honeycomb structures (e.g. Boucher [20]). Strek et al. [21] have presented that the sandwich composite structure with the two-phase auxetic core can exhibit auxetic properties even if all its are characterized by positive Poisson's ratio. The subject of research in the field of properties of auxetic composites was also the influence of temperature on the effective Poisson ratio (thermoauxeticity) of the composites made of solid matrix material with elliptical fibres [22]. Pozniak et al. [23] showed planar auxeticity of materials with various inclusions. Strek et al. [24] presented the simulations of torsional behaviour of a composite beam with auxetic phase.

Topology optimization method were used by many researchers to present new types of auxetic structures with optimal stiffness or damping characteristic of material or two or multi-phase auxetic materials, e.g. Gibiansky and Sigmund [25], Czarnecki and Wawruch [26], Lukasiak [27], Yi et al. [28], Whitty et al. [29], Schwerdtfeger et al. [30], Kaminakis et al. [31], Strek et al. [32].

In this work Finite Element Method (FEM) combined with Method of Moving Asymptotes (MMA) algorithm and Solid Isotropic Material with Penalization (SIMP) method is used to minimize the effective Poisson's ratio (PR) of two-phase anti-tetra-chiral structure. The domain of optimization is the shape of anti-tetra-chiral unit cell, which is characterized by the Poisson's ratio close to minus one. The Poisson's ratio of anti-tetra-chiral structure can be decreased by proper distribution of two different materials (two phases) inside the chiral geometry. Distribution of hard and soft material with given ratio of Young's modulus in anti-tetra-chiral domain influenced mechanical properties of structure. The calculations shows that the resultant anisotropic structure can exhibit greater values of negative Poisson's ratio than homogenous isotropic anti-tetra chiral structure made of one positive material. Topology optimization allows to find distribution of materials in anti-tetra-chiral unit cell creating original new structures with auxetic properties beyond the limits of isotropic one. The results are presented in the second section of this paper.

1.1. Method of evaluation of effective properties – SIMP interpolation

An analysis of effective properties of two-phase chiral shape uses the methods of evaluation of generalized parameters such as Young's modulus, Poisson's ratio and density. These generalized properties were written by means of Solid Isotropic Material with Penalization (SIMP) scheme. SIMP is very often used for material interpolation in geometry optimization. The main part of this method is the introduction of an interpolation function that expresses various generalized physical quantities as a function of continuous variable. The parameters in SIMP method fulfils the equations:

$$E(r) = E_1 + (E_2 - E_1) \cdot r^p, \tag{1}$$

$$\vartheta(r) = \vartheta_1 + (\vartheta_2 - \vartheta_1) \cdot r^p, \tag{2}$$

where $r = r(x)$ is the control variable, p is the penalization parameter, E_1 and E_2 are the Young's moduli and ν_1 and ν_2 are the Poisson's ratios for the first and the second material ($E_1 \leq E_2$ and $\nu_1 \leq \nu_2$). In all considered cases materials differ only with respect to the values of Young's modulus. All other parameters are the same for both materials. The density of material is represented by continuous variable $0 < r < 1$. The interpolating functions (1-2) are used to describe the distribution of two phases with different properties inside the analysed domain. The most important goal is to obtain such distribution for which the anti-tetra-chiral shape exhibit the minimal value of effective Poisson's ratio. The values of elastic properties of materials (E_1, E_2, ν_1, ν_2) of both phases were chosen as follows: structural steel (hard material) and computational artificial material (soft material). More detailed simulation would optimize the interpolation of bulk and shear moduli, but it isn't main subject of this work.

1.2. Computing of effective Poisson's ratio and FEM analysis

The commonly used definition of Young's modulus for a homogeneous, isotropic elastic solid domain is specified as the ratio of longitudinal stress to longitudinal strain. Respectively the Poisson's ratio for the same homogenous, isotropic elastic solid material is the negative ratio of transverse to longitudinal strain at every point in a body under longitudinal tension. In a material with two phases where the ratios may be changed from point to point both classic definitions of Young modulus and Poisson's ratio isn't still possible to use. The new definitions of the ratios were checked by many authors. The most common definition of Poisson's ratio is based on the assumption of small deformation and it is computed as a negative ratio of the average transverse to longitudinal strains [33]:

$$\nu_{eff} = -\frac{\overline{\varepsilon_t}}{\overline{\varepsilon_l}}, \tag{3}$$

where: $\overline{\varepsilon_t}$ is the average strain in transverse direction and $\overline{\varepsilon_l}$ is the average strain in longitudinal direction. If the force is applied along the y-axis, the average transverse strain is defined by equation:

$$\overline{\varepsilon_t} = \frac{\int_{G_1} u_1 dG}{L_x \cdot \int_{G_1} dG}, \tag{4}$$

where: G_1 is the boundary parallel ($x = L_x$) to the boundary with applied prescribed displacement and u_1 is displacement in x-axis. The average longitudinal strain is defined by equation:

$$\overline{\varepsilon_l} = \frac{\int_{G_2} u_2 dG}{L_y \cdot \int_{G_2} dG}, \tag{5}$$

where: G_2 is the boundary parallel ($y = L_y$), where a load is applied and u_2 is displacement in y-axis. Because of using SIMP scheme effective Poisson's ratio must be dependent on control variable *r*. In this way equation receive objective functions for the optimization problem of the two-phase chiral shape:

$$\vartheta_{eff}(r) = -\frac{\overline{\varepsilon_t(r)}}{\overline{\varepsilon_l(r)}}, \tag{6}$$

As a control variable function in SIMP scheme two constraints are applied - pointwise inequality (7) and integral inequality (8), which are given by formulas:

$$0 \leq r(x) \leq 1 \text{ for } x \in S, \tag{7}$$

$$0 \leq \int_S r(x) dS \leq A_f \cdot S, \tag{8}$$

where x is defined coordinate, A_f – is a percentage fraction of the second material in the domain S. The process of optimization is defined in following order: FEM – discretization, redefinition of function minimization with applied constraints as standard finite dimensional nonlinear programmable problem. Then the value of control variable is evaluated in every mesh node as:

$$r(x) = \sum_{i=1}^{N} r_i \cdot \phi_i(x), \tag{9}$$

where: $\phi_i(x)$ are shape functions and N is the number of an element node.

After discretization, pointwise inequality condition is expressed as follows (where: M – the number of all nodes of all mesh elements, where the value of control variable is derived):

$$0 \leq r_i \leq 1 \text{ for } x = 1, \ldots, M. \tag{10}$$

1.3. Definition of mechanical problem and boundary conditions

The analysis is made for anti-tetra-chiral (ATC) unit cell with circular nodes subjected to compression load from the top. Considered shape is defined by the size of the unit cell ($2L$ on $2L$), L – length of ligament, D – diameter of node and g – thickness of unit cell. In considered case $L_x = L_y = 2L$ (see figure 1b).

Due to symmetry of considered problem only quarter of the structure (the whole optimized geometry consists of four unit cells) is analysed. The boundary conditions (BC) used in computations (Figure 1B) are as follows: right boundary: x=L - free BC; left boundary (symmetry): x = -L - $\boldsymbol{n} \cdot \boldsymbol{u} = 0$; bottom boundary (symmetry): y =-L $\boldsymbol{n} \cdot \boldsymbol{u} = 0$; top boundary: y = L - prescribed displacement, $u_2 = w$, where: $w = 0.01$ and L=0.15.

The governing equations for the solid stress–strain mechanical problems by the displacement formulation are steady state Navier's equation with linear constitutive equation and zero body load.

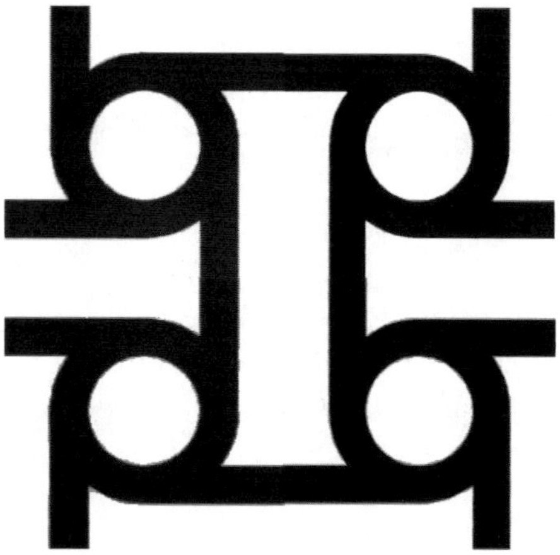

Figure 1A. Geometry of ATC structure.

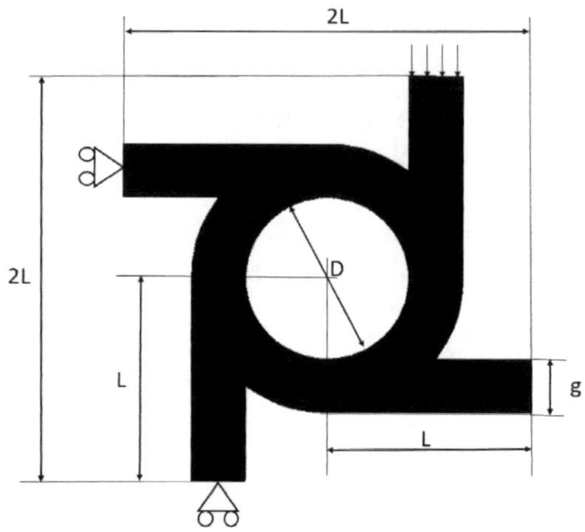

Figure 1B. Dimensions and boundary conditions.

2. Numerical results

In order to obtain the goal of topology optimization of a two-phase anti-chiral structure the correct and effective algorithm needs to be applied [29]. If the number of design variables is very large simple mathematical programming methods usually can't bring correct results. In this article the method of moving asymptotes (MMA) algorithm is used to find optimal distribution of two positive materials in anti-tetra-chiral shape. This algorithm was firstly used and described in 1986 by Svanberg [34, 35]. MMA method is constructed on the thing that it is used the changed Lagrangian function to make a convex approximation and solving a sequence of convex approximations. The base of MMA algorithm is the method of CONLIN (convex linearization) and it was used in a wide range of computational problems with the great successes. In this paper the MMA algorithm is used to solve the optimization problem, which distributes two materials in the chiral shape of single unit cell. As result it is created the whole structure with negative Poisson's ratio smaller than value of anti-chiral cell (close to -1). MMA algorithm minimizes the objective function which defines the effective PR of a structure with some appropriate constraints and finds the optimal distribution characterized by the minimal Poisson' ratio. All numerical computations were conducted by means of FEM software (Comsol Multiphysics). Control variable shape functions in optimizations were bilinear Lagrange polynomials. Quadratic Lagrange polynomials were defined as shape functions for displacement fields. The value of penalization parameter in SIMP method was p=3.

As described below, two phases were used characterized with positive Poisson's ratio and real Young moduli. One of these materials can be called "soft material" with the lower Young's modulus and second: "hard material" with the greater Young's modulus:

- soft material: Young's modulus is $E_1 = 10^7$ Pa and Poisson's ratio $\upsilon_1 = 0.33$;
- hard material (structural steel): Young's modulus is is $E_2 = 2 \cdot 10^{11}$ Pa and Poisson's ratio $\upsilon_2 = 0.33$.

In all simulations initial value of A_f is defined which means the percentage value of hard material in the whole structure and it changes between 20% and 50%.

The dimensions of optimized structure is L = 0.15 m, D = 0.1 m and thickness of ribs is g = 0.05 m and g = 0.025 m.

In the figure 2A the optimized distribution of materials in anti-tetra-chiral structure with A_f= 20% is presented. In figures 2-5 the green colour represents soft material and the blue one the hard material. The whole structure after optimization is characterized by the effective Poisson's ratio equal to -6.4188. The anti-tetra-chiral structure made of four unit cells is shown in the figure 2B.

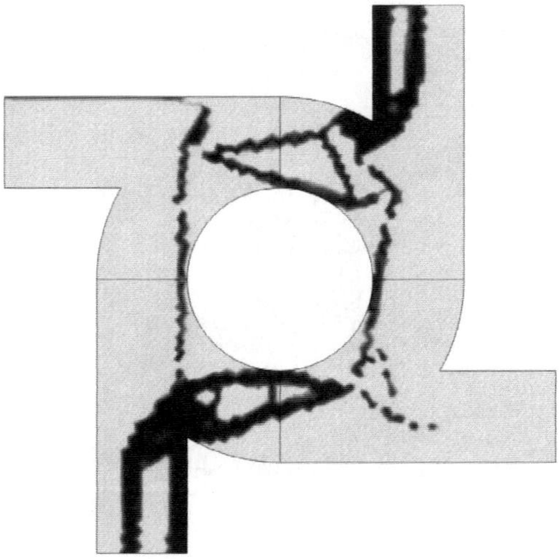

Figure 2A. Two-phase material with A_f=20%.

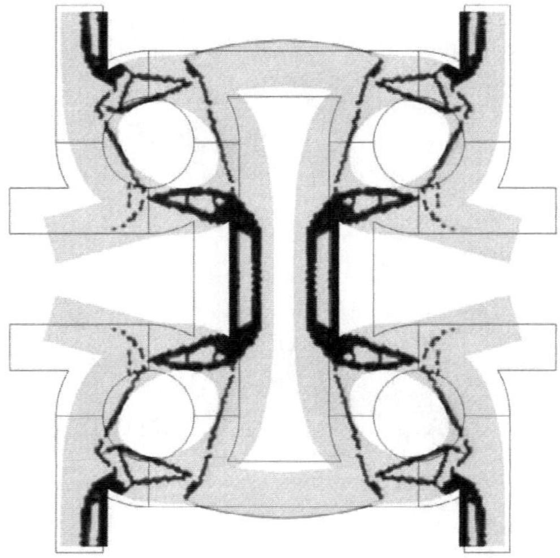

Figure 2B. Deformed shape of structure I.

In the figure 3A the distribution of materials in anti-tetra-chiral structure with A_f = 30% is shown. The whole structure after optimization exhibits the effective Poisson's ratio equal to -8.7017. In the figure 3B the deformed shape consisted of four unit cells is shown.

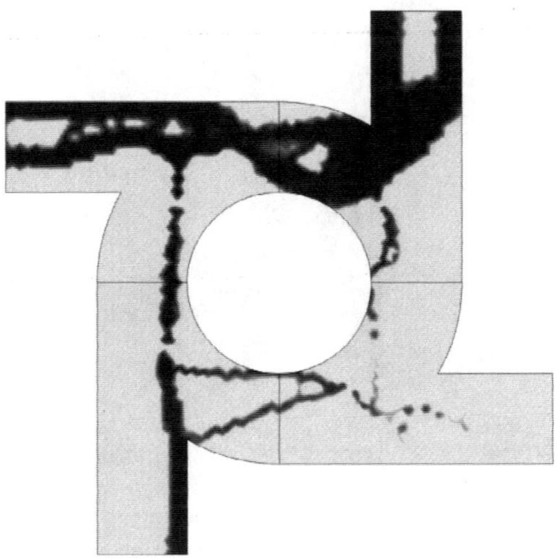

Figure 3A. Two-phase material with A_f=30%.

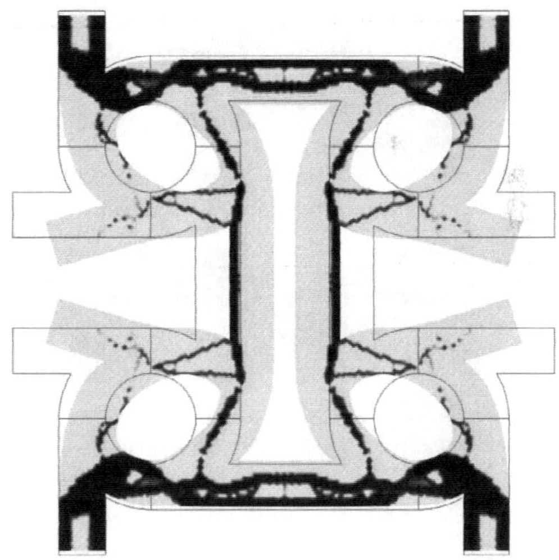

Figure 3B. Deformed shape of structure II.

In the figure 4A the next optimized distribution of materials is presented in anti-tetra-chiral structure with A_f = 40%. The whole structure after optimization exhibit the effective Poisson's ratio equal to -6.9386. In the figure 4B the network consisted of four optimized unit cells is presented.

In the figure 5A the next optimized distribution of materials in anti-tetra-chiral structure with A_f = 50% is presented. The whole structure after optimization is characterized by the effective Poisson's ratio equal -7.0516. In the figure 5B the network consisted of four optimized unit cells is presented.

IOP Conf. Series: Materials Science and Engineering **248** (2017) 012006 doi:10.1088/1757-899X/248/1/012006

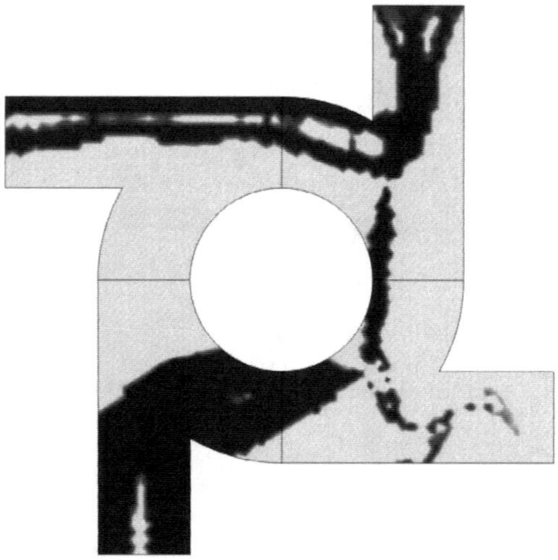

Figure 4A. Two-phase material with A_f=40%.

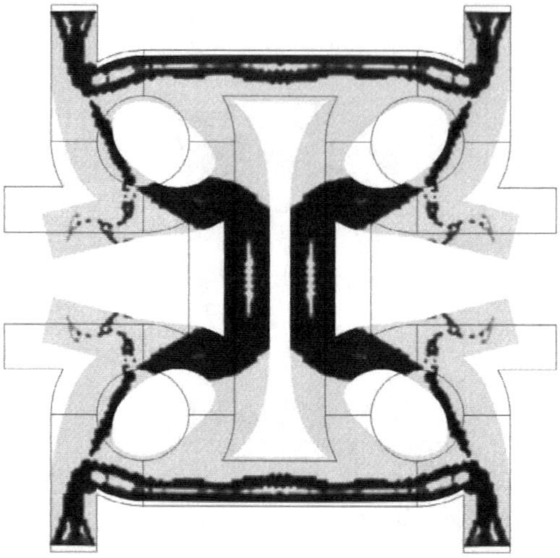

Figure 4B. Deformed shape of structure III.

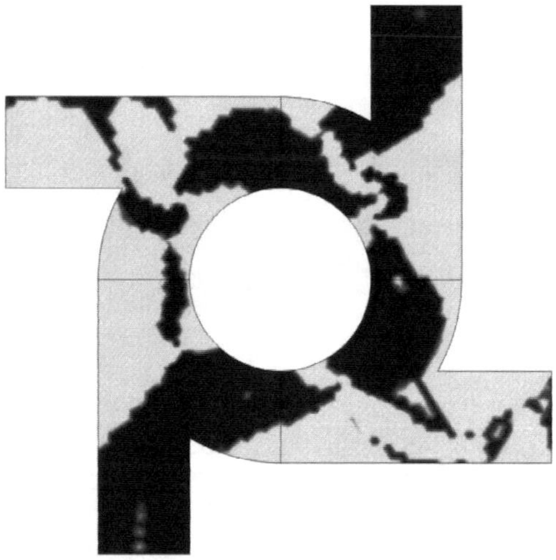

Figure 5A. Two-phase material with A_f=50%.

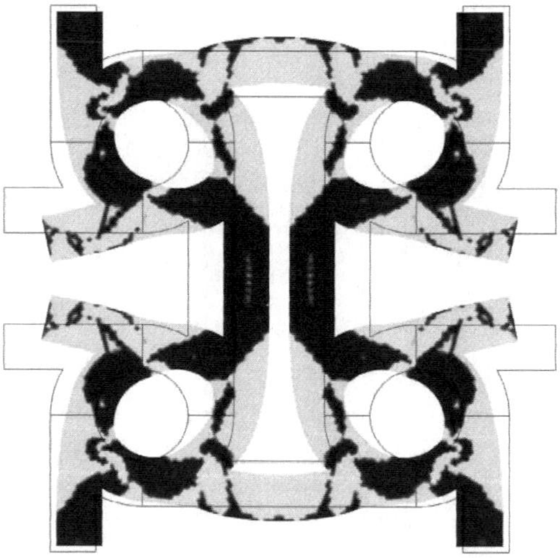

Figure 5B. Deformed shape of structure IV.

In the table 1 numerical values of Poisson's ratios for anti-tetra-chiral structures are presented for different values of A_f from 20% to 50% and g = 0.05 m. In table 2 numerical values of the Poisson's ratios for anti-tetra-chiral structures with smaller thickness of ribs g = 0.025 m are presented. All values of the Poisson's ratio are few times smaller than the value of the Poisson's ratio for homogenous isotropic structure.

Table 1. Poisson's ratio for anti-tetra-chiral structures for different values of A_f and g = 0.05 m.

Structure	I	II	III	IV
A_f	20%	30%	40%	50%
PR	-6.4188	-8.7017	-6.9386	-7.0516

Table 2. Poisson's ratio for anti-tetra-chiral structures for different values of A_f and g = 0.025 m.

Structure	V	VI	VII	VII
A_f	20%	30%	40%	50%
PR	-4.4994	-7.8173	-7.3032	-7.5293

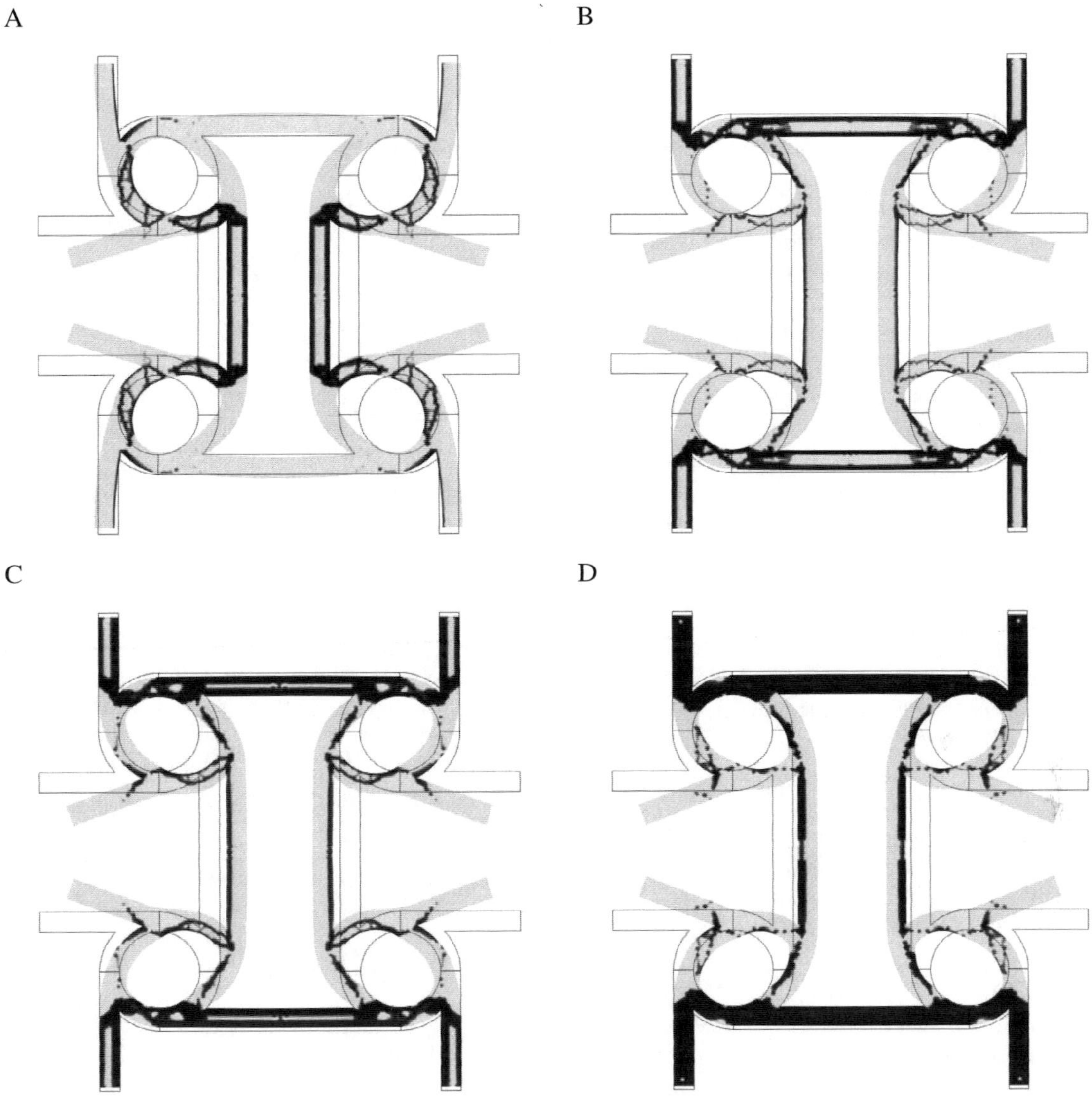

Figure 6. Deformed structures with g = 0.025 m and A_f (A) 20%, (B) 30%, (C) 40% and (D) 50%.

All presented optimized distributions of two-phase materials (Figures 2A-5A, 6) show that hard material (structural steel) represented by blue colour creates complicated shapes giving possibility to achieve anisotropic structures with characterized by great auxetic behaviour. One can find lattice-like structures created by hard material inside some parts of anti-tetra-chiral domain.

3. Conclusions

Normally, the value of the Poisson's ratio of anti-tetra-chiral structures made of one material approach to minus one and structure is isotropic. It is the lower bound for the Poisson's ratio of isotropic material. One-phase sample of considered structure is isotropic with Poisson's ratio equal -0.91099. However, the anti-tetra-chiral structure can be anisotropic if we use two materials to build the structure.

Optimisation of distribution of these materials allows to obtain structures characterized by the Poisson's ratio in the range from -4.5 to -8.7. The value depends on A_f parameter and mechanical properties of materials. The distribution of two materials causes that the whole structure is more auxetic and anisotropic. Also the distribution of reinforcing hard material inside soft matrix material in anti-tetra-chiral domain influenced mechanical properties of structure. This composite structure will exhibit great auxetic effect not known for recently built networks with anti-tetra chiral geometries of unit cells. The optimized structures may be in the future made by methods of 3-dimensional printing.

4. References

[1] Gibson L J and Ashby M F 1997 Cellular solids: structure and properties. *Cambridge University Press (2nd ed.).*

[2] Grima J N, Alderson A and Evans K E 2005 *Phys. Stat. Sol. B*, **242(3)** 561-75.

[3] Prall D and Lakes R S 1996 *Int. J. Mol. Sci.* **39(3)** 305-14.

[4] Alderson A, Alderson K L, Attard D, Evans K E, Gatt R, Grima J N and Zied K M 2010 *Comp. Sci. Tech.* **70(7)** 1042-8.

[5] Alderson A, Alderson K L, Chirima G, Ravirala N and Zied K M 2010 *Comp. Sci. Tech.* **70(7)** 1034-41.

[6] Lorato A, Innocenti P, Scarpa F, Alderson A, Alderson K L, Zied K M and Evans K E 2010 *Comp. Sci. Tech.* ***70*(7)** 1057-63.

[7] Miller W, Smith C W, Scarpa F, Evans K E 2010 *Comp. Sci. Tech.* **70(7)** 1049-56.

[8] Kolla A, Ju J, Summers J D, Fadel G and Ziegert J C 2010 *Design of Chiral Honeycomb Meso-Structures for High Shear Flexure.* In ASME 2010 International Design Engineering Technical Conferences and Computers and Information in Engineering Conference (pp. 43-49), American Society of Mechanical Engineers.

[9] Joshi H 2013 *Finite Element Analysis of effective mechanical properties, vibration and acoustic performance of auxetic chiral core sandwich structures,* Clemson University.

[10] Airoldi A, Bettini P, Panichelli P and Sala G 2015 *Phys. Stat. Sol. (b)*, **252(7)** 1446-54.

[11] Mousanezhad D, Haghpanah B, Ghosh R, Hamouda A M, Nayeb-Hashemi H and Vaziri A 2016 *Theoret. Appl. Mech. Lett.* **6(2)** 81-96.

[12] Allaire G 2002 *Shape optimization by the homogenization method* Springer Verlag, NewYork.

[13] Bendsøe M P and Sigmund O 2003 Topology optimization. Theory, Methods and Applications *Springer Verlag,* Berlin.

[14] Bendsøe M P and Sigmund O 1999 *Arch. Appl. Mech* **69** 635-54.

[15] Thomsen J 1992 *Struct. Optim.* **5** 108-15.

[16] Sigmund O 2000 *J. Mech. Phys. Solids* **48(2)** 397–428.

[17] Evans K E, Nkansah M A and Hutchinson I J 1992 *Acta Metall. Mater.* **40(9)** 2463-9.

[18] Milton G W 1992 *J. Mech. Phys. Solids* **40(5)** 1105-37.

[19] Lim T C 2012 *Phys. Stat. Sol. B* **249** 1366-72.

[20] Boucher M A, Smith C W, Scarpa F, Rajasekaran R and Evans K E 2013 *Comp. Struct.* **106** 1-14.

[21] Strek T, Jopek H and Nienartowicz M 2015 *Phys. Stat. Sol. B* **252(7)** 1540–50.

[22] Jopek H, Strek T 2015 *Phys. Stat. Sol. B*, **252(7)** 1551–58.

[23] Pozniak A A, Wojciechowski K W, Mizzi L and Grima J N 2016 *Comp. Part B-Eng* **94** 379.

[24] Strek T, Jopek H, and Fraska A 2016 *Phys. Stat. Sol. B* **10** 1359–68.

[25] Gibiansky L V and Sigmund O 2000 *J. Mech. Phys. Sol.* **48** 461-98.

[26] Czarnecki S and Wawruch P 2015 *Phys. Stat. Sol. B* **252(7)** 1620–30.

[27] Łukasiak P 2014 *Recent Adv. Comput. Mech.*, edited by: Łodygowski T, Rakowski J., Litewka P,(Taylor & Francis Group, London, 2014, pp. 213-222.

[28] Yi Y M, Park S H, Youn S K 2000 *Int. J. Sol. Struct.* **37** 4791-810.

[29] Whitty J P M., Alderson A, Myler P and Kandola B 2003 *Comp.: Part A* **34** 525–34.

[30] Schwerdtfeger J, Wein F, Leugering G, Singer R F, Körner C, Stingl M and Schury F 2011 *Adv. Mater.* **23** 2650–4.

[31] Kaminakis N T, Drosopoulos G A, and Stavroulakis G E 2014 *Arch. Appl. Mech.* **85** 1289.

[32] Strek T, Jopek H and Idczak E 2016 *Phys. Stat. Sol. B,* **253** 1387–94.
[33] Strek T, Jopek H and Wojciechowski K W 2016 *Smart Mater. Struct.* **25(5)** 054002.
[34] Svanberg K 1987 *Int. J. Numer. Meth. Eng.* **24(2)** 359–73.
[35] Svanberg K 2013, MMA and GCMMA – Fortran versions, *KTH, Royal Institute of Technology*, Stockholm.

Acknowledgments
This work was supported by grants of the Ministry of Science and Higher Education in Poland: 02/21/DS PB/3493/2017 and 02/21/DSMK/3498/2017. The simulations have been carried out at the Institute of Applied Mechanics, Poznan University of Technology.

IOP Conf. Series: Materials Science and Engineering **248** (2017) 012007 doi:10.1088/1757-899X/248/1/012007

Performance Study of Salt Cavern Air Storage Based Non-Supplementary Fired Compressed Air Energy Storage System

CHEN Xiaotao[1, 2], SONG Jie[3], LIANG Lixiao[3], SI Yang[2], WANG Le[3] and XUE Xiaodai[1, 2]

[1] Department of Electrical Engineering, Tsinghua University, 100084, Beijing, China.
[2] New Energy (photovoltaic) Industry Research Center, Qinghai University, 810036, Qinghai, China
[3] Research Institute of Global Energy Internet, Beijing, China.

Abstract. Large-scale energy storage system (ESS) plays an important role in the planning and operation of smart grid and energy internet. Compressed air energy storage (CAES) is one of promising large-scale energy storage techniques. However, the high cost of the storage of compressed air and the low capacity remain to be solved. This paper proposes a novel non-supplementary fired compressed air energy storage system (NSF-CAES) based on salt cavern air storage to address the issues of air storage and the efficiency of CAES. Operating mechanisms of the proposed NSF-CAES are analysed based on thermodynamics principle. Key factors which has impact on the system storage efficiency are thoroughly explored. The energy storage efficiency of the proposed NSF-CAES system can be improved by reducing the maximum working pressure of the salt cavern and improving inlet air pressure of the turbine. Simulation results show that the electric-to-electric conversion efficiency of the proposed NSF-CAES can reach 63.29% with a maximum salt cavern working pressure of 9.5 MPa and 9 MPa inlet air pressure of the turbine, which is higher than the current commercial CAES plants.

1. Introduction

With the pressure of environment and energy issues, renewable energy such as wind and solar have been developed rapidly [1, 2]. However, due to the intermittent and stochastic characteristic, large-scale integration of such renewable energies into power system is challengeable. Energy storage techniques provide a new insight for this phenomenon by capturing the features of load shifting, smoothing renewable energy output, and improving the safety operation of power grid [3, 4].

Compressed air energy storage (CAES) is one of the most prominent large-scale energy storage techniques [5, 6]. The CAES technique has no strict requirements on geographical condition and many researchers have been focused on the design [7, 8], control [9], and optimization [10]. The conventional CAES system is composed of compressor, air storage chamber, gas combustion, turbine, generator, etc.[11, 12]. Air storage chamber is the energy storage equipment while the compressor and the turbine are the energy conversion devices as well as the interface with power grid. The world-wide two commercial CAES power stations including the Huntorf plant and McIntosh plant are respectively built in 1978 and 1991[13, 14]. The electric-to-electric conversion efficiency of these two plants can reach 33%-46% and 54% on average, respectively. Both of them are still in operation until now [13].

The above two plants both adopted a supplementary fired CAES (SF-CAES) architecture [15]. However, it consumes fossil fuels, which brings about emission problems. Therefore, the non-supplementary fired CAES (NSF-CAES) technology is developed from the conventional SF-CAES [16]. In contrast, heat regeneration is the one that collected and stored thermal energy during

Content from this work may be used under the terms of the Creative Commons Attribution 3.0 licence. Any further distribution of this work must maintain attribution to the author(s) and the title of the work, journal citation and DOI.
Published under licence by IOP Publishing Ltd

compression, and then used to preheat inlet air of the turbine, which eliminates the dependence on fossil fuels, and enables the zero-carbon-emission during system operation in NSF-CAES.

Advanced adiabatic compressed air energy storage (AA-CAES) and TICC-500 [8] are representatives of NSF-CAES systems. By the cascade utilization of the collected heat to preheat inlet air of the turbine, its electric-to-electric conversion efficiency is expected to reach 41%. However, the high-pressured air in TICC-500 is stored in a high-pressure container, which both reduces the system capacity and results in high cost. Advanced high-pressure air storage techniques are needed for the further development of NSF-CAES technique [17].

This paper presents a novel NSF-CAES system scheme with salt cavern air storage based on our previous system TICC-500. The rest of this paper is organized as follows. Available salt cavern distribution and non-supplementary CAES system diagram based on salt cavern air storage is proposed in Section 2. Systematic thermodynamic is analysed in Section 3. Systematic performance analysis is elaborated in Section4. Conclusions are drawn in Section 5.

2. Salt cavern air storage based NSF-CAES

As indicated previously, air storage is a key factor for the implement of such compression heat feedback based NSF-CAES. As a mature technology, salt cavern has been used widely in the natural gas storage. It has merits such as large capacity, higher storage pressure, and lower construction cost, which meets the requirements of NSF-CAES. In this respect, we carried out researches on NSF-CAES based on the salt cavern air storage in this section.

2.1. System configuration

At present, the major underground gas storage methods includes depleted oil and gas reservoir, water aquifer air storages, discharge gas pit and salt cavern gas storage [19]. Salt cavern gas storage technology uses the artificial cavity in salt bed or salt dome to store high pressure gas. Compared with other existed gas storage methods, salt cavern gas storage captures the advantages of large capacity, stable reliability, low cost, good seal, and long service life. Salt cavern gas storage has become an important part of the natural gas gathering and transportation system in China.

The use of salt cavern for gas storage can date back to 1960s. In 1961, the world's first salt cavern natural gas storage station was built in Michigan. Subsequently, France, Germany, Britain and Denmark built many salt cavern gas storage stations. China also carried out the studies on salt cavern gas storage. The available salt caverns distribution in East China by 2015 illustrate in Figure 1 and corresponding available volume is shown in

Table **1**.

Figure 1. Available salt cavern distribution in East China by 2015.

Table 1. Available salt cavern volume in East China by 2015.

No.	Salt cavern name	Province	Available Volume ($\times 10^6 m^3$)
1	Jintan	Jiangsu	14.3
2	Huan'an	Jiangsu	10
3	Pingdingshan	Henan	4
4	Yingcheng	Hubei	8
5	Zhangshu	Jiangxi	10
6	Qianjiang	Hubei	4

Specially, a huge salt cavern storage system with the storage volume of $(10\text{-}15)\times 10^6 m^3$ was built in Jintan City, Jiangsu Province. This salt cavern is selected as a potential site for building a 10 MW/100 MWh NSF-CAES plant. Therefore, it is practical to develop salt cavern based NSF-CAES system in China [20].

The flow chart of NSF-CAES with 2-stage compression and 2-stage expansion architecture based on salt cavern air storage depicted in Figure 2. In this system, thermal energy storage is applied to recycle the compression heat. The compression heat stored in TES system is used to heat the inlet air of the turbine, which improves the energy storage efficiency of the system.

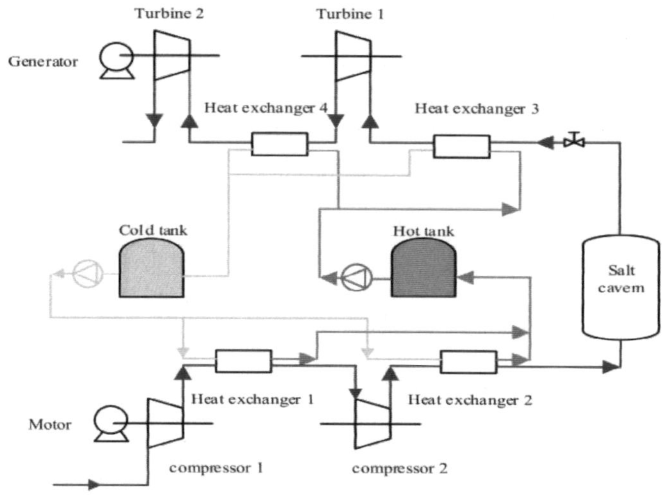

Figure 2. Diagram of salt cavern air storage based non-supplementary CAES system.

The proposed salt cavern air storage based NSF-CAES has the following advantages: First, salt cavern based air storage technique is widely used in natural gas storage fields supported by mature techniques. Second, salt cavern based air storage technique can guarantee a narrow pressure range, which is of great value for the stable and efficient operation of compressor and turbine, to yield a high-efficiency CAES. Third, salt cavern based air storage technique can save system cost, which is vital to the engineering application of this technique in smart grid and energy internet.

2.2. Operation mechanism

Operation of the proposed NSF-CAES system involves two basic processes: charging and discharging. In charging, ambient air is compressed to the high pressure by a 2-stage compressor. Then the high pressure air is stored in salt cavern. Meantime, the TES system recycles the compression heat generated with compression process. The electrical energy is converted into heat energy and potential energy of molecular [8]. While in discharging, the high pressure air is released from the salt cavern and heated by the heat energy stored in TES. The hot air expands in the first stage of the turbine and

IOP Conf. Series: Materials Science and Engineering **248** (2017) 012007 doi:10.1088/1757-899X/248/1/012007

generate electricity. After first stage expansion, both the pressure and temperature of the air decrease. The air is then heated again by the stored heat energy and expands in the second stage of the turbine to produce electricity.

The two-stage compression structure guarantees the compressor-discharged temperature less than 380℃, thus conventional industrial centrifugal compressors can satisfy the demand, which get rid of the limit of high temperature compressor. The TES system consists of cold tank, hot tank, working fluid, circulating pump, and heat exchanger. VP1, with high temperature characteristic, utilize as the heat storage medium. Maximum temperature of VP1 can reach to 398℃. To prevent the solidification of oil, the lowest temperature of heat conduction oil should higher than 30℃, and it driven by the oil-circulating pump. In heat exchanger 1 and 2, the compression heat is recycled and stored in the hot tank. In heat exchanger 3 and 4, the air is heated by high temperature oil, thus the reuse of compression heat is realized. Using the recycled heat stored in the TES to increase inlet air of the turbine temperature to 350℃, which greatly improves the system output power and the energy conversion efficiency. The pressure fluctuation of gas storage chamber is relatively small due to the huge volume of the salt cavern, so the system can operate more stable.

3. Systematic thermodynamic analysis

Thermodynamic analysis of the NSF-CAES based on salt cavern air storage is carried out in this section, following assumptions are made for the simplify of the analysis:

- The air is the ideal air, i.e., meets the ideal gas state equations;
- The mass flow rate of the compressor in charging and the mass flow rate of turbine in discharging are constant;
- The air storage chamber adopts the isothermal model, i.e., the air in air storage chamber can exchange heat sufficiently with the external environment, and the temperature of the air is equal to the ambient temperature during the whole charging and discharging process [7];
- The rated compression ratio is same in each stage, and the expansion ratio is also equal in each stage;
- The mechanical loss of the compressor, turbine and generator are considered with efficiency factor.

3.1. Compressor

The off-peak electricity, curtailed wind, solar and hydro power are used to compress the air to high pressure. In this process, with the increase of pressure in the salt cavern, the discharging pressure also increases, leading to the non-stationary operation of compressor [18]. The power consumed by compressor grows with the increase of the discharging pressure. Therefore, the first-stage compressor is working in rated operation conditions, while the second-stage is not.

During compression, the electric power consumed by i-stage compressor satisfies:

$$W_{c,i} = \frac{1}{\eta_{ci}}\frac{k}{k-1}R_g T_{ca,i}^{in} q_{m,c}\left[\left(\frac{p_{ca,i}^{out}}{p_{ca,i}^{in}}\right)^{\frac{k-1}{k}} - 1\right] \tag{1}$$

where $\eta_{c,i}$ is the adiabatic compression efficiency, k is the specific heat ratio, R_g is the gas constant, $T_{ca,i}^{in}$is the inlet air temperature, $q_{m,c}$is the mass flow rate of compressor, $p_{ca,i}^{in}$ and $p_{ca,i}^{out}$ are the inlet and outlet air pressure respectively. For simplicity, a two-stage compressor is utilized in the proposed salt cavern based NSF-CAES system. Thus, the total electricity consumption in charging can be determined by:

$$W_c = \eta_c \sum_{i=1}^{2} W_{c,i} t_c \tag{2}$$

where η_c is the efficiency of motor, t_c the compression time. The outlet air temperature of i-stage compressor is:

$$T_{ca,i}^{out} = T_{ca,i}^{in}\left\{1 + \frac{1}{\eta_{Ci}}\left[\left(\frac{p_{ca,i}^{out}}{p_{ca,i}^{in}}\right)^{\frac{k-1}{k}} - 1\right]\right\} \tag{3}$$

3.2. Salt cavern

Usually, the air in the salt cavern can exchange heat with the ambient through salt cavern surface, so the salt cavern adopts the isothermal model. The air temperature of salt cavern and the temperature of salt cavern surface keep constant during the whole charging and discharging process.

3.2.1. Air charging. The inner air pressure of salt cavern P_{cav} is equal to the high pressure air from the last stage exchanger, i.e., heat exchanger 2 in Figure 2 while the temperature is equal to the ambient $T_{cav} = T_0$. In charging, the air energy conservation equation in salt cavern is [12]

$$\mathrm{d}(mu) = h_{in}\delta m + \delta Q \tag{4}$$

In equation (4), $h_{in} = C_p T_{cav}^{in}$, $\mathrm{u} = C_V T_{cav}$, Thus:

$$\delta Q = \left(C_V T_0 - C_P T_{cav}^{in}\right)dm \tag{5}$$

Under the condition of $T_{cav}^{in} = T_0$, we have

$$\delta Q = -VdP_{cav} \tag{6}$$

The state equation for ideal air in salt cavern is

$$dP_{cav} = \frac{R_g T_0}{V}dm \tag{7}$$

Assuming that, the initial air pressure in storage room is$P_{cav,1}$. Air pressure in salt cavern increases gradually during charging. By integrating equation (7), we have the state of charge (SOC) of the salt cavern:

$$P_{cav} = P_{cav,1} + \frac{q_{m,c}R_g T_0}{V}t \tag{8}$$

3.2.2. Air discharging. Similar with equation (4), the energy conservation process in salt cavern during discharging can be illustrated as:

$$\mathrm{d}(mu) = -h_{out}\delta m + \delta Q \tag{9}$$

In equation (9), $h_{out} = C_p T_{cav}^{in}$, $\mathrm{u} = C_V T_{cav}$, Thus:

$$\delta Q = -R_g T_0 dm = -VdP_{cav} \tag{10}$$

From above analysis, heat exchange between the high pressure air in the salt cavern and the ambient is existed in charging and discharging.

3.3. Turbine and generator

To ensure a stable power output of the turbine, throttle valve is used to adjust the inlet air pressure of the turbine. Usually, the inlet air pressure of turbine is inlet air pressure of the first stage turbine, and equals to the minimum working pressure of salt cavern.

For each stage *i* of the turbine, the actual output shaft power is

$$W_{e,i} = C_p q_{m,e} T_{e,i}\eta_{e,i}\left[1 - \left(\frac{p_{ea,i}^{in}}{p_{ea,i}^{out}}\right)^{-\frac{k-1}{k}}\right] \tag{11}$$

The illustrated turbine in figure 2 adopts a two-stage architecture, thus, the actual electricity output of the whole turbine is

$$W_e = \eta_e \sum_{i=1}^{2} W_{e,i} t_e \tag{12}$$

where $\eta_{e,i}$ is the adiabatic efficiency of i-stage of turbine, η_e is the generator efficiency, q_{me} is the air mass flow rate of the turbine, $p_{ea,i}^{in}$ and $p_{ea,i}^{out}$ are the inlet and outlet air pressure of the turbine, respectively. $T_{ea,i}^{in}$ is the inlet air temperature and t_e is the turbine generation time.

3.4. Heat exchange

3.4.1. Recycling compression heat. The TES captures the function of recycling and reusing the compression heat, thus improves the system efficiency.

In charging, heat storage medium in heat exchanger 1 and 2 is heated by high pressure air from compressor, so the recycling of compression heat is fulfilled. The efficiency of heat exchanger is defined by [21].

$$\varepsilon = \frac{(c_p q_m)_1 (T_1^{in} - T_1^{out})}{(c_p q_m)_{min} (T_1^{in} - T_2^{out})} = \frac{(c_p q_m)_2 (T_2^{out} - T_2^{in})}{(c_p q_m)_{min} (T_1^{in} - T_2^{in})} \tag{13}$$

where subscript '1' stands for hot fluid, i.e., air, '2' cold fluid, i.e., heat conduction oil. Superscript 'in' is the inlet of heat exchanger, 'out' the outlet of heat exchanger.

For the i-stage cooler of adiabatic CAES system, the hot fluid inlet temperature is $T_{ca,i}^{out}$, outlet temperature is $T_{ca,i+1}^{in}$, the cold fluid inlet temperature is $T_{cw,i}^{in}$. The relationship between $T_{ca,i}^{out}$, $T_{ca,i+1}^{in}$ and $T_{cw,i}^{in}$ can be denoted as

$$T_{ca,i+1}^{in} = (1-\varepsilon) T_{ca,i}^{out} + \varepsilon T_{cw,i}^{in} \tag{14}$$

Through heat exchanger, the temperature of heat-storage medium is changed to

$$T_{cw,i}^{out} = (1-\varepsilon) T_{cw,i}^{in} + \varepsilon T_{ca,i}^{out} \tag{15}$$

The total heat absorbed by the heat-storage medium during charging is

$$Q_c = \sum_{i=1}^{2} \int_0^t \varepsilon C_p q_{m,c} (T_{ca,i}^{out} - T_{cw,i}^{in}) dt \tag{16}$$

3.4.2. Reusing Compression Heat. In discharging, the air in the heat exchanger 3 and 4 is heated by the heat storage medium, thus, the reuse of compression heat is achieved. After i-stage heat exchanger, the air temperature is

$$T_{ea,i}^{in} = (1-\varepsilon) T_{ea,i-1}^{out} + \varepsilon T_{tes} \tag{17}$$

The total heat released by the heat-storage medium during discharging is

$$Q_e = \sum_1^m \int_0^t \varepsilon C_p q_{m,e} (T_{tes} - T_{ca,i-1}^{out}) dt \tag{18}$$

3.5. Energy Storage Efficiency

The ratio between output electricity in discharging and the consumed electricity in charging is system energy storage efficiency, as shown in

$$\eta = \frac{W_e}{W_c} \tag{19}$$

Noting that, the efficiency in equation (19) is the electric-to-electric efficiency, which will be utilized to analyze the overall performance of the proposed salt cavern based NSF-CAES.

4. Systematic Performance Analysis

Under the condition of thermodynamic analysis, the performance of salt cavern based NSF-CAES including power consumption of compressor, temperature of TES, mass of heat-storage medium ,volume of salt cavern, energy storage density and system efficiency are studied with different

IOP Conf. Series: Materials Science and Engineering **248** (2017) 012007 doi:10.1088/1757-899X/248/1/012007

inlet air pressure of turbine P_{ea}^{in} and maximum working pressure of salt cavern P_{cav}^{max} . Key parameters of the proposed NSF-CAES system are shown in Table 2.

Table 2. Main parameters of the proposed NSF-CAES system.

Parameter	Capacity	Charging time	Discharge time
Value	50(MW)	8(h)	4(h)
parameter	Compressor stage	Turbine stage	Heat medium
value	2	2	VP1 conduction oil

4.1. Power Consumption of Compressor

The relationship between consumed electricity by the compressor W_c and inlet air pressure of the turbine P_{ea}^{in} and maximum working pressure of salt cavern P_{cav}^{max} is illustrated in Figure 3.

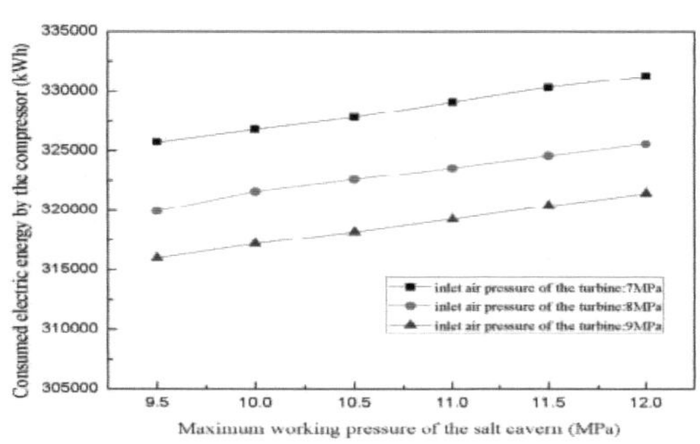

Figure 3. Influences on power consumption by the compressor.

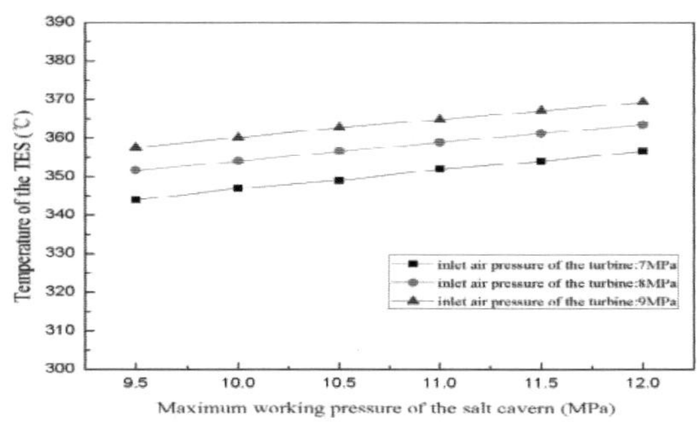

Figure 4. Influence on TES temperature.

With the growing of maximum working pressure of salt cavern P_{cav}^{max}, the compressor power consumption W_c increases under a certain inlet air pressure of turbine P_{ea}^{in}. The reason is that with the increase of P_{cav}^{max}, the discharge pressure of the compressor enhances, thus W_c gradually increases. On the other hand, the electricity consumption of compressor decreases as the turbine inlet air pressure increases under a constant working pressure of salt cavern. As the turbine inlet air pressure enhances, the work ability for per unit air improves. The total mass of air in need drops, thus the compression work goes down.

In this regard, the compressor can reach a 315.9 MWh minimum consumption under the condition of P_{ea}^{in} is 9 MPa and P_{cav}^{max} is 9.5 MPa. WhenP_{ea}^{in} is 7MPa and P_{cav}^{max} is 12 MPa, the compressor can reach a 331.2 MWh maximum consumption.

4.2. Temperature of TES

Temperature is an important index for TES system. The higher the temperature is, the higher the efficiency is. The relationship between temperature of TES and inlet air pressure of the turbine P_{ea}^{in} and the maximum working pressure of salt cavern P_{cav}^{max} is depicted in Figure 4.

At a certain inlet air pressure of the turbineP_{ea}^{in}, the temperature of TES system increase with the increase of P_{cav}^{max}. When P_{cav}^{max} increases, the discharge pressure of compressor increases and the discharge temperature of the also increases. Furthermore, the heat-storage medium can be heated to a higher temperature in the heat exchanger. Similarly, when the pressure of storage room is fixed, the increase of P_{ea}^{in} will also enhance pressure and the compressor discharge temperature.

In this regard, under the condition of P_{ea}^{in} is 7 MPa and P_{cav}^{max} is 9.5 MPa. The TES system can obtain a 344℃ lowest temperature while a 364℃ highest temperature can be reached whenP_{ea}^{in} is 9 MPa and P_{cav}^{max} is 12 MPa.

4.3. Mass of heat-storage medium

The proposed salt cavern based NSF-CAES system takes the heat conduction oil as the heat-storage medium. Mass of heat-storage medium is a key parameter for this kind of NSF-CAES system. The relationship between the mass of heat-storage medium and the inlet air pressure of the turbine P_{ea}^{in} and maximum working pressure of salt cavern P_{cav}^{max} is shown in Figure 5.

As shown in Figure 5, at a certain P_{ea}^{in} , with the raise of P_{cav}^{max}, the mass of heat storage medium needed by the TES system drops. The reason is that with the increment of P_{cav}^{max}, the temperature of heat-storage medium also increases. The specific heat capacity of oil enhances. Therefore, under a certain power generation, the quantity of heat in need drops, so does the mass of heat-storage medium. Similarly, with the increase of P_{ea}^{in}, the mass of the heat-storage needed by the system drops under a fixed operation pressure of salt cavern.

To be specific, when the inlet air pressure of turbine P_{ea}^{in} is 7 MPa and maximum working pressure of salt cavern P_{cav}^{max} is 9.5MPa, the TES system needs a 1644 t heat storage medium. While the TES system needs a 1485t minimum heat-storage medium with a 9 MPa P_{ea}^{in} and 12 MPa P_{cav}^{max}.

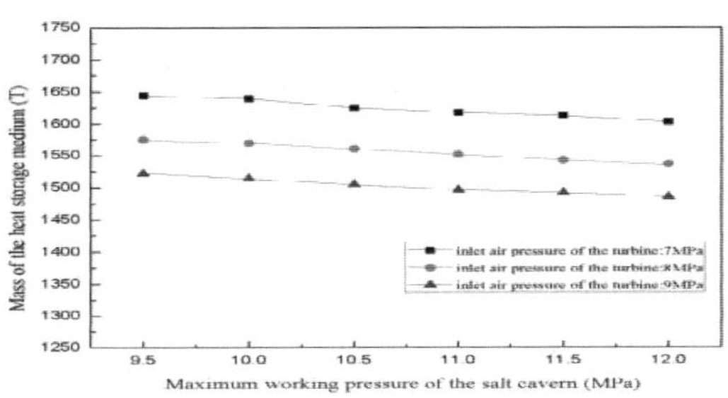

Figure 5. Influence on mass of heat storage medium.

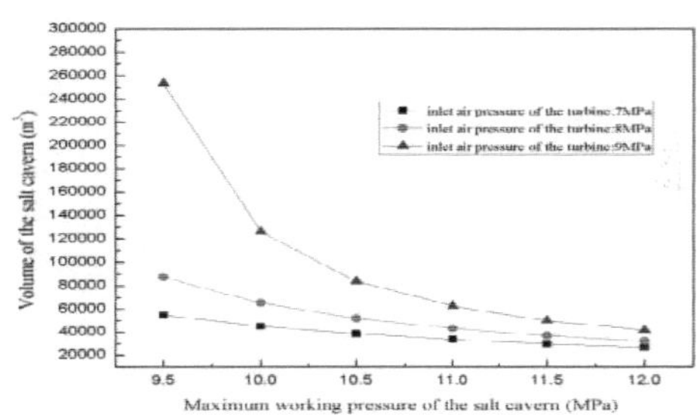

Figure 6. Influence on salt cavern volume.

4.4. Volume of salt cavern

CAES employs air as electric energy storage medium. A huge air storage space is needed to realize a high capacity ESS. As for the proposed salt cavern based NSF-CAES, the volume of salt cavern is vital to the system capacity and functionality. The relationship between volume of salt cavern and the inlet air pressure of the turbine P_{ea}^{in} and maximum working pressure of the salt cavern P_{cav}^{max} illustrates in Figure 6 shows that the volume of salt cavern decreases with the increment of P_{cav}^{max} under certainP_{ea}^{in}. Since the higher the storage room pressure is, the greater the air mass stored in per volume is. Under the same power generation, the mass of air required is the same, so the storage room that needed will drops.

When the maximum working pressure of the salt cavern is given, the volume of the salt cavern increases together with P_{ea}^{in}. This is due to the higher the turbine inlet pressure is, the smaller the available pressure interval of salt cavern is. For per unit air storage space, the mass of available air for power generation decreases. Therefore, for a given amount of power generation, the required storage space increases.

Definitely, for the proposed salt cavern based NSF-CAES with parameters in

Table **1**, the volume of salt cavern can reach the minimum value i.e. 26,500m^3, with a 7 MPa inlet air pressure of turbine and 12 MPa maximum working pressure of salt cavern. Alternatively, the volume of salt cavern can reach the maximum value, i.e. 253,000m^3, with a 9 MPa inlet air pressure of turbine and 9.5 MPa maximum working pressure of salt cavern. With different system parameters, the storage space differs greatly. The cost of storage room takes a great percentage on the whole cost. It is a key problem on how to choose adequate parameters to determine the storage space.

4.5. Energy Storage Density

Energy storage density is an important evaluation index for ESS. Take air storage chamber as an example, the quantity of electric energy that can be stored in per cubic meter of air storage space is analysed under different system parameters. The connection between the energy storage density with inlet air pressure of the turbine P_{ea}^{in} and maximum working pressure of salt cavern P_{cav}^{max} is given in Figure 7.We can learn from Figure 7 that energy storage density decreases with the increases of P_{ea}^{in} and P_{cav}^{max}. If the maximum working pressure of salt cavern P_{cav}^{max} is fixed, the available pressure interval of salt cavern goes down with the increase of inlet air pressure of turbine P_{ea}^{in}. The available air mass in unit storage space decreases, so does the power generation and energy storage density.

As for the NSF-CAES system shown in table 2. By fixing the inlet air pressure of the turbine, with increasing air pressure in storage room, the available air mass in unit storage room goes up, so does the power generation. The turbine inlet air pressure is 9 MPa, and air pressure in storage room is 9.5 MPa. The energy storage density can reach its least, i.e. 0.79kWhm^{-3}. When the turbine inlet air pressure is 7 MPa, and the air pressure in storage room is 12MPa, the energy storage density is at its most, which is 7.52kWhm^{-3}.

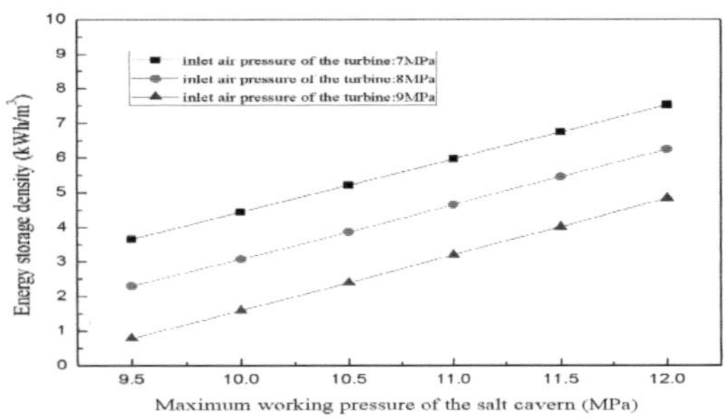

Figure 7. Influence on storage density.

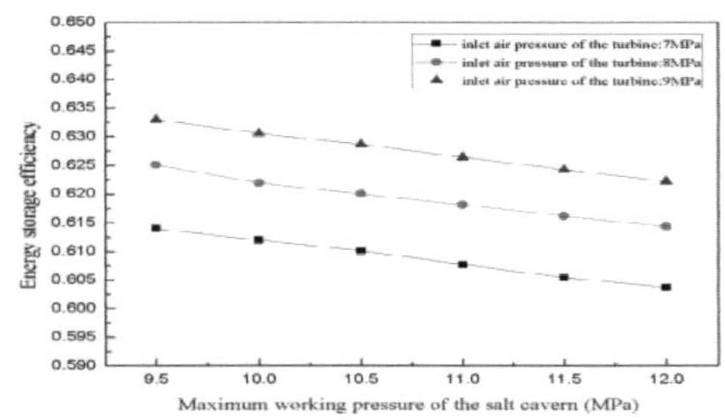

Figure 8. Influence on system efficiency.

4.6. System efficiency

Energy storage efficiency is the most important index for ESS. The influence of inlet air pressure of the turbine P_{ea}^{in} and maximum working pressure of salt cavern P_{cav}^{max} on the efficiency illustrates in Figure 8. We can draw the conclusion that the energy storage efficiency goes down as the maximum working pressure of salt cavern goes up under a fixed inlet air pressure of the turbine. When P_{cav}^{max} keeps constant, the energy storage efficiency goes up along with P_{ea}^{in}. At a given power generation, improving the turbine inlet pressure helps to reduce power demand of compressor, thus, improves the energy storage efficiency. However, if the working pressure of salt cavern goes up, the compressor power consumption will also become greater, thus, the efficiency will decrease. To achieve a higher energy storage efficiency, the inlet air pressure of turbine should be improved as much as possible, and the working pressure of salt cavern should be limited down as much as possible.

Specifically, for the proposed NSF-CAES system with parameters in table 2, the energy storage efficiency reach 63.29% with a 9 MPa inlet air pressure of turbine and 9.5 MPa maximum working pressure of salt cavern. In this system, the highest energy storage efficiency corresponds to the storage room volume of 253,000 m^3. It has been one of the key problems in system design about how to balance the energy storage efficiency and volume of salt cavern.

5. Conclusion

This paper proposes a novel CAES system based on salt cavern air storage technique and thermal energy techniques. It adopts mature technologies, with an easily applicable system architecture. Particularly, TES enables no gas combustion and zero-carbon emission in the whole energy storage process. Thermodynamic design and thermodynamic characteristics of the energy storage system are carried out.

Generally, with the increase of pressure and temperature of inlet air of turbine, power consumption of compressor decreases, and also the power loss of molten salt, so do the mass of heat conduction oil in the heat storage system. The volume of salt cavern decreases with the increase of inlet air temperature of the turbine. The energy storage density increases with the increasing turbine inlet temperature and decreasing turbine inlet pressure. The energy storage efficiency increases as the pressure and temperature of turbine inlet air. The electric-to-electric efficiency of the system can reach 63.29%, which is higher than the currently commercial CAES plants, under the condition of a 9.5 MPa salt cavern pressure, and 9 MPa inlet air pressure of turbine.

The findings in this paper are applicable to the design and construction of large-scale CAES power plants. To obtain more reasonable parameters configuration in the future, optimization and simulation are needed to realize a NSF-CAES with high energy efficiency, low construction cost, and high reliable technique.

6. References

[1] Xu J, He D and Zhao X. 2010 *Energ.* **35** 4439-44.
[2] Razykov T M, Ferekides C S, Morel D, Stefanakos E, Ullal H S and Upadhyaya H M 2011 *Sol Energy.* **85** 1580-608.
[3] Solomon A A, Kammen D M and Callaway D. 2014 *Appl. Energ.* **134** 75-89.
[4] Denholm P and Sioshansi R. 2009 *Energ. Policy.* **37** 3149-58.
[5] Ibrahim H, Ilinca A and Perron J. 2008 *Renew. Sust. Energ. Rev.* **12** 1221-50.
[6] Inage S I 2009 *Prospects for Large-Scale Energy Storage in Decarbonised Power Grids.* International Energy Agency Iea.
[7] Budt M, Wolf D, Span R and Yan J. 2016 *Appl. Energ.* **170** 250-68.
[8] Mei S W, Wang J J, Fang T, Chen L J, Xue X D, Qiang L, et al. 2015 *Sci. China Tech. Sci.* **58** 600-11.
[9] Saadat M, Shirazi F A and Li P Y. 2014 *Revenue maximization of electricity generation for a wind turbine integrated with a Compressed Air Energy Storage system.* American Control Conference - Acc 2014. pp. 1560-5.
[10] Rui L I, Chen L, Yuan T and Li C L 2016 *J. Mod. Pow. Syst. Clean Energ.* **4** 566-80.
[11] Chen L, Zheng T, Mei S, Xue X, Liu B and Lu Q. 2016 *J. Mod. Pow. Syst. Clean Energ.* **4** 1-13.
[12] Wang S, Zhang X, Yang L, Zhou Y and Wang J. 2016 *Energ.* **103** 182-91.
[13] Raju M and Khaitan S K. 2012 *Appl. Energ.* **89** 474-81.
[14] Swanekamp R. 2000 *Pow.*
[15] Luo X, Wang J, Dooner M, Clarke J and Krupke C. 2014 *Energ. Proc.* **62** 603-11.
[16] Karellas S and Tzouganatos N. 2014 *Renew. Sustain. Energ. Rev.* **29** 865-82.
[17] Hartmann N, Vöhringer O, Kruck C and Eltrop L. 2012 *Appl. Energ.* **93** 541-8.
[18] Liu B, Chen L, Mei S, Liu F, Wang J and Wang S. 2014 *J. Mod. Pow. Syst. Clean Energ.* **2** 422-30.
[19] Evans D J and Chadwick R A. 2014 *Underground gas storage* Springer Berlin Heidelberg.
[20] Chang X 2014 *Geological evaluation of underground salt in China and geological evaluation of salt hole construction* China University of Geosciences (Beijing).
[21] Fakheri A 2007 *J. Heat Transfer.* **129** 1268-76.

Acknowledgement

This work was supported in part by the National Natural Science Foundation of China (No. 36851621065, No.51567021, No.51577163) and state grid technology program (SGRI-DL-71-15-006).

IOP Conf. Series: Materials Science and Engineering **248** (2017) 012008 doi:10.1088/1757-899X/248/1/012008

Reuse of Coconut Shell, Rice Husk, and Coal Ash Blends in Geopolymer Synthesis

Tjokorde Walmiki Samadhi[1,*], **Winny Wulandari**[1], **Muhammad Iqbal Prasetyo**[1], **Muhammad Rizki Fernando**[1]

[1]Chemical Engineering Program, Faculty of Industrial Technology, Institut Teknologi Bandung, Labtek X Bldg., Jalan Ganesha 10 Bandung 40132 Indonesia

Abstract. Mixtures of biomass and coal ashes are likely to be produced in increasing volume as biomass-based energy production is gaining importance in Indonesia. This work highlights the reuse of coconut shell ash (CSA), rice husk ash (RHA), and coal fly ash (FA) for geopolymer synthesis by an activator solution containing concentrated KOH and Na2SiO3. Ash blend compositions are varied according to a simplex-centroid mixture experimental design. Activator to ash mass ratios are varied from 0.8 to 2.0, the higher value being applied for ash compositions with higher Si/Al ratio. The impact of ash blend composition on early strength is adequately modeled by an incomplete quadratic mixture model. Overall, the ashes can produce geopolymer mortars with an early strength exceeding the Indonesian SNI 15-2049-2004 standard minimum value of 2.0 MPa. Good workability of the geopolymer is indicated by their initial setting times which are longer than the minimum value of 45 mins. Geopolymers composed predominantly of RHA composition exhibit poor strength and excessive setting time. FTIR spectroscopy confirms the geopolymerization of the ashes by the shift of the Si-O-Si/Al asymmetric stretching vibrational mode. Overall, these results point to the feasibility of geopolymerization as a reuse pathway for biomass combustion waste.

1. Introduction

As Indonesia heads for an increased contribution of biomass-based energy to reach its planned national energy mix by 2025, the increased production of biomass ash as a waste of biomass thermal conversion is expected as well. Alternative reuse pathways for the ashes should help to ensure the economic sustainability of biomass-based energy.

Geopolymer is a generic term for a class of alkali-aluminosilicate solid materials produced via complex reactions between aluminosilicate solid materials and an activating solution. The activating solution is a concentrated alkali aqueous solution – typically NaOH or KOH – with or without the addition of soluble sodium silicate (Na_2SiO_3). Key feature in the geopolymeric structure is the presence of a solidified, crosslinked gel phase consisting of a random network of silica and alumina tetrahedra. The gel phase acts a continuous binder phase bridging unreacted aluminosilicate particles and aggregates into a coherent cementitious mass. Geopolymer may be applied as Portland cement substitute; when applied to concrete preparation, geopolymers exhibit early compressive strengths that are comparable to, and elastic moduli that are somewhat lower than ordinary Portland cement (OPC) concrete [1].

Geopolymer offers several environmental advantages. Unlike Portland cement which require massive amount of energy to sinter the raw minerals into clinker at temperatures in excess of 1400 °C, the aluminosilicate – alkali activator reaction may be conducted at ambient or very mildly elevated temperatures. Concrete made by using geopolymer has been argued to reduce the global warming

Content from this work may be used under the terms of the Creative Commons Attribution 3.0 licence. Any further distribution of this work must maintain attribution to the author(s) and the title of the work, journal citation and DOI.
Published under licence by IOP Publishing Ltd

potential relative to OPC concrete by approximately 45% [2]. Another advantage of geopolymer stems from the fact that various aluminosilicate waste materials have been proved to be good candidates for geopolymer synthesis. These include coal fly ash, metallurgical slag, building demolition waste, and several biomass ashes such as rice husk ash, palm oil fuel ash (POFA), and others [3-5].

This paper is a part of an ongoing study on the synthesis of geopolymers from biomass ashes that are abundant in Indonesia, with an emphasis on rice husk ash (RHA) and coconut (*Cocos nucifera*) shell ash (CSA). Coal fly ash (FA) is included, as it is expected that biomass combustion needs to incorporate the blending of coal with the biomass fuel to increase the ash fusion temperatures of the fuel, hence reducing its tendency to form slag in the combustion chamber [6]. The primary objective of this work is to determine the impact of biomass ash blend on key engineering properties of geopolymer, and the compliance of these properties to industrial standards.

Successful development of geopolymer from biomass ash is therefore of significant impact for Indonesia. Not only does it provide a pathway for better waste management of the biomass-based energy industry, it also reduces the national industrial CO_2 emission through partial substitution of the energy-intensive Portland cement.

2. Methodology

2.1. Materials

Starting materials for the biomass ashes are rice husk and coconut shell charcoal. The rice husk was obtained from a local farmers' cooperative in the greater Bandung area. The coconut shell charcoal was purchased from a local market in Bandung. The coal fly ash is an ASTM Class F fly ash produced by the Suralaya powerplant in Banten province. Based on thermogravimetric analyses of the biomass materials, they are combusted at 550 ^{o}C for 3 hours in an electric furnace for ashing. The activator solution is an aqueous solution of 10.0 M KOH with the addition of technical-grade Na_2SiO_3 at a sodium silicate to KOH solution mass ratio of 2.0. Ash characterizations include phase analyses using X-ray diffractometry (XRD) and oxide composition determination using X-ray fluorescence (XRF).

2.2. Experimental Design

Key engineering properties examined in this work are early strength as measured by compressive strength after 7-day setting of geopolymer-sand mortars, and initial setting time of geopolymer paste.

To prepare the geopolymer paste, dry ash blends are mixed with the activator solution in a low-speed planetary mixer until a smooth paste is obtained (approximately 2 minutes). The activator solution to ash mass ratios are varied from 0.2 to 2.0, with higher values assigned to higher ash Si/Al ratios. Ternary ash blend compositions consisting of RHA, CSA, and FA are varied according to a conventional simplex centroid mixture experimental design. Statistical analysis of the measured data is undertaken by the ANOVA method using the Minitab software package.

For initial setting time measurement, the pastes are directly hand-shaped after mixing into balls for testing using a Vicat apparatus. For the compressive strength measurement, the paste is further blended with a graded sand at a sand to geopolymer mass ratio of 2.75. The mortar is then cast into a 5x5x5 cm steel mold, tapped by hand, and transferred to an electric oven for a quick curing at 60 ^{o}C for 24 hours. The filled molds are then taken out of the oven, the mortars removed from the mold, and left to stand at ambient conditions for 6 additional days. The 7-day compressive strength is then measured using a universal uniaxial load tester.

3. Results and Discussion

The oxide compositions of the RHA, CSA, and FA used to formulate the ash blends for geopolymer synthesis are presented in table 1. Figure 1 compiles the X-ray diffractograms of the three ashes. RHA is characterized by a very high SiO_2 content of 89.6 %-mass and Si/Al molar ratio of 544, and a largely amorphous structure. A diffuse hump in the diffractogram in the vicinity of $2\theta = 21^{o}$ is a fingerprint of a random network structure consisting predominantly of SiO_4 tetrahedra. The CSA is characterized by a high alkali content (particularly K_2O), relatively low Al_2O_3 and SiO_2 contents, and Si/Al molar ratio of 0.49. Crystalline phases observed in CSA include quartz, sylvite (KCl), and calcite ($CaCO_3$). Oxide composition of the Suralaya powerplant FA indicates that it belongs to the ASTM Class F category. The Si/Al molar ratio is 2.04, and the observed crystalline phases are quartz and hematite. The presence of these crystalline phases in biomass ashes has been reported in the literature [7].

Table 1. Oxide compositions of RHA, CSA, and FA.

Ash	%-mass of oxide component								
	Al_2O_3	SiO_2	CaO	Fe_2O_3	MgO	Na_2O	K_2O	P_2O_5	SO_3
RHA	0.14	89.6	1.14	0.18	0.37	0.14	5.26	1.57	1.29
CSA	5.71	3.28	13.6	13.6	15.4	5.89	30.4	6.93	3.83
FA	21.1	50.7	15.8	3.71	2.29	1.24	2.09	0.52	1.91

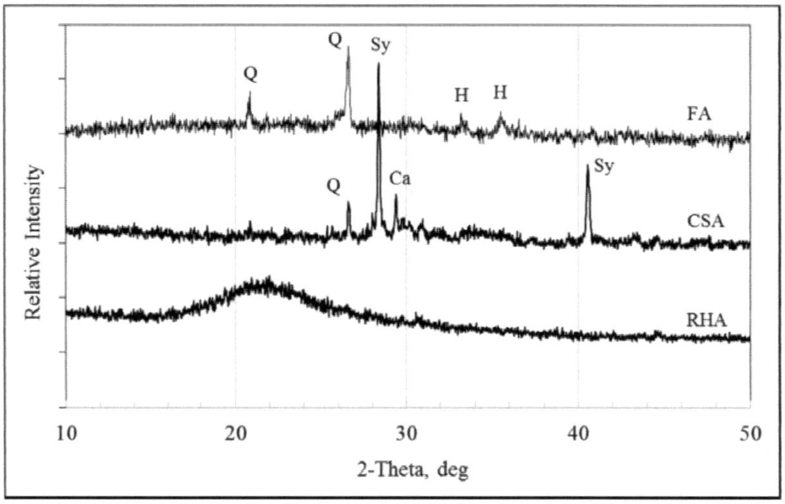

Figure 1. X-ray diffractograms of rice husk ash (RHA), coconut shell ash (CSA), and coal fly ash (FA) (Q = quartz, H = hematite, Sy = sylvite, Ca = calcite).

Table 2 compiles the early strength measurement results of the geopolymer mortars after a quick curing at 60 °C for 24 hours followed by setting at ambient conditions for 6 days. The mean strengths vary from 1.40 to 15.7 MPa, with the lowest strength exhibited by the RHA-only ash composition, and the highest by the FA-only composition. Compared to the SNI 15-2049-2004 Indonesian national standard Portland cement mortar minimum early strength value of 2.0 MPa, RHA-only geopolymer formulation is unfavorable, while the CSA-only composition produces an acceptable 3.15 MPa early strength. Most (5 out of 7) ash blend compositions in this work produces satistfactory early strength with respect to the SNI industrial standard. As a comparison, the early strength of the Portland cement mortar specimens lie approximately in the middle of the geopolymer mortar strength range.

The decrease of geopolymer strength with increasing Si/Al molar ratio at Si/Al above 1.90 has been reported in the literature. The RHA has Si/Al ratio that is two orders of magnitude higher than CSA and FA. It has been argued that at high Si/Al ratios, the geopolymer mechanical strength decreases with increasing Si/Al due to the higher content of unreacted aluminosilicate raw materials [8].

To quantify the effect of ash blend composition on the geopolymer mortar early strength, the data in table 2 are analyzed using the Minitab statistical software package. Table 3 summarizes the ANOVA data treatment results using a special cubic mixture model. At a confidence level of 95%, the table indicates that the RHA*CSA quadratic blending and the RHA*CSA*FA cubic blending terms are not statistically significant. By dropping these terms, the geopolymer mortar early strength is adequately represented by an incomplete quadratic mixture model as described in equation 1.

IOP Conf. Series: Materials Science and Engineering **248** (2017) 012008 doi:10.1088/1757-899X/248/1/012008

$$CS(MPa) = 1.40RHA + 3.15CSA + 15.65FA - 20.30(RHA \cdot FA) - 7.60(CSA \cdot FA) \quad (1)$$

Table 2. Early strength (7-day compressive strength) measurement results of the geopolymer mortar specimens.

Run no.	Ash blend composition (mass fractions)			Compressive strength (Mpa)			Initial setting time (mins)
	RHA	CSA	FA	Rep. I	Rep. II	Mean	
1	1.0	0	0	1.4	1.4	1.40	> 300
2	0	1.0	0	3.0	3.3	3.15	90.0
3	0	0	1.0	16.5	14.8	15.7	51.8
4	0.5	0.5	0	2.1	1.6	1.85	> 300
5	0.5	0	0.5	3.5	3.4	3.45	> 300
6	0	0.5	0.5	6.7	8.3	7.50	57.5
7	0.33	0.33	0.33	2.6	2.3	2.45	271.7
8	Portland cement			6.2	5.9	6.05	

Table 3. ANOVA treatment results of the geopolymer mortar early strength data.

Source	Deg. of freedom	Sum of squares	Mean squares	F	P-value
Linear	2	241.58	120.79	287.11	0.000
Quadratic					
RHA*CSA	1	0.241	0.241	0.57	0.474
RHA*FA	1	34.34	34.34	81.63	0.000
CSA*FA	1	4.81	4.81	11.44	0.012
Special cubic					
RHA*CSA*FA	1	1.21	1.21	2.88	0.133

Table 2 also compiles the initial setting times of the geopolymer paste specimens measured by the standard Vicat apparatus method. The SNI 15-2049-2004 Indonesian national standard calls for a minimum initial setting time for Type I Portland cement of 45 minutes. The initial setting time is a measure of the workability of the cement paste, in which longer setting time translates to better workability. Data in table 2 suggest that the geopolymers in general exhibit good workability. However, RHA-containing formulations exhibit excessively long setting time, in accordance with their low early strengths. Of the three ashes, FA produces the shortest setting time. The presence of excessively long setting times of RHA-containing formulations prevent the ANOVA treatment of the data.

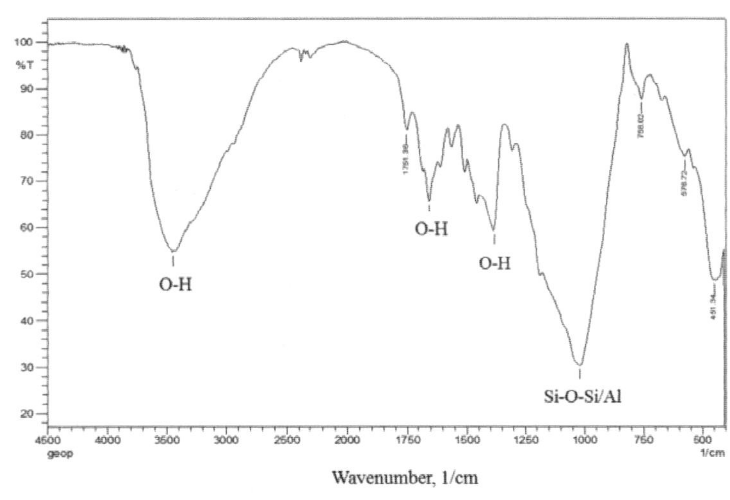

IOP Conf. Series: Materials Science and Engineering **248** (2017) 012008 doi:10.1088/1757-899X/248/1/012008

Figure 2. FTIR spectra of the RHA-CSA-FA ternary ash blend geopolymer paste.

To verify the reactivity of the biomass ashes to geopolymerization, chemical bonds in the geopolymer paste specimens are identified by FTIR spectroscopy. Figure 2 presents the FTIR spectra of the ternary ash blend (RHA-CSA-FA) geopolymer. The silanol group asymmetric stretching vibrations at 1018 cm-1 is shifted with respect to its position in the spectra of unreacted aluminosilicate geopolymer precursor materials such as fly ash and metakaolin, which typically lie in the 1070 – 1080 cm-1 range. This shift is an indication of the formation of the geopolymer gel phase [9, 10], thereby confirming the reactivity of the RHA and CSA biomass ashes to geopolymerization.

4. Conclusions

The technical feasibility of the reuse of rice husk ash (RHA) and coconut shell ash (CSA) for geopolymer synthesis has been demonstrated, with and without the addition of ASTM Class F coal fly ash (FA). A shift in the ash blend composition towards RHA reduces the geopolymer compressive strength, while a shift towards FA increases the strength. RHA-only geopolymer does not comply with the Indonesian industrial standard for early strength, and produces excessively long initial setting time, while CSA-only geopolymer produces acceptable said properties. Low strength and excessive setting time at high ash Si/Al molar ratio is likely to be due to higher content of unreacted aluminosilicate raw materials. The presence of FA in the ash blend can ensure that the strength and setting time are satisfactory. Biomass ash blends reactivity towards geopolymerization has been proved by the shift in the silanol group stretching vibrational mode, which is an IR-absorption fingerprint of the geopolymeric gel phase.

5. References

[1] Noushini A, Aslani F, Castel A, Gilbert R I, Uy B and Foster S 2016 *Cem. Concr. Res.* **73** 136-46.
[2] Habert G, d'Espinose J B and Roussel N 2011 *J. Clean. Prod.* **19** 1229-38.
[3] Xu H and van Deventer J S J 2003 *Ind. Eng. Chem.* **42** 1698-706.
[4] Mijarsh M J A, Johari M A M and Ahmad Z A 2014 *Constr. Build. Mater.* **52** 473-81.
[5] Detphan S and Chindaprasirt P 2009 *Int. J. Min. Met. Mater.* **16** 720-6.
[6] Samadhi T W, Wulandari W, Gumilar, F G and Ramadani K F 2016 *Key Eng. Mat.* **709** 38-41.
[7] Vassilev S V, Baxter D, Andersen L K and Vassileva C G 2013 *Fuel* **105** 40-76.
[8] Duxson P, Mallicoat S W, Lukey G C, Kriven W M, van Deventer J S J 2007 *Coll. Surf. A* **292** 8-20.
[9] Belmokhtar N, Ammari M, Brigui J and Allal L B 2017 *Constr. Build. Mater.* **146** 621-9.
[10] Singh G V P B and Subramaniam K V L 2017 *Cem. Concr. Comp.* **81** 122-32.

Acknowledgment

This research is funded by the Republic of Indonesia Ministry of Research, Technology, and Higher Education via the Higher Education Distinguished Research 2017 (PUPT DIKTI 2017) grant program.

IOP Conf. Series: Materials Science and Engineering **248** (2017) 012009 doi:10.1088/1757-899X/248/1/012009

On the Active and Passive Flow Separation Control Techniques over Airfoils

Tohid Moghaddam[1,*] and Nafiseh Banazadeh Neishabouri[2]

[1]Department of Mechanical Engineering, K. N. Toosi University of Technology, Iran
[2]Department of Mechanical Engineering, University of Tulsa, Tulsa, OK 74104 USA

Abstract. In the present work, recent advances in the field of the active and passive flow separation control, particularly blowing and suction flow control techniques, applied on the common airfoils are briefly reviewed. This broad research area has remained the point of interest for many years as it is applicable to various applications. The suction and blowing flow control methods, among other methods, are more technically feasible and market ready techniques. It is well established that the uniform and/or oscillatory blowing and suction flow control mechanisms significantly improve the lift-to-drag ratio, and further, postpone the boundary layer separation as well as the stall. The oscillatory blowing and suction flow control, however, is more efficient compared to the uniform one. A wide range of parameters is involved in controlling the behavior of a blowing and/or suction flow control, including the location, length, and angle of the jet slots. The oscillation range of the jet slot is another substantial parameter.

1. Introduction

Potential flow theory describes various phenomena in the realm of fluid mechanics, but it is not adequate for most boundary layer cases, e.g. flow field around an airfoil. The boundary layer over a rigid surface keeps growing as long as the pressure gradient remains zero along the surface. For an adverse pressure, the thickness of boundary layer increases drastically. This adverse pressure and shear (viscous) forces reduce momentum in the boundary layer, and if any of these will be effective along a considerable length of the surface, the growth of boundary layer is halted. This is called separation. If separation over an airfoil can be avoided, the boundary layer remains thin and the pressure drop in the trailing edge would be prevented, and consequently, the drag force decreases to a minimum value [1].

A significant number of recent studies have been focused on the effects of the boundary layer on the lift and drag forces (e.g. [2-4]) since the existence of the boundary layer produces multiple designing and engineering challenges, particularly in the aerodynamic field [5-7]. Methods developed for the boundary layer manipulation are termed as the boundary layer separation control or flow separation control [8]. The ultimate goal of the boundary layer control over an airfoil is to increase the lift and decrease the drag, and therefore, to enhance the airfoil performance by increasing the lift-to-drag ratio along with stall-delay.

In the current paper, we provide a short review of recent studies associated with the blowing and suction flow separation control techniques over finite and infinite wings. The paper starts with an introduction to the passive and active flow control techniques and continues with the latest contributions in the field. It wraps up by discussing possible further research areas in the flow control field and ends with presenting some ongoing thoughts.

Content from this work may be used under the terms of the Creative Commons Attribution 3.0 licence. Any further distribution of this work must maintain attribution to the author(s) and the title of the work, journal citation and DOI.
Published under licence by IOP Publishing Ltd

2. Boundary layer control methods

Controlling the behavior of mechanical systems is an integral concept that inevitably arises in entire engineering research fields from dynamics and robotics [9-11] to fluid mechanics [12-14] and turbulent boundary layers [4-7]. Boundary layer can be controlled through a passive method, which not needs a controller, and/or an active method, which requires a controller and controlling sensors [8]. In some literature, boundary layer management or flow management are used instead of the boundary layer control (e.g. [15]). As schematically shown in figure 1, the active flow control can be divided into two categories, predetermined and reactive. Predetermined active flow control includes uniform and non-uniform energy expenditure, regardless of the state of the flow. Reactive flow control is a particular active flow control in which the power expenditure of controllers is regularly tuned by measured parameters via in situ sensors [8, 16-17].

The classification of boundary layer control methods into active and passive categories is one of the most common schemes. Vortex generator, distributed roughness, streamlining, and uniform blowing and suction are among various devices that are employing for the passive flow control technique. On the other hand, some of the active flow control methods are heating wall, movement of surface elements, oscillatory blowing and suction, and synthetic jets.

2.1. Blowing and suction

Suction is the act of extracting fluid from the boundary layer before the fluid separates upstream. Blowing, on the other hand, can be defined as entering fluid into the boundary layer in order to increase its momentum. The blowing and suction flow control can be accomplished by using a uniform or an oscillatory (periodic) process. In the case of uniform blowing, the boundary layer momentum is increased by the injection of a high-energy fluid adjacent to the surface. If the fluid is injected tangentially, the boundary layer momentum will be increased directly. In the normal blowing, however, the fluid is injected perpendicular to the surface. In the case of oscillatory blowing and suction, periodical blowing (suction) to (from) the surface is applied which is more effective than the uniform blowing and suction. One way of achieving the oscillatory blowing and suction is by using a synthetic jet with zero mass flux as illustrated in figure 2 [18].

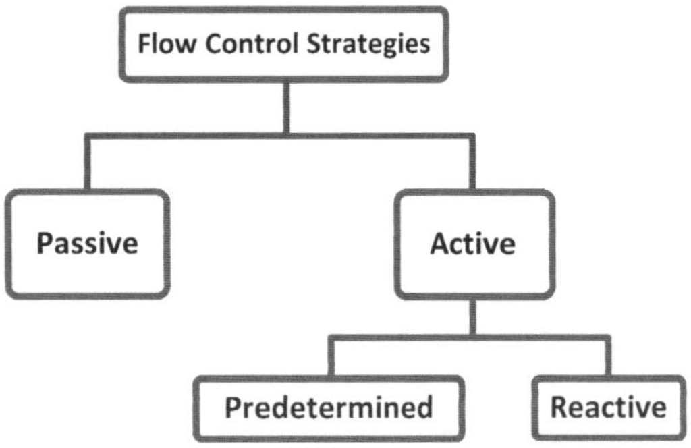

Figure 1. Classification of flow control strategies.

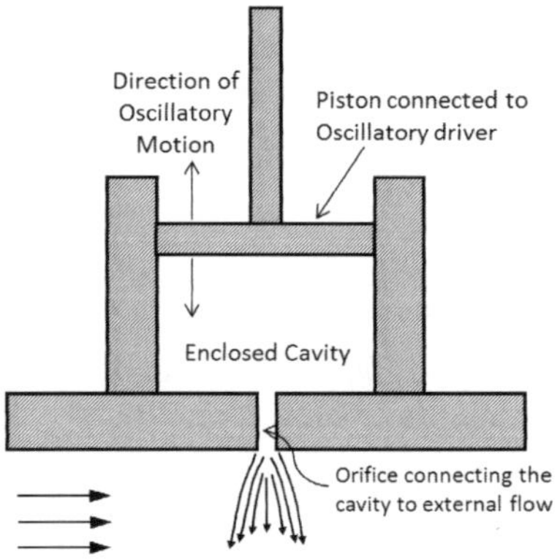

Figure 2. Synthetic jets with zero mass flux.

3. Recent contributions

Many investigations have been recently conducted on the active flow separation control, either experimentally or computationally. The fundamentals of the active flow control using synthetic jets over a NACA 0015 airfoil were studied by Parthasarathy and Das [19]. The chord Reynolds number was considered to be 896,000 at the angle of attack of 20 degrees wherein the flow was separated. The analysis of the phase-averaged pressure and streamlines clarified that the synthetic jet is strongly effective in increasing the lift coefficient, which is more pronounced for larger jet amplitudes and at smaller angles of jet injection. In a computational study, Montazer et al. [20] investigated the fundamental parameters of a synthetic jet actuator for controlling the flow separation around a NACA

0015 airfoil. By applying the response surface optimization method, it was presented that the synthetic jet is more effective for post-stall angles of attack, which can enhance the lift-to-drag ratio up to 66%. Feero et al. [21] experimentally examined the effectiveness of the active flow control through the synthetic jet technique to re-attach the stalled flow on a NACA 0025 airfoil at a relatively low Reynolds number. It was observed that the boundary layer was reattached in both low- and high-frequency actuation in a time-averaged sense.

In recent years, the study of the synthetic jets as an active flow control method over wind and water turbines have become more popular. Velasco et al. [22] simulated the flow field around a cross-flow vertical-axis water turbine favoring synthetic jets. It was revealed that the net torque and power output of the turbine can be considerably improved by employing synthetic jets over the extrados and intrados of the airfoil, which leads to an increase in the global efficiency of the turbine. The effects of the synthetic jet were also studied numerically by Moshfeghi and Hur [23] where they consider a finite span wing with the S809 airfoil at different angles of attack ranging from 0 to 25 degrees. The results for an attached flow showed that the synthetic jet causes early separation, and consequently, declines the lift coefficient, while for a separated flow the lift coefficient considerably enhances. They also observed that the lift does not follow a distinguished pattern for varying jet angles at a large angle of attack.

The passive flow control has also been the point of interest in many researches. Yousefi and Saleh [24] conducted a three-dimensional suction flow control study to investigate the aerodynamic characteristics of a wing with a NACA 0012 airfoil section. Two different suction slot distributions were considered at the leading edge of the wing, center suction and tip suction. The results indicated that the lift-to-drag ratio increases by the suction jet length for both center suction and tip suction distributions. The center suction distribution, however, was more effective in the enhancement of the overall aerodynamic performance. It was concluded that the tip suction is more efficient when the jet length is less than half the wingspan, and consistently, the center suction provides better aerodynamic performance when the jet length is greater than half the wingspan. The effects of center suction and tip suction on the streamlines behind the wing are shown in figures 3 and 4 [24], respectively. In another study, Yousefi et al. [25] and Yousefi and Saleh [17] investigated the effects of suction and blowing jet width on the aerodynamic characteristics of a NACA 0012 airfoil. It was found that the jet width of about 3.5% ~ 4% of the chord length is optimum for tangential blowing, while for perpendicular blowing smaller jet widths are more efficient. Consistent with the literature, the suction jet width of 2.5% of the chord length is the value that maximizes the lift-to-drag ratio.

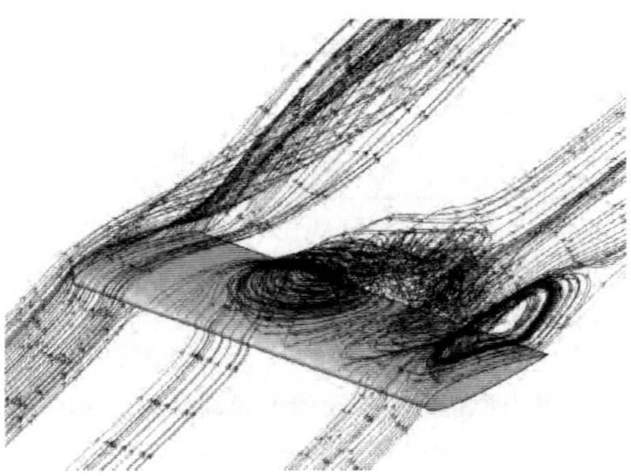

(a)

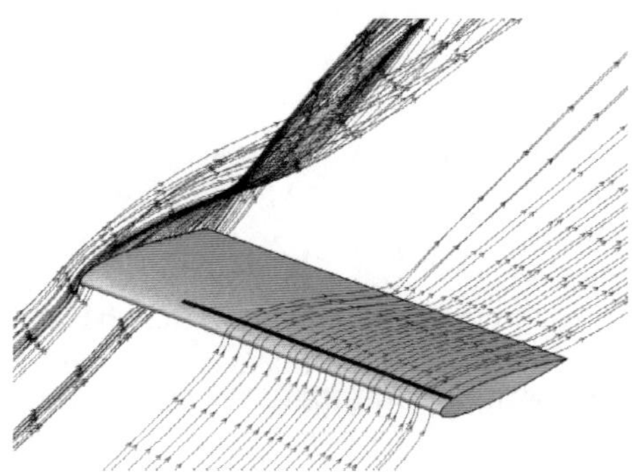

(b)

Figure 3. The effects of suction jet length on streamlines over the finite wing at angle of attack of 18° for center suction, (a) no control, (b) suction jet length of 1.5C [24].

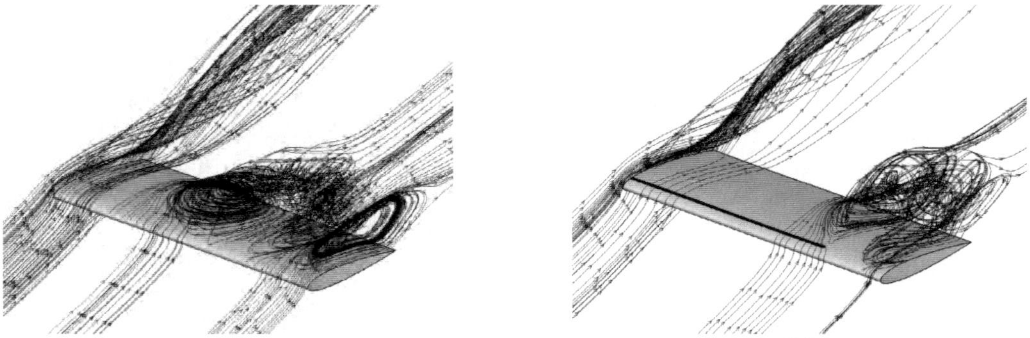

(a) (b)

Figure 4. The effect of suction jet length on streamlines over the finite wing at angle of attack 18° for tip suction, (a) no control, (b) suction jet length of 1.5C [24].

The effects of steady blowing flow separation control technique were analyzed by Svorcan et al. [26]. Three different flow fields were considered including subsonic flow past an Aerospatiale A airfoil, transonic flow past a NACA 0012 airfoil, and transonic flow in linear compressor/turbine cascade. It was found that lift coefficients and lift-to-drag ratios are improved for all controlled cases. The steady blowing was also found to be of significant importance in controlling the boundary layer separation in the study of the flow field around NACA 0012 and LA203A airfoils performed by James et al. [27]. A secondary blowing jet was positioned on the suction side of the airfoils at 60% of the chord length. The results of this study demonstrated that the secondary blowing causes the separation point to move downstream, the lift coefficient to increase, and the stall of airfoils to delay. Zhang et al. [28] computationally studied the suction flow control over a NACA 0012 airfoil at different angles of attack. The results indicated that the flow separation point travels toward the trailing edge by increasing the suction coefficient as defined in [29], and the vortex shedding can be fully suppressed when the suction coefficient is greater than 0.01 for a steady flow. Furthermore, the increase in suction coefficient resulted in the increase of lift-to-drag ratio. Hafien et al. [30] considered self-adapting flexible flaps on the upper side of the NACA 0012 airfoil. The results indicated that the size and intensity of the vortex-shedding can be considerably reduced, which explains the increase of the lift-to-drag ratio.

Consistent with the active flow control, the research field of the aerodynamic performance of wind and water turbines undergoes many studies investigating the effects of passive flow control techniques. Shehata et al. [31] investigated the effects of the suction flow control on the stall angle of Wells turbine. It was found that with the use of suction flow separation control, the torque coefficient can be increased by more than 17% before stall and by almost 40% within the stall regime. They further extended their work by employing multi-suction slots over the turbine blade [32] to enhance the performance of the turbine in the stall regime. It was observed that the airfoil with three suction slots located at 40%, 55% and 90% of the chord from the leading edge can improve the torque coefficient by 26.7% and 51% before and after the stall, respectively. In another study, Moshfeghi et al. [33] further investigated the influences of a passive flow control method on the aerodynamic performance of S809 airfoil by splitting the airfoil along the span. The split location was considered as the integral parameter and the effects of split location on power coefficient, low-speed shaft torque, and flow patterns were numerically studied. It was deduced that the torque is quite sensitive to the split location and injected flow angle for an attached flow. This sensitivity significantly declines for a highly separated flow in the stall region. Moreover, for a partially separated flow, splitting may impose either positive or negative effect on the torque generation.

4. Closing remarks

In the current paper, we briefly reviewed the latest advances in the field of active and passive flow control where we focused on the blowing, suction, and synthetic jets flow control methods. From a flow control point of view, blowing is fundamentally different from the suction. The primary advantage of suction is that it maintains the boundary layer attached to the surface, and consequently, drag decreases. The blowing, on the other hand, is much more sensitive, and in most cases, it reduces the aerodynamic performance. The blowing at the leading edge increases the lift by producing additional vortices but causes pressure to significantly improve such that the boundary layer detaches from the surface, and therefore, drag increases. The blowing at the trailing edge, however, can enhance the lift, particularly by using smaller oscillation range. Hence, perpendicular suction at the leading edge and tangential blowing at the trailing edge are the best locations for suction and blowing jet slots [17, 24-25].

The use of synthetic jets not only stabilizes the boundary layer but also increases the mixing level between the inner and outer portions of the boundary layer by adding (extracting) momentum to (from) the boundary layer. The optimum jet width is determined to be 2.5 percent of the chord length [25] in order to maximize the aerodynamic performance. However, the effects of the jet dissipate before it reaches the trailing edge of the airfoil if the jet width is not adequate. This is due to the strong momentum of the free-stream flow.

References

[1] Schlichting H 1968 *Boundary Layer Theory* (New York: McGraw-Hill).
[2] Homann H, Bec J and Grauer R 2013 *J. Fluid Mech.* **721** 155-79.
[3] Ma D, Zhao Y, Qiao Y and Li G 2015 *Chinese J. Aeronaut.* **28** 1003-15.
[4] Ma D, Li G, Yang M and Wang S 2017 *Proc. Inst. Mech. Eng., Part G: J. Aeros. Eng.* 0954410017694057.
[5] Yousefi K, Saleh R and Zahedi P 2013 *Int. J. Eng.* **7** 10-24.
[6] Yousefi K, Saleh R and Zahedi P 2013 *Int. J. Mater. Mech. Manuf.* **1** 136-42.
[7] You D and Moin P 2008 *J. Fluid. Struct.* **24** 1349-57.
[8] Gad-el-hak M 2000 *Control Flow: Passive, Active and Reactive Flow Management* (United Kingdom: Cambridge University Press).
[9] Korayem M H, Zehfroosh A, Tourajizadeh H and Manteghi S 2014 *Nonlin. Dyn.* **76** 1423-41.
[10] Korayem M H, Tourajizadeh H, Zehfroosh A and Korayem A H 2014 *Nonlin. Dyn.* **78** 1515-43.
[11] Korayem M H, Tourajizadeh H, Zehfroosh A and Korayem A H 2015 *Robotica* **33** 933-52.
[12] Bozorgnezhad A, Shams M, Kanani H, Hasheminasab M and Ahmadi G 2016 *Int. J. Hydr. Energ.* **41** 19164-81.
[13] Bozorgnezhad A, Shams M, Kanani H, Hasheminasab M and Ahmadi G 2015 *Int. J. Hydr. Energ.* **40** 2808-32.
[14] Bozorgnezhad A, Shams M, Kanani H and Hasheminasab M 2015 *J. Dispers. Sci. Tech.* **36** 1190-7.
[15] Fiedler H E and Fernholz H H 1990 *Prog. Aeros. Sci.* **27** 305-87.
[16] Gad-el-Hak M 1996 *Appl. Mech. Rev.* **49** 365-80.
[17] Yousefi K and Saleh R 2014 *J. Theoret. Appl. Mech.* **52** 165-79.
[18] Rosas C R 2005 *Numerical Simulation of Flow Separation Control by Oscillatory Fluid Injection* (Doctoral dissertation, Texas A&M University) pp 146.
[19] Parthasarathy T and Das S P 2017 *J. Phys.: Conf. Ser.* **822** 012009.
[20] Montazer E, Mirzaei M, Salami E, Ward T A, Romli F I and Kazi S N 2016 *IOP Conf. Ser.: Mater. Sci. Eng.* **152** 012023.
[21] Feero M A, Lavoie P and Sullivan P E 2016 *ASME 2016 Fluids Engineering Division Summer Meeting (Washington, DC, USA, 10-14 July 2016)* pp V01AT13A011.
[22] Velasco D, López O and Laín S 2017 *Renew. Energ.* **113** 129-40.
[23] Moshfeghi M and Hur N 2017 *J. Mech. Sci. Tech.* **31** 1233-40.
[24] Yousefi K and Saleh R 2015 *Meccanica* **50** 1481-94.
[25] Yousefi K, Saleh R and Zahedi P 2014 *J. Mech. Sci. Tech.* **28** 1297-310.
[26] Svorcan J M, Fotev V G, Petrović N B and Stupar S N 2016 *Therm. Sci.* **20** Suppl. 6 S1-S14.
[27] James S E, Suryan A, Sebastian J J, Mohan A and Kim H D 2017 *Comput. Fluid.* In Press.
[28] Zhang W, Zhang Z, Chen Z and Tang Q 2017 *European J. Mech. -B/Fluid.* **65** 88-97.
[29] Wahidi R and Bridges D H 2012 *AIAA J.* **50** 523-39.

[30] Hafien C, Bourehla A and Bouzaien M 2016 *J. Appl. Fluid Mech.* **9** 2569-80.
[31] Shehata A S, Xiao Q, Saqr K M, Naguib A and Alexander D 2017 *Ocean Eng.* **136** 31-53.
[32] Shehata A S, Xiao Q, Selim M M, Elbatran A H and Alexander D 2017 *Renew. Energ.* **113** 369-92.
[33] Moshfeghi M, Shams S and Hur N 2017 *J. Wind Eng. Ind. Aerodyn.* **167** 148-59.

Edge Trimming Analysis for Surface Quality of Hybrid Composite - CFRP/Al2024

J Ismail[1], S B Mohamed[1], M Mohamad[1], W N F Mohamad[1], A S Mohamed[2] and A Mohd[3]

[1]Universiti Sultan Zainal Abidin, 21300 Kuala Terengganu, Terengganu, Malaysia
[2]Universiti Teknikal Malaysia Melaka, 76100 Durian Tunggal, Melaka, Malaysia
[3]TATi University College, 24000 Kemaman, Terengganu, Malaysia

Abstract. The application of hybrid composite materials has increased due to their strength and light weight ratio. Edge trimming of these materials was extremely difficult due to the anisotropic and non-homogeneous structures of CFRP and ductile nature of aluminium. This research was focused on edge trimming process via down milling operation. Three cutting parameters were examined namely spindle speed, feed rate and depth of cut. The extension of two level full factorial design, Centre Composite Design (CCD) was used to plan systematic experimental methodology. The analysis of the influence and the interaction factors associated to surface quality was studied. The objective was to obtain machined surface quality of CFRP/Al2024 between 0.4 µm to 0.6 µm. The depth of cut was the most significant factor for Al2014 and meanwhile the spindle speed and feed rate were significant factors of CFRP. The validation experiment was conducted at optimum level of recommendation control factors setting to compare the deviation of predicted value from actual/measured value. Surface roughness of CFRP was found to be at 0.594 micron at the setting of spindle speed, 11750 rpm, feed rate, 750 mm/min and depth of cut, 0.255 mm. For Al2024, the surface roughness was found to be at 0.32 micron.

1. Introduction

The use of hybrid composite materials has significantly increased especially in aerospace, naval and automotive industries due to their strength and light weight ratio [1, 2]. Trimming the edges is the first step to be carried out after the composite parts are taken out from the mold. This is normally done by conventional machining [3, 4] with the aims to satisfy the surface integrity requirement. Hence, investigations on the significant factors of machining parameters had been performed by some researchers which associated with machined surface of drilling [5], turning [6], milling [7] and trimming [8-11].

This research project described the edges trimming process of CFRP/Al2024 via down milling operation. The objectives of this study is to optimize cutting parameters in achieving surface quality of below 1 μm. The down milling method for trimming operation where the feeding direction of specimen is in the direction of cutting tool rotation was carried out to achieve optimum surface quality. This is to prevent the cutting edges form rubbing and burnishing against the surface of the specimen before engaging the cut. Every trimming operation was carried out randomly to avoid any systematic errors.

Content from this work may be used under the terms of the Creative Commons Attribution 3.0 licence. Any further distribution of this work must maintain attribution to the author(s) and the title of the work, journal citation and DOI.
Published under licence by IOP Publishing Ltd

2. Material preparation and experimental set-up

The stacking of 16 layers of Carbon Fibre Reinforced Plastic (CFRP) with 4.0mm thickness and Aluminium alloy 2024 (Al2024) with a thickness of 1.2 mm was used. The edge trimming operation was carried out on Mori Seiki NV4000 DCG CNC milling machine using KENNAMETAL Burr-Style Routers Helix 15^0, KCN05 of 10mm in diameter as shown in figure 1. The value of control parameters was set according to the table 1.

Figure 1. Experimental set-up for edge trimming process applying down milling approach.

Table 1. Cutting Parameters (control factors) for Trimming Process.

Cutting Parameters (Control Factors)	Unit	Min Value	Max Value
Spindle Speed	N (rpm)	1000	13500
Feed Rate	F_r (mm/min)	500	1000
Depth of Cut	d_c (mm)	0.01	0.5

The surface roughness for both CFRP and Al2024 was measured separately using Mitutoyo SJ-301. Five readings of 100 mm length machined surface were measured, averaged and recorded in order to encounter for unknown error. Because of composition of the composite specimen consists of metal and non-metal, therefore the machined surfaces were treated and recorded as two separate responses. By doing this, it is more convenient to observe and analyze each response individually. Moreover understanding and identifying of the interaction and the influential factors can be easily done statistically hence, the optimal formulation of the controlled factors can be predicted and formulated.

3. Results and discussion

Design-Expert V9 software was used to predict the cutting parameters. The results of the previous two level full factorial design ($2^3 = 8$ experiments) showed a significant lack of fit in the design. It was found that addition of curvature terms in the model would improve the model significantly. The software suggested to do replication of the current design (8 experiments x 2 = 16 experiments). Adding of 3 centre points was recommended in order to detect the present of any curvature in the response in the experiment. Thus, the total of 19 experiments were conducted using two level full factorial design. Table 2 shows the surface roughness (Ra) of CFRP and Al2024.

The analysis of variance (ANOVA) was performed to analyse the factors that affect the surface roughness of CFRP and Al2024 separately due to their different responses. Statistically, the quadratic model F-value of 6.29 implied that the model was significant. P-value is less than 0.05, which indicated that the model was significant. In this case A, B, and BC were significant model terms. The analysis showed that spindle speed and feed rate have a significant effect on surface roughness of CFRP. The depth of cut has no significant effect, but there was a significant interaction effect between feed rate and depth of cut as shown in table 3.

Table 2. Experimental Results of Surface Roughness (Ra) of CFRP and Al2024.

Experiment No.	Spindle Speed, N (rpm)	Feed Rate, f_r (mm/min)	Depth of Cut, d_c (mm)	Surface Roughness, Ra (μm) CFRP	Al2024
1	13500	1000	0.5	0.56	0.34
2	10000	500	0.5	0.66	0.35
3	10000	1000	0.5	0.73	0.54
4	11750	750	0.25	0.49	0.22
5	13500	500	0.01	0.45	0.34
6	10000	1000	0.01	1.04	0.21
7	13500	500	0.5	0.42	0.73
8	10000	500	0.01	0.44	0.28
9	11750	750	0.25	0.49	0.22
10	13500	1000	0.01	0.89	2.02
11	11750	750	0.25	0.50	0.20
12	11750	1000	0.25	1.03	0.55
13	11750	750	0.25	0.75	0.37
14	11750	750	0.5	0.54	0.26
15	11750	500	0.25	0.50	0.42
16	11750	750	0.01	0.43	0.47
17	13500	750	0.25	0.53	0.80
18	11750	750	0.25	0.74	0.36
19	10000	750	0.25	0.82	0.74

Table 3. ANOVA for Surface Roughness of CFRP.

Source of Variation	Sum of Square (SS)	Degree of Freedom (d.o.f)	Mean Square = SS/d.o.f	F Value	P Value	Remarks
Model	0.063	9	0.070	6.29	0.0182	Significant
A-Spindle Speed, N	0.091	1	0.091	8.19	0.0287	Significant
B-Feed Rate, f_r	0.25	1	0.25	22.72	0.0031	Significant
C-Depth of Cut, d_c	0.055	1	0.055	4.99	0.0669	
AB	1.513E-003	1	1.513E-003	0.14	0.7243	
AC	0.017	1	0.017	1.49	0.7243	
BC	0.11	1	0.11	9.74	0.0206	Significant
Residual	0.066	6	0.011			
Lack of Fit	0.066	4	0.017	497.29	0.0020	Significant

Table 4 showed the quality of surface roughness of Al2024 was influenced by the interaction between A-spindle speed and B-feed rate. The results of the model P-Value of 0.0384 and lack of fit of 0.0088 are significant. C-depth of cut of 0.0114 has a significant effect on surface roughness of Al2024 but A-spindle speed of 0.1068 and B-feed rate of 0.2531 are not significant.

Table 4. ANOVA for Surface Roughness of Al2024.

Source of Variation	Sum of Square (SS)	Degree of Freedom (d.o.f)	Mean Square = SS/d.o.f	F Value	P Value	Remarks
Model	0.42	9	0.046	4.60	0.0384	Significant
A-Spindle Speed, N	0.036	1	0.036	3.59	0.1068	
B-Feed Rate, f_r	0.016	1	0.016	1.60	0.2531	
C-Depth of Cut, d_c	0.13	1	0.13	12.94	0.0114	Significant
AB	0.061	1	0.061	6.11	0.0484	Significant
AC	3.966E-004	1	3.966E-004	0.040	0.8489	
BC	4.561E-004	1	4.561E-004	0.045	0.8382	
Residual	0.060	6	0.010			
Lack of Fit	0.060	4	0.015	112.32	0.0088	Significant

4. Optimization of Edge Trimming Parameters

Design expert allows criteria setting for all variables including factors and desired responses. So in this case the researcher left the setting of spindle speed, feed rate and depth of cut without any changes. The target value response criteria was set to below 1 μm for both. Since the responses of the experiment involved two responses, thus the setting of numerical criteria optimization was done separately.

For CFRP, the setting for lower and upper bound were 0.42 and 1.04 respectively. Meanwhile, for Al2024, the settings for the lower and upper bound were 0.2 and 0.8 respectively. These limits indicated that it was the most desirable range to achieve the targeted value of less than 1 μm. The design expert provided optimal designs with the different desirability factors ranging the least desirable to the most desirable (0 to 1). The goal of optimization is to obtain a secure set of conditions that will satisfy all the goals. The program randomly picked a set of conditions from which to begin its search for desirable outcomes. After grinding through cycles of optimization, Design Expert sorted the results as shown in table 5.

Table 5. Experimental Results of Surface Roughness (Ra) of CFRP and Al2024.

No	Spindle speed	Feed Rate	Depth of cut	Ra CFRP	Ra Al2024	Desirability
1	11750.001	750.000	0.255	0.607	0.335	0.771
2	11775.572	750.011	0.255	0.606	0.336	0.769
3	11749.985	758.312	0.255	0.612	0.334	0767
4	11717.068	750.005	0.254	0.609	0.334	0.767
5	11749.981	762.408	0.255	0.615	0.333	0.767

The validation experiment conducted at optimum level of predicted control factors setting. The deviation of predicted value from the actual (measured) value is 3.11% for CFRP and 3.43% of Al2024.

5. Conclusion

Edges trimming analysis for surface quality of hybrid composite using Design of Experiment methodology, two level full factorial design has beed carried out. The influence and interaction of cutting parameters namely, spindle speed, feed rate and depth of cut on machined surface quality of

CFRP/Al2024 has been experimentally studied via trimming process. The surface quality of CFRP was influenced by the interaction effect between feed rate and depth of cut. The spindle speed and feed rate were the most significant parameters on surface quality of Al2024.

6. References

[1] Andrei G, Dima D & Andrei L 2006 *J. Optoelectr. Adv. Mater.* **8(2)** 726–30.
[2] Mangalgiri P D 1999 *Compos. Mater. Aeros. Appl.* **22(3)** 657–64.
[3] Davim J P and Reis P 2005 *J. Mater. Proc. Tech.* **160(2)** 160–67.
[4] Kasim M S, Che Haron C H, Jaharah A, Ghani Hadi M A, Izamshah R, Anand T J S and Mohamed S B 2016 *Trans. IMF* **94(4)** 175-81.
[5] Krishnaraj V, Zitoune R and Collombet F 2012 *Usak Univ. J. Mater. Sci.* **2** 95–109.
[6] Surinder K, Meenu Satsangi P S and Sardana H K 2012 *Indian J. Eng. Mater. Sci.* **19** 163–74.
[7] Selvam M D, Karuppusami G and Dawood A K S 2012 *Eng. Sci. Tech.: Int. J.* **2(4)** 544–8.
[8] Mohamed S B, Mohamad W N F W, Kasim M S, Ibrahim Z and Musanih R 2015 *Appl. Mech. Mater. J.***789-790** 105-10.
[9] Mohamed S B, Mohamad W N F, Muhamad M, Ismail J, Yew B S, Mohd A and Musanih M R 2016 *Mater. Sci. Forum* **863** 111-115.
[10] Agarwal N 2012 *MIT Int. J. Mech. Eng.* **2(1)** 55–61.
[11] Haddad M, Zitoune R, Eyma F and Castanié B 2013 *Int. J. Mach. Mach. Mater.* **13(2-3)** http://doiorg/101504/IJMMM2013053229

Acknowledgement
The authors would like to thank Universiti Sultan Zainal Abidin for funding of this research project (UniSZA/2015/DPU/29) and TATi University College for allowing us to utilize equipment and laboratory.

IOP Conf. Series: Materials Science and Engineering **248** (2017) 012011 doi:10.1088/1757-899X/248/1/012011

Preparation and thermal properties of graphene nanosheet/Ti composites

Z Y Zhang[1, 2, 3], Y Zhu[1, 2] and Y L Liang[1, 2*]
[1]College of Materials and Metallurgy, Guizhou University, Guiyang 550025, China;
[2]Key Laboratory for material structure and strength of Guizhou province, Guiyang 550025, Guizhou, China;
[3]School of Electronic and Inform Engineering, Anshun University, 516000, Anshun, Guizhou, China.

Abstract. Fully dense graphene nanosheet (GNS)/Ti composites with homogeneously distributed GNSs of thicknesses ranging from 5 to 20 nm have been fabricated from ball milled expanded graphite and Ti by hot pressing sintering (HPS). The phase and morphology of as-synthesized GNS–Ti composites are characterized by FESEM and TEM. This new composite outperforms most of carbon nanotube/Ti in thermal conductivity. The authors focused, in this study, on finding the reason of the increasement of thermal conductivity in the as-synthesized GNS–Ti composites. In addition, graphene with titanium substrate has been clearly shown and their growth model from experiment data of thermal conductivity given at the same time.

1. Introduction

Graphene has a high thermal conductivity of 5000W.m^{-1}.K^{-1}[1], so it can be used as an ideal thermal performance reinforcement for composite materials. While graphene nanosheets (GNSs) composed of a few graphene layers possess properties similar to that of single-layer graphene but are much easier to produce and handle [2, 3], GNS is a good candidate for the reinforcement of metal matrix to enhance the thermal conductivity [4]. However, pure titanium has poor thermal performance, which greatly limit its applications. In order to enhance the thermal conductivity of pure titanium, we used GNS as a reinforcement to fabricate the GNS/Ti composites. To ensure the as-prepared GNS/Ti composites have a good thermally structure, we employed ball milling technology to disperse the GNS and titanium powders. Then the GNS/Ti powders mixtures have been obtained through the above way. Finally, the dense GNS/Ti bulk composites were prepared by hot pressing sintering (HPS) using the GNS/Ti powders mixtures. In this paper, we got a clear GNS-Ti interface. Relative to pure titanium, the thermal conductivity of the composite are improved obviously. At the same time, thermal conductivity enhancement mechanism has also been a more reasonable explanation.

2. Experimental details

Firstly, GNSs (99% purity) were added to absolute ethyl alcohol and subjected to an ultrasonic dispersion for 40 min until the solution became brown in color, implying that the graphene was exfoliated into several-layered or even single-layered nanosheets. The spherical Ti powders (100 g, 45 μm in diameter, 99% purity) were transformed into 10-μm-thick Ti flakes through ball milling in an attritor at 325 rpm. Then, the refined Ti flakes were added to the solution containing graphene to form a powder slurry. The mixed slurry was mechanically stirred until its color changed from brown to transparent, before being filtered and rinsed with absolute ethyl alcohol to obtain GNSs–Ti composite powders.

Content from this work may be used under the terms of the Creative Commons Attribution 3.0 licence. Any further distribution of this work must maintain attribution to the author(s) and the title of the work, journal citation and DOI.
Published under licence by IOP Publishing Ltd

Secondly, samples of the GNS/Ti powder were loaded into a 20mm inner diameter and 100mm long graphite die. A sheet of graphite paper was placed between the punch and the powder as well as between the die and the powder for easy removal. The bulk composites were prepared using HPS apparatus. The samples were sintered in a vacuum (residual cell pressure $\leq 3.2\times10^{-2}$Pa).In the sitering process, the heating rate was set up to 10℃/min. An uniaxial pressure of 50MPa was applied from 1000℃ upwards and maintained during the dwell at 1300℃ for 2hours.The final product took the form of cylindyical samples with a height of 5mm and diameter of 25mm.Five compositions of powder mixtures were prepared with 0.5%, 1%, 3%, 5%, 10% (vol%) of GNS.

Thirdly, density measurements were conducted using Archimedes's method. Thermal diffusivities of the samples are measured by a light flash apparatus (LFA 447 NanoFlash). The morphology and microsture of the samples were characterized by field emission scanning electron microscopy (FESEM) (SUPRA40-41-90) and transmission electron microscopy (TEM). And the thin foils, which were further mechanically grinded to below 80 mm in thickness from about 200 mm and ion thinning, were examined in a Tecnai G2 F20 S-TWIN (200 kV) high-resolution transmission electron micro-scopy (TEM) to observe the boundary between graphene and Ti matrix.

3. Results and discussion

3.1 Microstructure of the as-prepared samples

The image of expanded graphite presents a loose structure containing many flake (figure 1). Magnified image shows the thickness of an individual flake is about 100 nm (figure 1), which indicates expanded graphite is a good start for fabricating GNSs compared with natural flake graphite.

Figure 1. Images of original material.

The SEM observations of the composites reveal no obvious defects in the specimens and the micrographs were featureless, as shown in figure 2, confirming their fully dense structure. To evaluate the distribution of GNSs in the Ti matrix, fracture surface of the composite containing 3 vol.% GNSs was investigated (figure 2b). By comparing with the pure Ti (figure 2a), the growth of the grain are found to be restrained remarkably with the presence of GNSs. Moreover, it is shown that the grain sizes of Ti are homogeneous, which verify the good dispersion of GNSs in the matrix (figure 2c). Compared with the pure Ti, in GNS/Ti composite the phenomena of transgranular fractures increase, suggesting the high strength of GNSs. The SEM images show no thin GNSs (below 20nm) but a few pulled-out sheets dispersed in the matrix (marked by yellow rings in figure 2b), but most of these pulled-out sheets are as thick as 50nm (figure 2d). This phenomenon can be explained as follows: when the sheets are thin (less than 20nm) and vertical to the fracture surface, they are easy to rupture under the shearing stress. Besides, when the thin sheets are located on the fracture surface, they cannot be identified in a SEM image due to the small thickness of GNSs. On the contrary, when we get thick GNSs in the grain boundary, they are strong enough to bear the shearing stress and expose themselves when fractures occur.

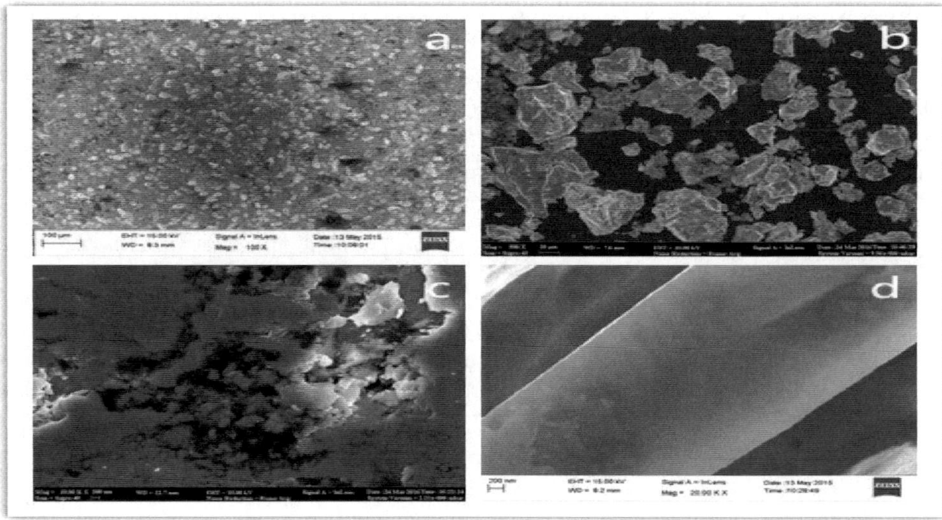

Figure 2. SEM images of fractured surfaces of (a) pure Ti and (b-d)GNS/Ti composite containing 1vol% GNSs;(c) and (d) are magnified parts of (b).

To investigate the morphology of as-prepared composites,FESEM and TEM measurements were carried out. Figure 3 shows the FESEM, TEM, HRTEM images and the SAED pattern of GNS/Ti composite. As shown in figure 3(a), it can be seen that a large amount of Ti nanoparticles are loaded on the surface of GNS h a crumpled and rippled structure of GNS-Ti composite. The spherical Ti nanoparticles are attached on GNSs and are wrapped by GNSs, which could be beneficial to prevent Ti from agglomerating and enables a good dispersion over the graphene sheets, as shown in figure 3(b). Further characterization of the microstructure of the composites was carried out by TEM. Figure 3 are representative TEM and HRTEM images of a composite containing 5 vol.% GNSs. The image reveals that the GNSs semi-wrap around the Ti grains and connect with each other to form a network structure (figure 3a). It is observed that more than half of the GNSs are below 10nm in thickness and sheets with the thick- ness of more than 20 nm have rarely been observed. Some of the GNSs are as thin as 5 nm, which means it comprises of only seven layers of graphene, as shown in figure 3c. It can also be seen that only very small fraction of overlap exist in the Ti grain boundary (figure 3b), indicating the effectiveness and success of our strategy in preparing and dispersing of GNSs.

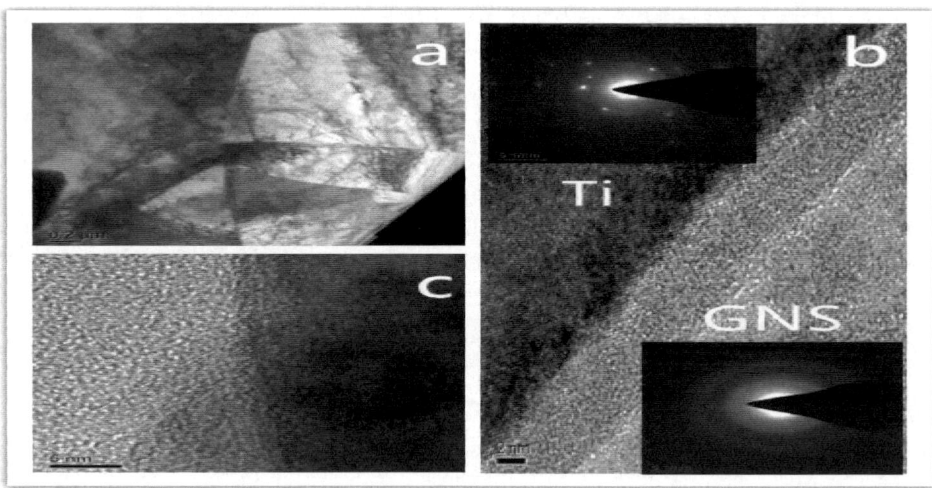

Figure. 3 TEM and HRTEM images of a GNS/Ti composite containing 5 vol.% GNSs: (a) GNSs surrounding Ti nanoparticles; (b) the interface of GNSs between Ti nano-particles and the SAED pattern of GNS/Ti composite.(c) GNSs with a thickness of 5 nm.

There are two key factors of preparing and dispersing GNSs well in our method. The first one is the usage of planetary ball mill. The possibility of preparing tailored carbon nanomaterials by mechanical deformation has been widely investigated before [5], and the mostly used experimental method is mechanical milling in a ball milling device. The two most common types of ball milling equipments, planetary and vibratory mills, have been both experimentally tested. In this study we exploit planetary ball mill as the main implement to exfoliate graphite into GNSs because the planetary mill favors the cleavage of the particles [5]. One of the effective methods for preparing graphene is called laborious drawing or micromechanical exfoliation [6].

The other important factor is the selection of dispersal media. It is well-known that the presence of a liquid in the milling process leads to lubrication, moderates the violent shocks and avoids large agglomerates of the particles. For instance, Inam et al. have reported dimethylformamide as an effective dispersant for making CNT/ceramic composites. Furthermore, it has been reported recently that graphite [7] could be exfoliated in certain solvents to achieve defect-free monolayer.

In order to obtain fully dense composites and retain nano-crystalline Ti grain size without damaging the GNSs during sintering, HPS was employed. The relative density of HPSed samples is all above 94%.

3.2. Thermal properties of the as-prepared samples

3.2.1 Model. Thermal diffusivities of the samples are measured by a light flash apparatus (LFA 447 NanoFlash). The details of the laser flash measurement procedures have been reported elsewhere [8]. The thermal conductivity (λ) was calculated from the thermal diffusivity (α) using the equation (1).

$$\lambda = \rho\alpha C_p \tag{1}$$

where ρ and Cp are the density and specific heat of the composites, respectively. Density was calculated from sample weight and dimensions. Heat capacity was tested by a differential scanning calorimeter (TA Instruments, 2920) with a temperature increase rate of 10℃/min from room temperature to 250 ℃.

Through the calculation of formula (1), the data of GNS/Ti's thermal conductivity are shown in figure 4. At the same time ,we obtain that thermal conductivities of the composite matrix and GNS are 14.82 and 2985.42$Wm^{-1}K^{-1}$ respectively This behavior can be described an effective medium theory [9] reported by Nan et al. The effective thermal conductive of the composite (λ_e) is given by the equation (2).

$$\frac{\lambda_e}{\lambda_m} = 1 + \frac{(V_f a)}{3} \frac{\lambda_f / \lambda_m}{a + \left(2R_k \lambda_f / d \right)} \tag{2}$$

where λ_m and λ_f are thermal conductivities of the composite matrix and GNS, respectively; V_f, d and a are the volume fraction, diameter and aspect ratio of GNS, respectively; and R_k is the thermal interface resistance. For a perfect interface without any thermal resistance (R_k=0) and a>1, the thermal conductivity enhancement is further simplified as:

$$\frac{\lambda_e}{\lambda_m} = 1 + \frac{V_f \lambda_f}{3\lambda_m} \tag{3}$$

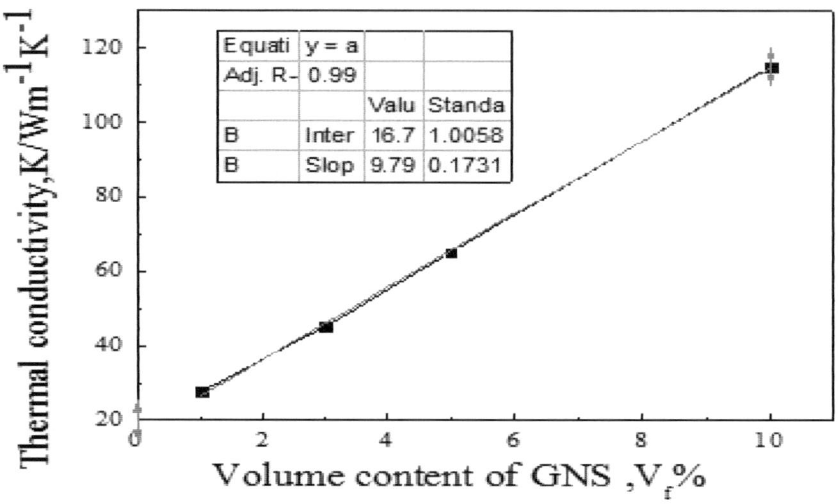

Figure 4. Thermal conductivity vs GNS content for GNS/Ti composites.

Figure 4 shows the plots of thermal conductivity vs GNS content for GNS/Ti composites. This result show that the incorporation of graphene sheet into metals can significantly improve their thermal conductivity. Two beneficial factors for the effective thermal conductivity improvement achieved are the homogeneous dispersion of GNSs in the GNSs/Ti matrix and the low thermal boundary resistance (TBR) at GNS-Ti interface. The GNS clusters or stacks cause the phenomenon of reciprocal phonon vector, which acts like a heat reservoir and restrict the heat flow diffusion [10, 11], while a homogeneous GNS dispersion effectively increases the heat flow region and enhance the diffusion of phonon in the composites.

Through mathematical calculation, we can draw that the rule of thermal conductivity in the prepared-samples can accord with the law of the third formula .At the same time, we can roll out the thermal interface resistance may be attributed to the zero. As shown in the figure 4(b) this conclusion proves that the composite interface bonding is good.

3.2.2 Theory. GNS in composite structures were placed at grain boundaries in agreement with their positions in the sintered mixtures of titanium and GNS powders. According to the rule of mixture, thermal expansion of composites constitutes of proportional contributions from both titanium and graphene according to their volume percentages. Therefore, the good approximations of experimental values and rule of mixture predictions are representative of good interface bonding between the two phases. Because titanium and graphene flakes are both soft in nature, the two phases can expand freely along the directions parallel to the basal planes. Since the expansion of titanium and graphene is additive by proportion, the shear stress at the titanium-GNS interfaces must also be minimal with-out restraining either phase to expand.

4. Conclusions

GNSs/Ti composites were successfully fabricated by HPS. In the fabrication process the graphene reinforcement was applied in the form of graphene /alcoholcolloid for a safer and simpler manufacturing process. The GNSs after ball milling are 5-20 nm in thickness and its composite interface bonding integrates well with Ti matrix. This simple and effective process suggested that preparation of GNSs /Ti MMCs with enhanced thermal conductivity may provide benefits in the development of compact and light-weight heat exchangers. The suggested process may be applied to the sheet material, which will be formed into a plate or cylindrical shape heat exchanging unit.

IOP Conf. Series: Materials Science and Engineering **248** (2017) 012011 doi:10.1088/1757-899X/248/1/012011

5. Reference

[1] Beall G W. 2014 *Method and system for producing graphene and functionalized graphene*: US, US8865307.
[2] Wang J, Li Z, Fan G, et al. 2012 *Scrip. Mater.* **66(8)** 594–7.
[3] Fan Y, Wang L, Li J, et al. 2010 Int. Conf. Geosci. Remote Sens. **6(2)** 190-3.
[4] Stankovich S, Dikin D A, Dommett G H B, et al. 2011 *Nature* **442(7100)** 282-6.
[5] Janot R, Guerard D 2002 *Carbon* **40(15)** 2887–96.
[6] Novoselov K S, Jiang D, Schedin F, Booth T J, Khotkevich V V, Morozov S V, et al. 2005 *Proc. Natl. Acad. Sci. USA* **102(30)** 10451–3.
[7] Inam F, Yan H, Reece M J, Peijs T. 2008 *Nanotech.* **19(19)** 195710.
[8] Hernandez Y, Nicolosi V, Lotya M, et al. 2008 *Nat. Nanotech.* **3(9)** 563–8.
[9] Tjong S C. 2013 *Mater. Sci. Eng. R Rep.* **74(10)** 281-350.
[10] Li D, Müller M B, Gilje S, et al. 2008 *Nat. Nanotech.* **3(2)** 101-5.
[11] Chen L Y, Konishi H, Fehrenbacher A, et al. 2012 *Scrip. Mater.* **67(67)** 29-32.

Acknowledgment

The financial support of Scientific and technological cooperation of major projects of Guizhou Province (No.[2014]6013) and Scientific is gratefully acknowledged.

IOP Conf. Series: Materials Science and Engineering **248** (2017) 012012 doi:10.1088/1757-899X/248/1/012012

Parameter optimization for selective laser melting of TiAl6V4 alloy by CO_2 laser

R M Baitimerov[1], P A Lykov[1], L V Radionova[1], E V Safonov[2]

[1]South Ural State University, Lenin avenue 74, Chelyabinsk, Russian Federation
[2]Moscow State University of Mechanical Engineering, Bolshaya Semenovskaya 38, Moscow, Russian Federation

Abstract. TiAl6V4 alloy is one of the widely used materials in powder bed fusion additive manufacturing technologies. In recent years selective laser melting (SLM) of TiAl6V4 alloy by fiber laser has been well studied, but SLM by CO_2-lasers has not. SLM of TiAl6V4 powder by CO_2-laser was studied in this paper. Nine 10x10x10 mm cubic specimens were fabricated using different SLM process parameters. All of the fabricated specimens have a good dense structure and a good surface finish quality without dimensional distortion. The lowest porosity that was achieved was about 0.5%.

1. Introduction

Selective laser melting (SLM) is powder bed additive manufacturing technique that makes it possible to fabricate near net shape parts from metal powder. Now SLM is widely used in aerospace industry [1-3] and medicine [4, 5] for production of parts with complex shapes such as fuel injectors, cooling combustion chambers, and surgical and dental implants [6-8]. TiAl6V4 is one of the most used materials for SLM fabrication due to its high strength, corrosion resistance, low density, and high biocompatibility [9-11]. In recent years, SLM of TiAl6V4 alloy by fiber laser has been well studied [12-16], but SLM by CO_2-lasers has not. Laser energy absorption by metal powder material is strongly wavelength dependent, for titanium alloy absorptance of CO_2-laser energy is lower than for fiber laser [17]. The aim of this study is determine of SLM parameters for the fabrication of high density TiAl6V4 alloy parts by CO_2-laser.

2. Experimental setup

The Sinterstation® Pro DM125 SLM system from 3D Systems was used within this investigation for TiAl6V4 porosity optimization. The Sinterstation® Pro DM125 SLM system is equipped with a 200 W CO_2-laser with 35 μm beam diameter on the powder surface, an automatic powder spreading system, an inert argon gas protection system, and a powder bed preheating system.

Nine 10x10x10 mm cubic specimens were manufactured for porosity optimization in an argon atmosphere by using different process parameters. Point distance (PD), hatch space (HS), and layer thickness were fixed to 50 μm, the oxygen content in building chamber was limited to 500 ppm, and powder bed preheating was not used. Laser power (P) and exposure time (ET) values are presented in table 1. All specimens were fabricated using an "alternating" scanning strategy, wherein each new layer is rotated by 90° in relation to the preceding layer [18].

Content from this work may be used under the terms of the Creative Commons Attribution 3.0 licence. Any further distribution of this work must maintain attribution to the author(s) and the title of the work, journal citation and DOI.
Published under licence by IOP Publishing Ltd

IOP Conf. Series: Materials Science and Engineering **248** (2017) 012012 doi:10.1088/1757-899X/248/1/012012

Table 1. Parameters of SLM.

№	1	2	3	4	5	6	7	8	9
P, (W)	200	200	200	150	150	150	100	100	100
ET, (μs)	100	50	150	100	50	150	100	50	150

TiAl6V4 powder with 20-63 μm particle size from MTT Technologies GmbH was used within this investigation. The mean particle size of TiAl6V4 powder is d_{50}=44μm. The mean ISO Roundness of powder particles by volume is 63.74%. Optical granulomorphometer Occhio 500nano was used for determination of powder particle size and ISO Roundness. The flowability of powder is 19 s/50 g, the apparent density is 2482 kg/m^3.

An SEM image of TiAl6V4 powder is presented in figure 1. As can be seen, most powder particles have a regular spherical morphology and there are some irregular particles formed in atomization.

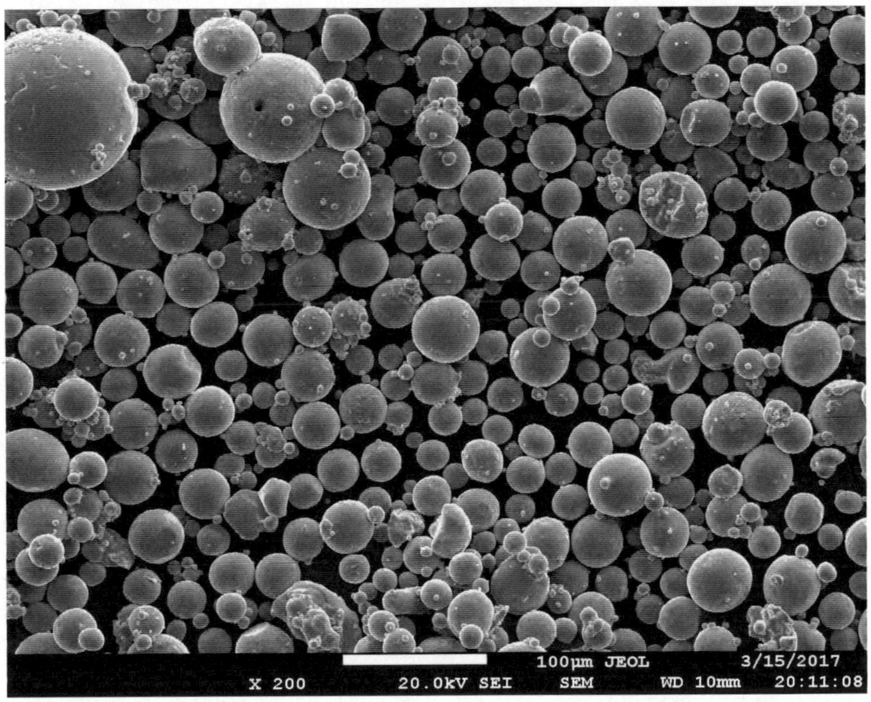

Figure 1. Morphology ofTiAl6V4 powder.

The porosity of the fabricated 10x10x10 mm cubes was determined through microscopy analysis by the cross-sections method [19]. For each specimen, two cross-sections were prepared (parallel and perpendicular to the building direction) by cutting. Each specimen was mounted in bakelite, then grinded and polished by standard procedures. Images of the cross-sections were analyzed by GIMP 2.8.22 software to determine porosity.

3. Results and discussion

SLM fabricated TiAl6V4 specimens are shown in figure 2. All of the specimens have a dense structure and a good surface finish quality without dimensional distortions.

IOP Conf. Series: Materials Science and Engineering **248** (2017) 012012 doi:10.1088/1757-899X/248/1/012012

Figure 2. SLM-fabricated TiAl6V4 specimens.

In figure 3, porosity evolution in TiAl6V4 samples was processed using different combinations of laser power and exposure time. The 8th specimen has the highest porosity - 22.6%, its cross-section is located in the top left corner of figure 3 on low laser power and low exposure time zone. It can be seen that the pores of specimen №8 have an irregular shape and large size. Such pores are called keyhole pores and arise due to insufficient energy for the full melting of powder layer [16].

With increasing of exposure time (7th specimen) and laser power (5th specimen), the number of pores reduced, but the morphology and size of pores almost was not changed. This demonstrates that laser energy is still insufficient for full melting of the powder layer. With further increase of exposure time and laser power (1st, 2nd, 4th, 6th, and 9th specimens), the number of keyhole pores further reduced, however, there were other type of pores with spherical shape and small size. Such pores are called metallurgical pores and arise from gases trapped within melt pool [18].

The 3rd specimen has the lowest porosity (about 0.5%) and is characterized only by metallurgical pores. Since the metallurgical pores are created due to gases dissolved in powder for the fabrication of low porosity materials by SLM, it is necessary to use a powder with low dissolved gases content.

4. Conclusion

Selective laser melting of TiAl6V4 powder using a CO_2-laser was studied in this paper. Nine 10x10x10 mm cubic specimens were fabricated using different SLM process parameters. All of the fabricated specimens have a good dense structure and a good surface finish quality without dimensional distortion. The lowest porosity that was achieved is about 0.5%.

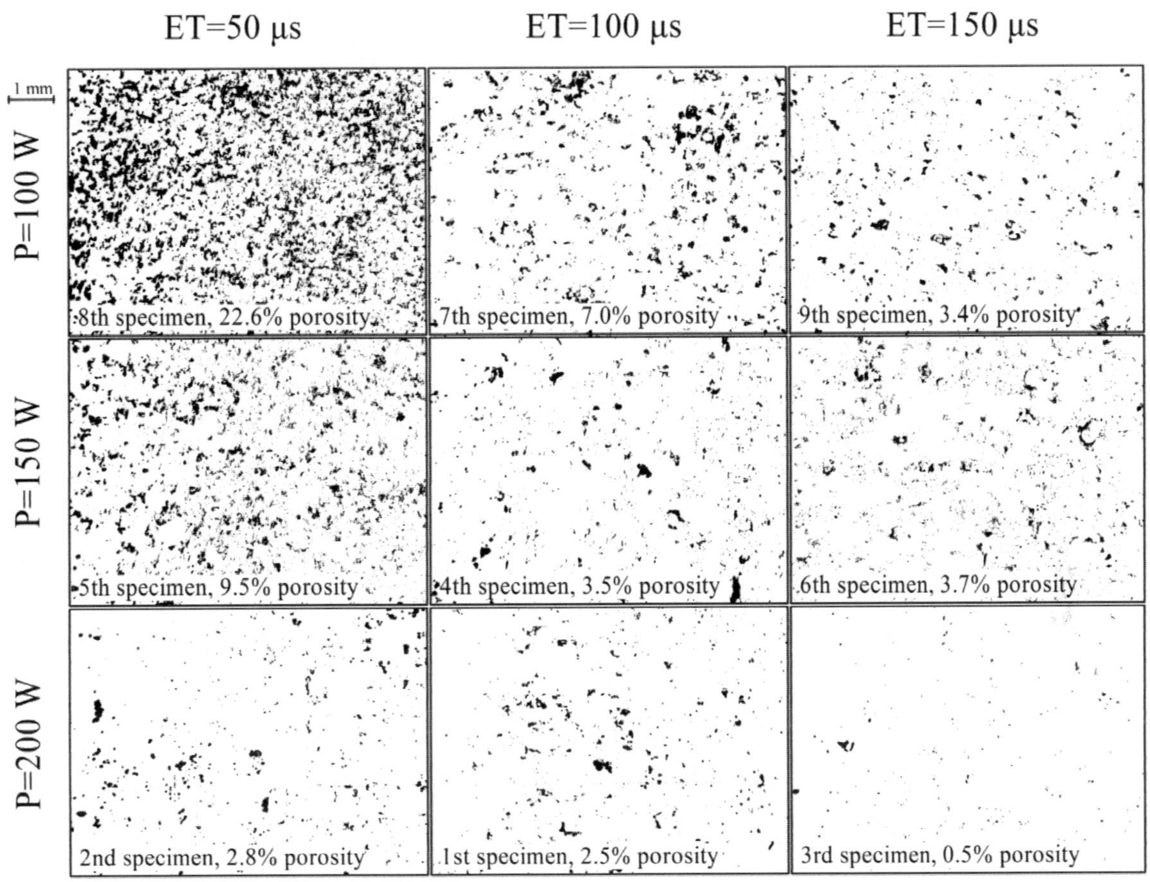

Figure 3. Porosity evolution in TiAl6V4 samples processed using different combinations of laser power and exposure time.

Acknowledgements
This research was financially supported by a grant of the President of the Russian Federation (No. MK-1881.2017.8).

References

[1] Uhlmann E, Kersting R, Klein T B, Cruz M F and Borille A V 2015 *Proc. CIRP* **35** 55.
[2] Haynes J 2014 *Proc. of the International Astronautical Congress, IAC* vol 9 (International Astronautical Federation, IAF) p 6120.
[3] Wagner S M and Walton R O 2016 *Prod. Plan. Control.* **27** 1124.
[4] Wauthle R, van der Stok J, Amin Yavari S, Van Humbeeck J, Kruth J P, Zadpoor A A, Weinans H, Mulier M and Schrooten J 2015 *Acta Biomater.* **14** 217.
[5] You J, Fang L, Zhang Q, Gao Y and Peng W 2015 *Chinese J. Biomed. Eng.* **34** 315.
[6] 2015 *Met. Powder Rep.***70** 196.
[7] Popovich A A, Sufiiarov V S, Polozov I A, Borisov E V, Masaylo D V, Vopilovskiy P N, Sharonov A A, Tikhilov R M, Tsybin A V, Kovalenko A N and Bilyk S S 2016 *Biomed. Eng. (NY).* **50** 202.
[8] Li J, Chi S, Xu J, Wang Y, Zhan D 2014 *West China J. Stomatol.* **32** 115.
[9] Bandyopadhyay A, Espana F, Balla V K, Bose S, Ohgami Y and Davies N M 2010 *Acta Biomater.* **6** 1640.
[10] Van Hooreweder B, Moens D, Boonen R, Kruth J P and Sas P 2012 *Adv. Eng. Mater.* **14** 92.
[11] Bartolomeu F, Faria S, Carvalho O, Pinto E, Alves N, Silva F S and Miranda G 2016 *Mater. Sci. Eng. A.* **663** 181.

IOP Conf. Series: Materials Science and Engineering **248** (2017) 012012 doi:10.1088/1757-899X/248/1/012012

[12] Sun J, Yang Y and Wang D 2013 *Opt. Laser Technol.* **49** 118.
[13] Dhansay N M, Tait R and Becker T 2014 *Adv. Mater. Res.* **1019** 248.
[14] Yang J, Yu H, Yin J, Gao M, Wang Z and Zeng X 2016 *Mater. Des.* **108** 308.
[15] Thijs L, Verhaeghe F, Craeghs T, Van Humbeeck J and Kruth J P 2010 *Acta Mater.* **58** 3303.
[16] Ali H, Ma L, Ghadbeigi H and Mumtaz K 2017 *Mater. Sci. Eng. A.* **695** 211.
[17] Tolochko N K, Khlopkov Yu V, Mozzharov S E, Ignatiev M B, Laoui T and Titov V I 2000 *Rapid Prototyp. J.* **6** 155.
[18] Aboulkhair N T, Nicola M E, Ashcroft I and Tuck C 2014 *Addit. Manuf.* **1-4** 77.
[19] Spierings A B, Schneider M and Eggenberger R 2011 *Rapid Prototyp. J.* **17** 380.

IOP Conf. Series: Materials Science and Engineering **248** (2017) 012013 doi:10.1088/1757-899X/248/1/012013

Evaluation of welding performance of 20 kHz and 40 kHz ultrasonic metal welding

W H Kim[1], E J Kang[2] and D S Park[1, 3]

[1]Department of Mechanical Engineering, Incheon National University, Incheon, 22012, Korea
[2]R&D Center, Yushin Precision Industrial Co., Incheon, 21999, Korea
[3]Author to whom any correspondence should be addressed.

Abstract. In this study, ultrasonic horns are designed by using vibration equations, vibration modal analysis and harmonic response analysis in order to compare welding performance when ultrasonic welding is performed at resonance frequencies of 20 kHz and 40 kHz. For the weldability evaluation of the manufactured horn for 20 kHz and 40 kHz, welding strength between Ni specimens with a thickness of 0.1 mm using tensile test are compared and analyzed. The lengths of horns with resonance frequencies of 20kHz and 40kHz were calculated as 130mm and 68mm respectively. As a result of vibration modal analysis, the optimum longitudinal vibration modes of 19,584Hz and 39,794Hz are obtained in 10th mode, and the frequency response of the two horns are 19,600 Hz and 39,800 Hz respectively. As the welding conditions are changed to welding pressure 2 bar, 3 bar and 4 bar, vibration amplitude of horn 60%, 80% and 100%, tensile strengths of welded specimens are observed. The welding strength was smaller at 40 kHz than at 20 kHz even at the same amplitude. This is because diffusion action of Ni in the weld interface is facilitated at 20 kHz than at 40 kHz.

1. Introduction

Due to recent environmental regulations, carbon emissions are controlled and interests in energy conservation are increasing. Therefore, it is important to develop low-cost and high-efficiency production technologies. Ultrasonic welding has come to the forefront as eco-friendly welding method for welding plastics and metals, which does not require special gases and welding materials used in conventional welding techniques.

Ultrasonic welding is characterized by high mechanical welding strength and almost no electrical resistance between the weld materials, because two metal surfaces are welded by physical diffusion actions removing the contaminants and deposit oxide film existed on the metal surfaces. It has been applied to the welding of secondary battery foils, can bottom, electronic and communication device parts, cell phones, speakers, vacuum glasses, and precision bonding of flip chips.

Figure 1 shows main parts of a typical ultrasonic welder. The converter or transducer converts alternating current of 50 to 60 Hz into a mechanical energy of 20 to 40 kHz. The booster amplifies the vibrational amplitude of the converter, and the horn transfers the mechanical vibration energy to the weld materials. The process of ultrasonic metal welding is such that the electric energy input through the power supply is converted into the mechanical vibration energy through the transducer, and then the vibrational amplitude is amplified by the booster. While the amplified ultrasonic vibration energy is transmitted to the weld through the horn, strong joining is achieved by the forced diffusion in the weld interfaces [1].

Content from this work may be used under the terms of the Creative Commons Attribution 3.0 licence. Any further distribution of this work must maintain attribution to the author(s) and the title of the work, journal citation and DOI.
Published under licence by IOP Publishing Ltd

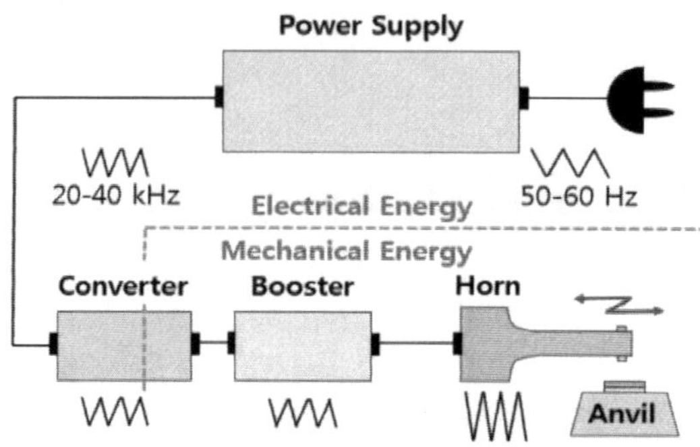

Figure 1. The schematic diagram of ultrasonic metal welding process.

Many researches on the ultrasonic welding have been conducted so far. Specially, analysis and design of horns with working frequency of 20 kHz and 40 kHz, welding experiment and weldability evaluation have been widely conducted [2-7], but the study on comparing the welding performance of two horns have not been presented. It is necessary to determine the optimal frequency of the horn to improve the welding performance when welding various types of metal. However, since there is no research data on these, there are many difficulties in manufacturing and utilizing the horn.

In this study, horns with resonance frequencies of 20 kHz and 40 kHz are designed and manufactured based on the modal analysis and harmonic response analysis, and the frequency characteristics of the manufactured horns are experimentally analyzed. In additions, the welding performance and characteristics of the two horns are investigated by analyzing the welding strength of the Ni-Ni thin plate.

2. Design of horns for ultrasonic welding

2.1. Horn design process

The material of the horn is selected according to the weld material and the weight of the horn, and the strength and hardness are considered to be the most important parameters. If the material of the welding member becomes harder than the horn, it is preferable to select a material having higher hardness than the material of the weld. The shapes of the horn are step, exponential and conical, and the amplitude amplification ratio is large in this order. In general, the step type is widely used because of high amplitude amplification. However, there is a disadvantage that the stress distribution increases as the amplitude amplification ratio increases due to a radical change in diameter [8].

The first process of the horn design is to calculate the length of the horn roughly using the vibration equations, and sequentially determine the shape and dimensions of the horn based on the mode shape and natural frequency analysis. Since the performance of the horn designed through the theoretical analysis produces some errors, the design and fabrication are completed through tuning processes.

2.2. Material and shape of horns

Figure 2 shows a 2D drawing of the horn which is finally modified through vibration analysis of 20 kHz and 40 kHz horns. Table 1 shows the physical properties of titanium alloys for finite element analysis of 20 kHz and 40 kHz horns.

3. Results of vibration analysis of horns

3.1. Vibration mode analysis

The number and shape of meshes for finite element analysis are automatically generated by the analysis tool. The boundary conditions for the analysis are cylinder restraint in the fastening part which is fastened with the booster and the bolt, and the cylinder is set so that only the axial displacement is freely generated. Horns are designed and manufactured based on modal analysis and harmonic response analysis with tuning processes.

Figure 3 shows the result of the modal analysis of 20 kHz horn. In the 10th mode, the natural frequency of 19,584 Hz close to 20 kHz is obtained and the maximum amplitude occurred at the tip of the horn. Figure 4 shows the result of the modal analysis of the horn for 40 kHz. As a result, it is found that the natural frequency of 39,794 Hz close to 40 kHz is obtained in the 10th mode, and the maximum amplitude of the transverse vibration is found in the tip of the horn.

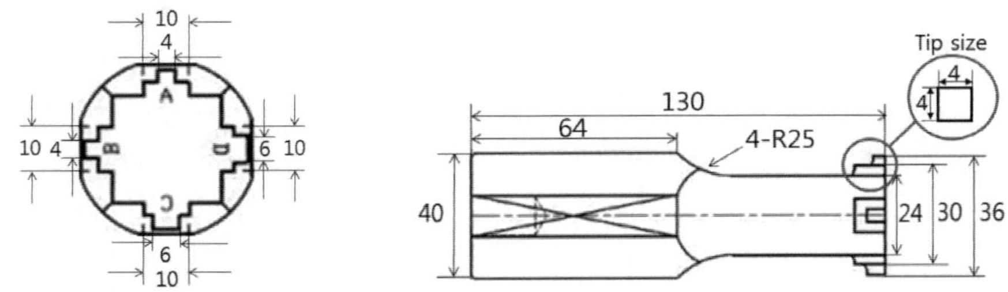

(a) 20 kHz horn

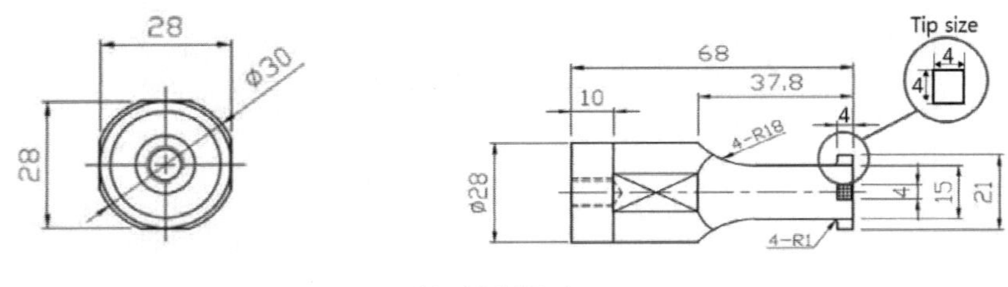

(b) 40 kHz horn

Figure 2. 2D Drawings of the horns.

Table 1. Mechanical properties of horn material.

Horn frequency	Density (g/m^3)	Poission's ratio	Young's modulus (GPa)
20kHz	4.5	0.34	116
40kHz	8.14	0.3	207

3.2. Harmonic response analysis

Figure 5(a) shows the result of the harmonic response of the horn tip when excited with a vibration frequency of 10 kHz to 30 kHz on the input side (left end of the horn) of the optimised 20 kHz horn. It is found that the frequency response of the horn's output surface (right end of the horn) is similar to the modal analysis result of 19,600 Hz.

Figure 5(b) shows the result of the harmonic response of the horn tip when excited with a vibration frequency of 30 kHz to 50 kHz on the input side (left end of the horn) of the optimised 40 kHz horn. The frequency response of the output of the horn (right end of the horn) is 39,800 Hz, which is similar to the result of the modal analysis.

4. Evaluation of the welding performance

4.1. Experimental method

Figure 6 shows the shape of the Ni specimen used in the weldability test using a horn of 20 kHz and 40 kHz. Test specimens of 50 mm x 10 mm x 0.1 mm are fabricated. Welding tests are carried out by overlapping 10mm width of each material. In this experiment, the welding pressure, the vibration amplitude and the welding time are selected as the welding parameters in order to evaluate the weldability of the thin Ni plate. Table 2 shows the experimental conditions for welding tests.

Seven Ni specimens are welded for each experimental condition. The welding strength is calculated by the maximum tensile force when welded Ni specimens are separated in the weld interface by using a tensile tester. In order to increase the reliability of the data, the average force of five tensile forces excluding the maximum and the minimum tensile forces was obtained.

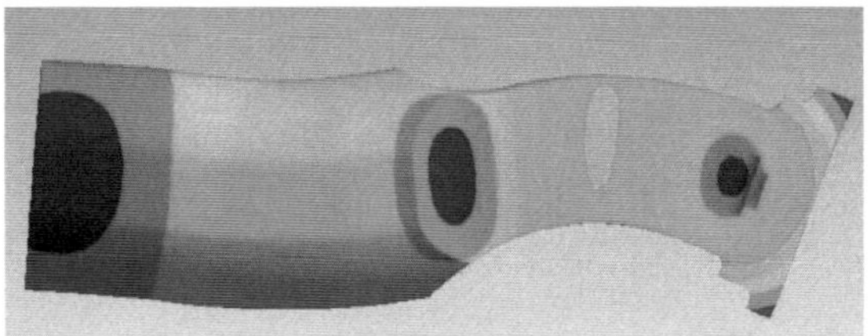

(a) 9th mode – 18,024 Hz

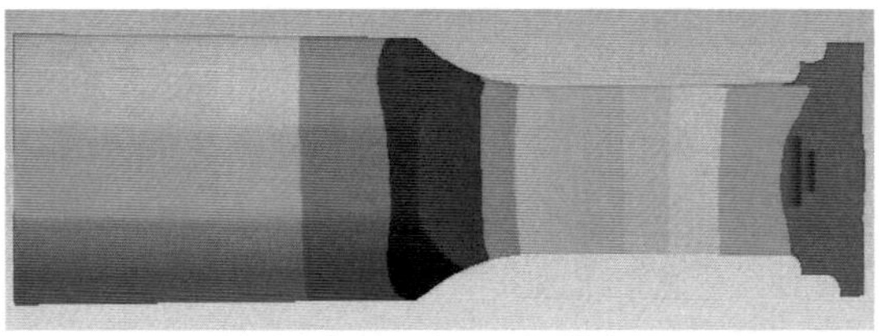

(b) 10th mode - 19,584 Hz

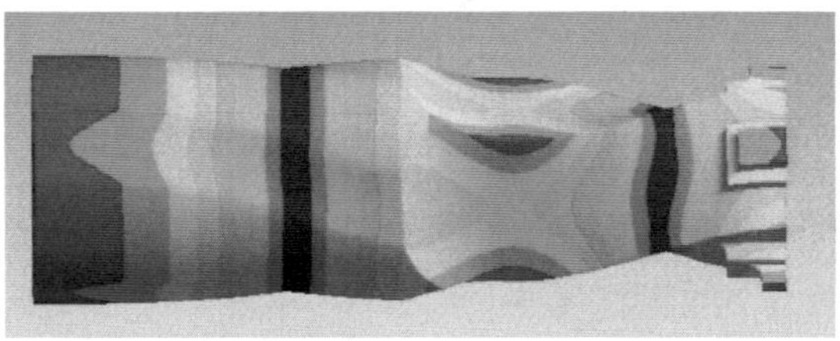

(c) 11th mode – 28,159 Hz

Figure 3. Mode shape and natural frequency of the 20 kHz horn.

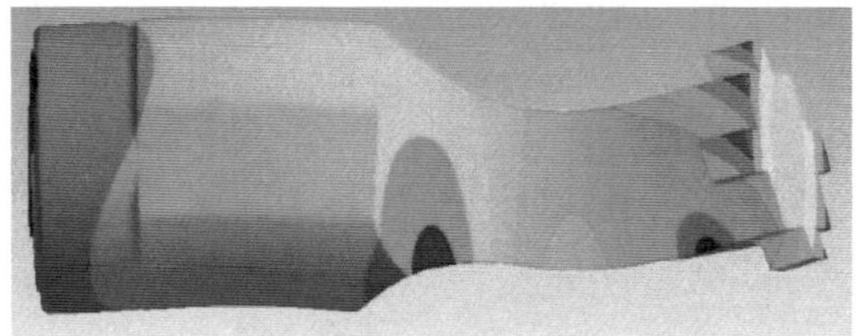

(a) 9th mode – 39,373 Hz

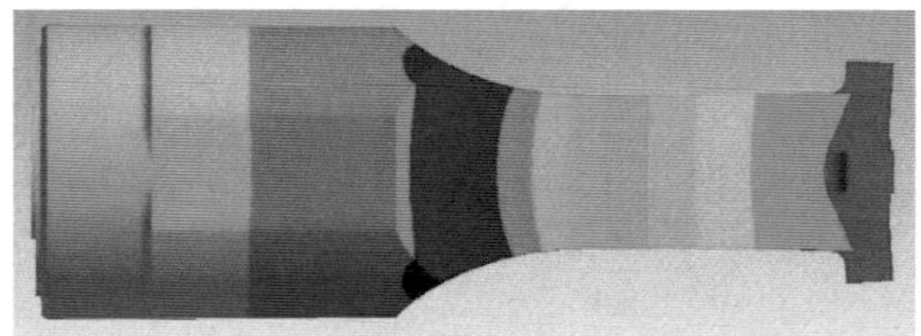

(b) 10th mode – 39,794 Hz

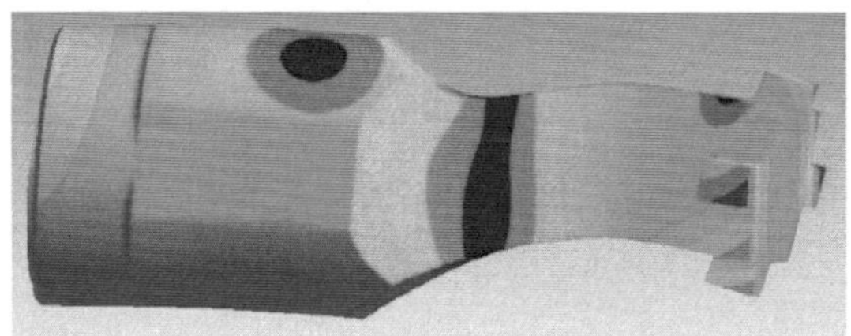

(c) 11th mode – 53,401 Hz

Figure 4. Mode shape and natural frequency of the 40 kHz horn.

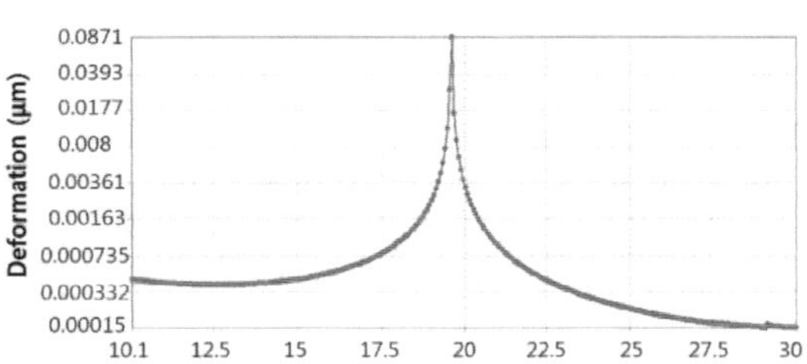

(a) 20 kHz

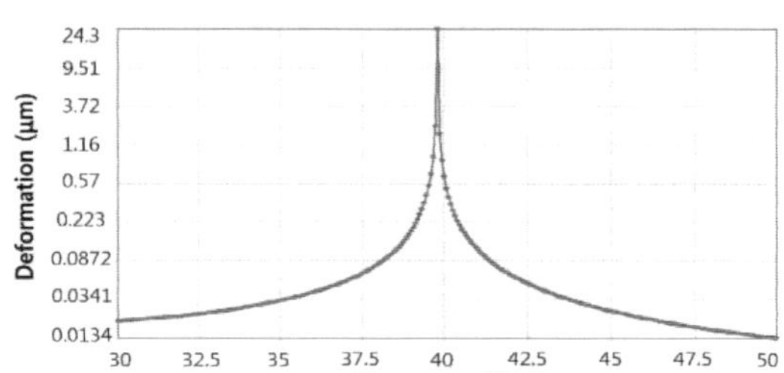

(b) 40 kHz

Figure 5. Harmonic response analysis results of the designed horn.

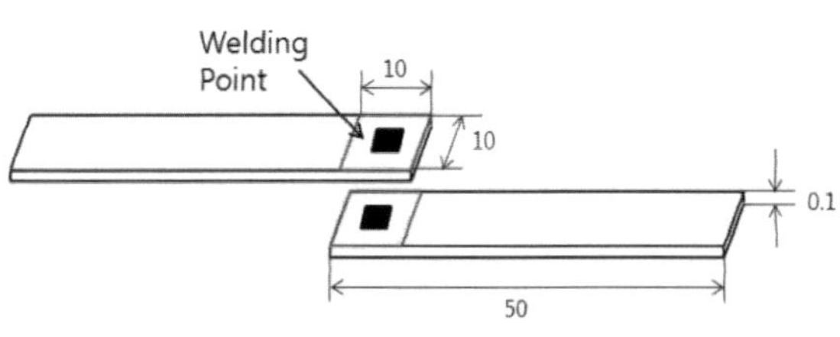

Figure 6. Welding specimen and welding point.

Table 2. Welding conditions.

Welding pressure (bar)	2.0, 3.0, 4.0
Amplitude (%)	60, 80, 100
Welding time (sec)	0.10 ~ 0.40

5. Experimental result and discussion

Figure 7 shows the tensile strength, ie, weld strength, of specimens welded by ultrasonic metal welders of 20 kHz and 40 kHz. Overall tendency is that the welding strength increases as the welding time increases under any conditions, and the welding strength is smaller at 40 kHz than at 20 kHz at the same vibration amplitude. This is because the ultrasonic friction heat between the welding specimens is concentrated in the horn tip, and diffusion of Ni between the welding parts is facilitated at 20 kHz than 40 kHz.

The maximum welding strength is 226.75 N at 20 kHz under the condition of pressure of 2 bar, amplitude of 100%, and a time of 0.38 second. At 40 kHz, it is confirmed that the pressure is 2bar, the amplitude is 100%, and the time is 0.38 second at 226.3 N. It is confirmed that the weld strength is higher at 2 bar than at 4 bar, and tensile strength increases with increasing amplitude, but there is not a tendency that the welding strength increases with increasing pressure.

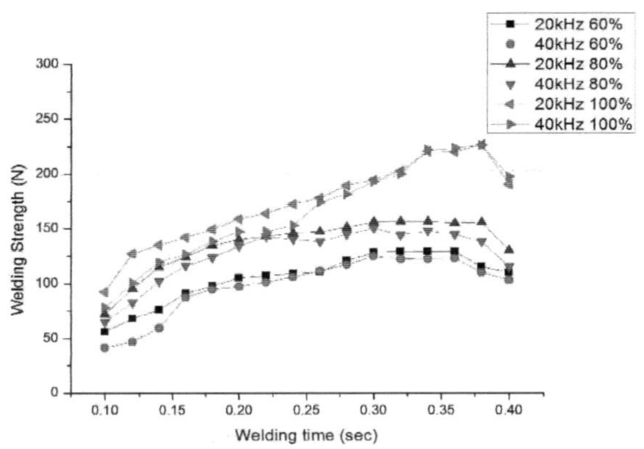

(a) welding pressure 2 bar

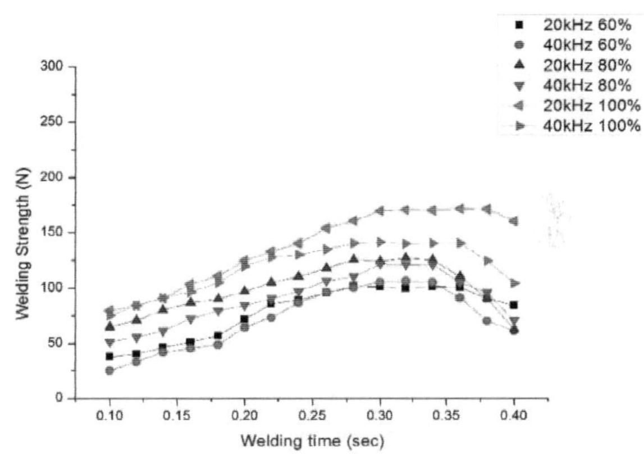

(c) welding pressure 4 bar

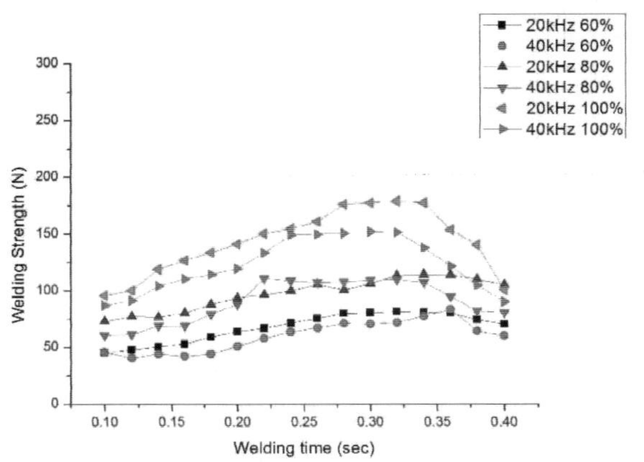

(b) welding pressure 3 bar

Figure 7. Welding strength variations according to welding conditions of welding time, welding pressure, vibration amplitude, and welding frequency of 20 kHz and 40 kHz horns.

6. Summary

In this study, to develop horns with resonance frequency of 20 kHz and 40 kHz, modal analysis and harmonic response analysis are performed by finite element method. In addition, for the evaluation of the weldability of the fabricated 20 kHz and 40 kHz horns, the weld strength between Ni specimens with a thickness of 0.1mm are compared and analysed. As for the overall tendency of welding strength according to the welding condition change, it is obvious that the welding strength increases with the increase of welding time under any welding conditions. It is confirmed that the welding strength is smaller at 40 kHz than at 20 kHz even at the same vibration amplitude. This is because the ultrasonic friction heat between the welding specimens is concentrated in the horn tip, and diffusion of Ni in the weld interface is facilitated at 20 kHz than at 40 kHz.

References

[1] John W *Ultrasonic metal welding for wire splicing and termination* Branson Ultrasonic Corp.
[2] Sooriyamoorthy E, Prakasan K and Jaiganesh V 2010 *Int. J. Adv. Manuf. Technol.* **51** 163–71.
[3] Kim E M, Jang H S and Park D S 2010, *J. Korean Soc. Mach. Tool Eng.* **19** 790-96.
[4] Jang H S, Park W Y and Park D S 2013 *J. Korean Soc. Manuf. Proc. Eng.* **12** 55-60.
[5] Park W Y, Lee C H and Park D S 2014 *Adv. Mater. Res.* **941-944** 2053-57.
[6] Volkov S S 2012 *Weld. Int.* **26** 796–99.
[7] Seo J S, Jang H S and Park D S 2015 *Mater. Manuf. Proc.* **30** 1069-73.
[8] Amin S G, Ahmed M H M and Youssef H A 1995 *J. Mater. Proc. Tech.* **55** 254-60 .

Acknowledgement

This work was supported by Incheon National University Research Grant in 2013.

IOP Conf. Series: Materials Science and Engineering **248** (2017) 012014 doi:10.1088/1757-899X/248/1/012014

The Variation of Work Productivity and Muscle Activities at Different Levels of Production Target

Nurhayati Mohd Nur[1], Siti Zawiah Md Dawal[2], Mahidzal Dahari[2] and Nurul Zuhairah Mahmud Zuhudi[1]

[1]Aerospace Department, Universiti Kuala Lumpur, Malaysian Institute of Aviation Technology, Jenderam Hulu, 43800 Dengkil, Selangor, Malaysia
[2]Department of Mechanical Engineering University of Malaya, Lembah Pantai, 50603 Kuala Lumpur, Malaysia

Abstract. This paper aims to investigate the variation of work productivity and muscle activities among workers performing industrial repetitive tasks at four different levels of production target. The work productivity and muscle activities data were recorded from twenty workers at four levels of production target corresponding to "participative (PS1)", "normal (PS2)", "high (PS3)" and "very high (PS4)". The results showed that worker productivity was found to increase at higher production target and there was a significant change ($p < 0.005$) in work productivity across the four different production targets. The muscle activities were found to increase at higher production target and correspond to more discomfort and a higher rate of muscle fatigue. The results indicated that working with a higher production target results in higher worker productivity, but could lead to higher risk of WMSDs.

1. Introduction

A number of quality management systems (QMS) applied and implemented in the manufacturing industries such as ISO 9001, ISO/TS 16949 and AS 9000. It is clearly stated in all the QMS to measure customer's satisfaction. One of the elements in the customer's satisfaction measurement is on time delivery. There are many improvement activities implemented by the industry in order to achieve the target such as KAIZEN, lean manufacturing and JIT (just in time). The improvement activities are well known in the industry and remarkable as appropriate tools to satisfy the customer and achieve high productivity. Instead of fulfilling QMS requirements and satisfy customers, manufacturing industry also tends to set high production target to get higher returns on their investments [1]. High production targets produce higher work productivity compared to low targets [2], but the effect on the worker's safety and health is unknown.

Productivity is related to the ability of the individual to produce. Thus, work productivity is referred as the ability of the workers to produce or deliver products that are expected from their jobs [3]. In the industrial-repetitive task, productivity is measured through observed work productivity (OWP) [3]. The OWP measure was based on the production target [4, 5]. The production target is obtained from the production standard time of the tasks. The standard time is inherently linked to the frequency of repetitive movements [6]. In the industrial-repetitive task, the workplace is not self-chosen, and the worker must follow a specified production standard time [7]. Work productivity is important to achieve customer's satisfaction, higher returns, more profits and more business opportunity. Therefore, it is frequently viewed from economic perspectives while the workers' role and workers' productivity state at any specific time is often underrated [3]. A few studies found that workers performing industrial-repetitive tasks are exposed to the risk of work-related musculoskeletal

Content from this work may be used under the terms of the Creative Commons Attribution 3.0 licence. Any further distribution of this work must maintain attribution to the author(s) and the title of the work, journal citation and DOI.
Published under licence by IOP Publishing Ltd

disorders (WMSDs) [8, 9]. WMSDs is caused by the accumulation of muscle fatigue [10] and it has a great impact on task performance [11].

Social Security Organization (SOCSO) reported that WMSDs cases are increasing every year [12]. The compensation cost paid due to this disease is also higher compared to other industrial diseases [13, 14]. WMSDs represents approximately one-third of workers' compensation costs in the Malaysian private industry [15]. Workers performing industrial-repetitive tasks may expose to the WMSDs risks. In addition, a high intention of the industry to achieve high production targets may provide an opportunity for the increase of the WMSDs cases reported. However, a study on the variation of work productivity, muscle activities and WMSDs risks at different levels of production target is still scarce. Therefore, it is timely to investigate the relationship and find ways to optimize work productivity with minimum risk of WMSDs.

2. Methodology

2.1 Experimental Task
An experimental task is conducted to collect data on the work productivity and muscle activities at different levels of production target. The task selected is an industry based, repetitive and involved upper limb muscles.

2.2 Subjects
Twenty healthy subjects, 10 females and 10 males were involved in this study. The subjects were between the ages of 22 and 45. All subjects were right-handed dominance and have no previous history of musculoskeletal injuries.

2.3 Production Target
The experimental tasks were performed in four levels of production target. The targets are participative target (PS1) which is referred to participative standard time, normal target (PS2) (referred to 100% normal standard time), high production target (PS3) (referred to 126% normal standard time) and very high production target (referred to 140% normal standard time). A high production target which is equivalent to 126% of the normal standard time is a realistic work pace in the manufacturing industry [16]. A very high production target is 140% of the normal standard time [2]. In the participative production target, no standard time is given to the subjects. They are allowed to perform the task at their own pace.

2.4 Experimental Task Procedure
Before the experimental task, the data on weight and height of the subject are recorded. Surface Electromyography (sEMG) was used to measure forearm muscles activity throughout the experiment. The surface electrodes were placed over the belly of the following muscles bilaterally [17]:-

1. Flexor Carpi Radialis – Left and Right
2. Extensor Carpi Radialis – Left and Right

The surface electrodes and lead electrode cable which are placed on the belly of the muscles are connected to the Noraxon Telemyo 2400 (Noraxon, INC USA), using small signal transmitters carried on a belt by the subjects. Muscles activity signals picked up by the electrodes were transmitted telemetrically to a receiver, which was connected to a computer. The task involved repetitive assembly actions similar to actual industrial assembly task. The task is a process of assembling foam ring to the plastic clip as presented in figure 1. The subjects performed the repetitive task for 1 hour for each production target.

IOP Conf. Series: Materials Science and Engineering **248** (2017) 012014 doi:10.1088/1757-899X/248/1/012014

Figure 1. Assembly Task.

3. Results

3.1 Work Productivity

The result of mean work productivity at four levels of production target is shown in table 1. The work productivity was found to increase at higher production target. A one-way repeated measures ANOVA was conducted to compare the average of work productivity with production targets (PS1, PS2, PS3 and PS4). There was a significant effect of production targets on the work productivity (Wilk's Lambda=0.188, F(3,17)=24.51, $p < 0.0005$, multivariate partial eta squared = 0.812).

Table 1. Mean work productivity.

Production Target	Work productivity (mean)
PS1	774
PS2	851
PS3	890
PS4	928

3.2 Muscle Activities

In this study, the root mean square (RMS) value of the EMG data was analysed. The mean and standard deviation of the normalized RMS for all muscles is presented in table 2. The mean values of normalized RMS represent muscle activities in the study.

Table 2. Muscle Activities (normalized RMS).

Production Target	Mean (Standard Deviation)			
	FCRR	FCRL	ECRR	ECRL
PS1	8.319(1.824)	10.311(2.825)	10.107(2.811)	10.248(3.349)
PS2	8.824(1.827)	11.010(2.809)	10.377(2.344)	11.409(3.416)
PS3	9.596(2.257)	11.890(2.496)	11.035(2.552)	11.974(3.431)
PS4	9.925(2.402)	12.475(2.921)	11.904(3.154)	12.702(3.486)

The RMS values were found to increase at higher production targets. The very high production target (PS4) showed the highest value of RMS for all muscles. The highest RMS was found at Extensor Carpi Radialis-Left (ECRL). A repeated measures ANOVA results were summarized in table 3 and it can be seen that there was a significant effect of productivity targets on the muscle activities for all muscles ($P < 0.05$).

Table 3. ANOVA results for muscle activities.

Muscle	Wilk's Lambda	F	P	Partial Eta Squared
FCRR	0.478	6.179	0.005	0.522
FCRL	0.410	8.142	0.001	0.590
ECRR	0.420	7.819	0.002	0.580
ECRL	0.222	19.830	0.000	0.778

The muscle fatigue rate for all muscles was summarized in table 4. Muscle fatigue rate is found higher at higher production target.

Table 4. Muscle fatigue rate.

Production Target	Muscle Fatigue Rate			
	FCRR	FCRL	ECRR	ECRL
PS1	0.044	0.082	0.083	0.053
PS2	0.045	0.084	0.088	0.059
PS3	0.046	0.086	0.092	0.105
PS4	0.050	0.096	0.103	0.115

4. Discussion

The aim of this study is to investigate the variation of work productivity and muscle activities while performing industrial repetitive tasks at different levels of production targets. This study was conducted in the ergonomics laboratory under a controlled environment. The subjects were industrial workers and performing the industrial repetitive task. The work productivity and muscle activities of the workers were recorded during the experimental tasks. The results showed that worker productivity was found to increase at higher production targets. The results are aligned with the findings of a previous study [2]. The results also showed that the workers work harder in order to achieve the higher output.

In participative production target, no production target is assigned to the workers and they are free to decide their work productivity. The results showed that they are able to achieve work productivity with 5% higher compared to the normal target. In addition, the work productivity for the normal target is found 18% higher than expected. However, the workers are not able to achieve the target for higher production target (126% and 140% normal standard), even though the results are higher than the normal production target. Based on the ANOVA result, there is a significant effect ($p < 0.005$) of production target to work productivity and the result suggests that there was a change in work productivity across the four different production targets.

The muscle activities for all muscles were found to increase at higher production target and the muscles are most active at very high production target, followed by high and normal production target. The lowest muscle activities are found in participative production target. It can be observed that there was a steady increase in the muscle activities while the production target is higher and showed that there is a significant difference ($p < 0.05$) in the muscle activities across the four production target.

Further analysis of the muscle activities over time for all muscles involved was found to increase over time at four levels of production target. All muscles gave the positive slope of muscle activities and indicate the sign of muscle fatigue. The higher muscle activities should correspond to more discomfort [18] and a higher rate of muscle fatigue. The most active muscle in this study is left Extensor Carpi Radialis (ECRL) followed by left Flexor Carpi Radialis (FCRL), right Extensor Carpi Radialis (ECRR) and right Flexor Carpi Radialis (FCRR). The results showed that the left muscles are more active compared to the right muscles. The reason for this is due to the task performed required more left muscles movement compared to the right muscles.

The overall results indicated that WMSDS risk is higher in higher production target and workers tend to slow down due to WMSDs risks which then results to lower productivity. The result is consistent with the findings of a previous study [19]. The results provide a significant reference to the industries where industrial engineers and managers should more concern with the production target assigned and the risks of WMSDs to the workers.

5. Conclusion

In conclusion, work productivity was found to increase at higher production target and there was a significant change in work productivity across the four different production targets. However, the muscle activities were found to increase at higher production target and correspond to more discomfort and a higher rate of muscle fatigue. The results indicated that working with a higher production target

IOP Conf. Series: Materials Science and Engineering **248** (2017) 012014 doi:10.1088/1757-899X/248/1/012014

results in higher worker productivity, but could lead to higher risk of WMSDs. The results can be used as a reference in work productivity planning to optimize work productivity with minimum risk of WMSDs to the workers.

6. References

[1] Erik B 2005 *Opt. Mag* **4(5)** 27-35.
[2] Shikdar A and Das B 2003 *Ergonom.* **46** 466–81.
[3] Escorpizo R 2008 *Int. J. Ind. Ergon.* **38** 291–7.
[4] Burton W N, Conti D J, Chen C Y, Shultz A B and Edington D W 1999 *J. Occup. Environ. Med.* **41** 863–77.
[5] Lerner D, Amick B C, Lee J C, Rooney T, Rogers W H and Chang H 2003 *Med. Care* **41** 649–59.
[6] Andersen J H, Kaergaard A, Mikkelsen S, Jensen U F, Frost P, Bonde J P, Fallentin N and Thomsen J F 2003 *Occup. Environ. Med.* **60** 649–54.
[7] Sundelin G and Hagberg M 1992 *Scand. J. Work Environ. Health* **18** 262–8.
[8] Moore A and Wells R 2005 *Ergonom.* **48** 859–73.
[9] Chung M K, Lee I and Kee D 2005 *Ergonom.* **48** 331–41.
[10] Ma L, Chablat D, Bennis F, Zhang W 2009 *Int. J. Ind. Ergon.* **39** 211–20.
[11] Joana S, Joao S B, Pedro R R M, Alberto S M, Rubmin S and Mario A P V 2016 *Int. J. Ind. Ergon.* **52** 78-91.
[12] SOCSO 2014 *Social Security Organisation (SOCSO)Annual Report 2014.*
[13] Joseph B, Naveen B, Suguna A and Surekha A 2016 *J. Health Manag.* **18** 545–54.
[14] SOCSO 2011 *Social Security Organisation (SOCSO)Annual Report 2011*
[15] Mohamed Azman A M 2007 *Occupational disease in Asian countries* World Soc. Secur. Forum, Moscow.
[16] Bosch T, Mathiassen S E, Visser B, de Looze M P and van Dieën J H 2011 *Ergonom.* **54** 154–68.
[17] Lehman K R, Psihogiost. J P and Meulenbroek R G J 2001 *Ergonom.* **44** 719–38.
[18] Kuijt-Evers L F M, Bosch T, Huysmans M A, de Looze M P, Vink P 2007 *Appl. Ergon.* **38** 643–54.
[19] Resnick M L and Zanotti A 1997 *Comput. Ind. Eng.* **33** 185–88.

IOP Conf. Series: Materials Science and Engineering **248** (2017) 012015 doi:10.1088/1757-899X/248/1/012015

The Mediating Effects of Muscle Activities on the Relationship of Production Standard Time and Work Productivity

Nurhayati Mohd Nur[1], Siti Zawiah Md Dawal[2], Mahidzal Dahari[2] and Nur Faraihan Zulkefli[1]

[1]Aerospace Department, Universiti Kuala Lumpur, Malaysian Institute of Aviation Technology, Jenderam Hulu, 43800 Dengkil, Selangor, Malaysia
[2]Department of Mechanical Engineering University of Malaya, Lembah Pantai, 50603 Kuala Lumpur, Malaysia

Abstract. The aim of this study is to investigate the mediation effects of muscle activities on the relationship between production standard time and work productivity. The work productivity and muscle activities data are collected from twenty workers (10 males, 10 females) while performing industrial repetitive tasks at three different levels of production standard time corresponding to "normal", "hard" and 'very hard". Mediation test was performed on the data and the results showed that muscle activities act as a mediator in the relationship between production standard time and work productivity. The result indicates the importance of assessing muscle activities in relation to work productivity at different levels of production standard time in order to optimize work productivity and reduce WMSDs risks.

1. Introduction

Work productivity can be referred as the worker's ability to produce goods that are expected from his or her job [1]. The work productivity is commonly measured based on the quantity produced per hour during specified production target in a standardized and short cycle time. With this approach, it is insufficient to identify the effect of the functional incapacity of workers associated with the work productivity [2]. Thus, workers may be exposed to a higher risk of contracting WMSDs while carrying out tasks in high work productivity targets or hard production standard times. WMSDs are recognized for their adverse impact on worker productivity and health [3, 4]. It has been suggested that work exposure such as task demands and work capacity may influence the occurrence of muscle fatigue, discomfort and injuries [5] which are associated with WMSDs. WMSDs cases are increasing every year [6] and the occurrence rate is still high. WMSDs provide a significant impact on the performance of the workers [7] and involved higher compensation cost [8]. Hard production standard time is a form of work exposure related to task demands. In general, tasks become more repetitive with hard production standard times, which will expose workers to higher muscle fatigue rate and WMSDs risks.

Muscle fatigue is related to muscle activity. Hence, muscle fatigue refers to a decline in muscle performance which is associated with intense muscle activity [9]. An increase in muscle activity signifies the development of muscle fatigue [10]. Thus, muscle fatigue is defined as the point at which the muscles are no longer able to sustain the required force or work output level [11, 12]. Muscle fatigue begins with the contraction of muscles and manifests itself eventually as a failure to proceed with the activity at the initial amount of force [12]. Muscle fatigue and muscle injury are interrelated to one another [9]. Workers with muscle fatigue may be at higher risk of developing WMSDs [13]. Based on the above argument, there is very high possibility workers are dealing with high muscle

Content from this work may be used under the terms of the Creative Commons Attribution 3.0 licence. Any further distribution of this work must maintain attribution to the author(s) and the title of the work, journal citation and DOI.
Published under licence by IOP Publishing Ltd

activities while performing the task with high work productivity target. However, it is unknown either muscle activities is the mediator for work productivity and it is timely to investigate the relationship.

2. Methodology

2.1 Measurement of Work Productivity and Muscle Activities

Work productivity and muscle activities data are collected from twenty workers (10 males, 10 females) while performing industrial repetitive tasks at three different levels of production standard time. The subjects were between 22 and 45 years old. The production standard times are listed as follows:

1. Normal (100% of the normal standard time)
2. Hard (126% of the normal standard time)
3. Very Hard (140% of the normal standard time)

2.2 Mediation Test

Mediation is a hypothesized causal chain in which one variable affects a second variable which in turn affects the third variable. The intervening variable (M) is the mediator and it mediates the relationship between a predictor (X) and an outcome (Y). This relationship is represented graphically in figure 1. In this study, M represents muscle activity, whereas X and Y represent production standard time and work productivity, respectively.

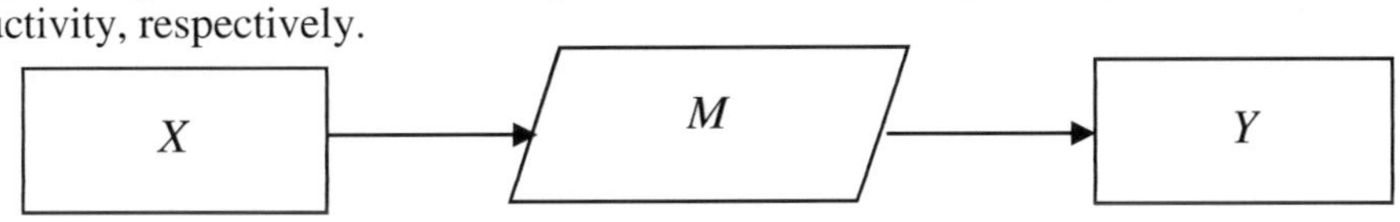

Figure 1. Graphical representation of the mediation model.

3. Results

3.1 Work Productivity and Muscle Activities

The result of mean work productivity at three levels of production standard time is shown in table 1.

Table 1. Work Productivity and Muscle Activities.

Production Standard Time	Work Productivity (mean)	Muscle Activities (mean)			
		FCRR	FCRL	ECRR	ECRL
Normal	851	8.824(1.827)	11.010(2.809)	10.377(2.344)	11.409(3.416)
Hard	890	9.596(2.257)	11.890(2.496)	11.035(2.552)	11.974(3.431)
Very Hard	928	9.925(2.402)	12.475(2.921)	11.904(3.154)	12.702(3.486)

3.2 Mediation Analysis Results

The mediation effect of muscle activity on the work productivity is tested. The relationship between X and Y refers to the total effect of X and Y and is given by the value of c, as shown in figure 2.

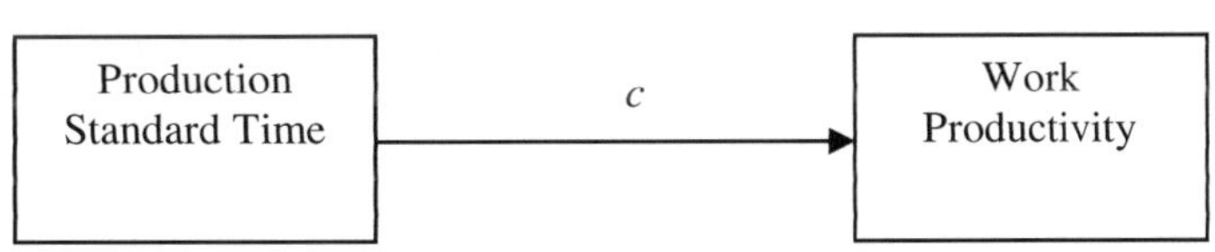

Figure 2. Total effect of X (production standard time) on Y (work productivity).

The direct effect of *X* on *Y* after controlling *M* is given by *c'*, as shown in figure 3.

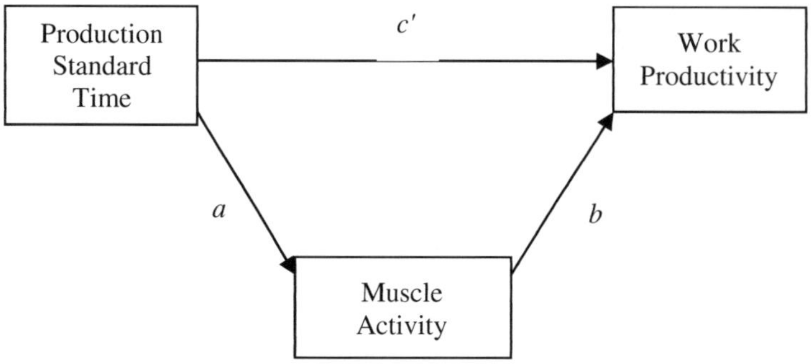

Figure 3. Direct and indirect effects of the mediation model.

Mediation analysis was carried out according to the procedure outlined by Baron and Kenny [14]. The results of the mediation analysis are summarized in table 2. It can be seen that the muscle activity acts as a mediator based on the following observations:

1. Production standard time significantly predicts work productivity
($c = 0.571$, $c \neq 0$).
2. Production standard time significantly predicts muscle activity
($a = 0.402$, $a \neq 0$).
3. Muscle activity significantly predicts work productivity while the production standard time is controlled ($b = 0.475$, $b \neq 0$).

Table 2. Summary of the mediation analysis results.

	Dependent Variable (Work Productivity)		**Conclusion**
	Without Mediator	**With Mediator**	
Production standard time	0.571	0.374	Mediation occurs
Muscle activity	-	0.475	

The unstandardized coefficient should either be zero or reduced significantly in the presence of mediation effects [14]. The results show that the value of the unstandardized coefficient decreases, which indicates the occurrence of partial mediation. However, according to Hayes [15], the terms 'partial mediation' and 'full mediation' are no longer applicable in the 21st century. The complete mediation model is shown in figure 4.

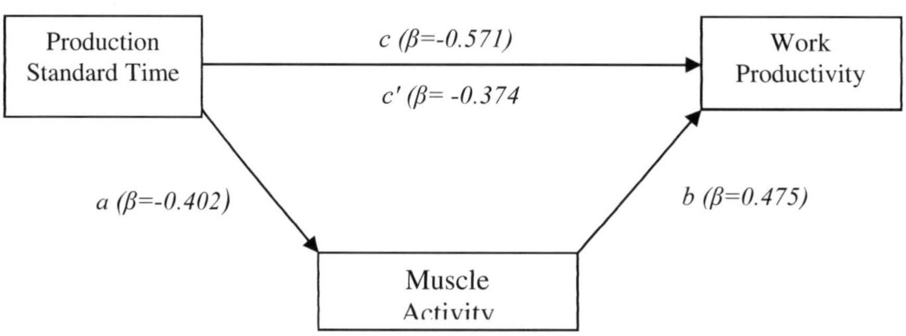

Figure 4. Complete mediation model.

The mediation model was validated using PROCESS add-on macro [16] which is available in the SPSS software and the results are summarized in table 3. The results reveal that there are no zeroes within the bias-corrected bootstrap confidence intervals, which indicates that the null hypothesis (i.e. H_0: mediation effects do not occur) can be refuted.

Table 3. Mediation results generated by PROCESS add-on macro.

Effect	Estimate	SE	t	p	Bias-Corrected Bootstrap Interval
c	0.1235	0.0242	5.0979	0.0000	(0.0139, 0.0790)
a	1.8590	0.5593	3.3236	0.0015	
b	0.0230	0.0061	3.7385	0.0004	
c'	0.0808	0.0241	3.3566	0.0014	
a*b	0.0427				

4. Discussion

Muscle activities are found to vary at different levels of production standard time. In general, muscle activities increases in harder production standard times, which reduces work productivity in terms of the percentage of normal production standard time. The increment of muscle activities over time signifies the development of muscle fatigue [17]. Muscle fatigue is an initiating factor of WMSDs [18] and it has been suggested that the accumulation of muscle fatigue causes WMSDs [12]. Previous studies suggested that WMSDs cause a reduction in work productivity [19].

The mediating test was conducted. The results show that muscle activities act as a mediator in this study. The muscle activities was used to predict work productivity while the production standard time is controlled ($\beta = 0.475$, $p < 0.05$ and $\beta = 0.374$, $p < 0.05$). The mediation test was validated using the procedure by Hayes [16]. The validation test shows that there are no zeroes within the bias-corrected bootstrap confidence intervals which indicate the presence of mediation. Hence, it is deemed necessary that the muscle activities of workers are assessed in relation to work design, particularly during the work productivity planning stage. Knowledge of the muscle activities will assist organizations to design tasks which will sustain work productivity and reduce the risk of WMSDs among workers.

5. Conclusion

In conclusion, muscle activities act as a mediator for production standard time and work productivity. The result indicates the importance of assessing muscle activities in relation to work productivity at different levels of production standard time in order to optimize work productivity and reduce WMSDs risks concurrently. A mediation model has been developed and validated. The model can also be used as a reference in other studies pertaining to muscle activities, work productivity and production standard time.

References

[1] Escorpizo R 2008 *Int. J. Ind. Ergon.* **38** 291–7.
[2] Allesina S, Azzi A, Battini D and Regattieri A 2010 *Int. J. Prod. Res.* **48** 2297–321.
[3] Bevan S 2015 *Best. Pract. Res. Clin. Rheumatol.* **29** 356–73.
[4] Buchbinder R, Blyth F, March L, Brooks P, Woolf A and Hoy D 2013 *Best. Pract. Res. Clin. Rheumatol.* **27** 575–89.
[5] Dempsey P 1999 *Int. J. Ind. Ergon.* **24** 405–16.
[6] SOCSO 2014 *Social Security Organisation (SOCSO) Annual Report 2014.*
[7] Joana S, Joao S B, Pedro R R M, Alberto S M, Rubmin S and Mario A P V 2016 *Int. J. Ind. Ergon.* **52** 78-91.
[8] Joseph B, Naveen B, Suguna A and Surekha A 2016 *J. Health Manag.* **18** 545–54.
[9] Allen D G, Lamb G D and Westerblad H 2008 *Phys. Rev.* **88** 287–332.
[10] Basmaijan J, DeLuca C J 1985 *Muscles Alive : Their functions revealed by electromyography*

5th ed. Williams and Wilkins Baltimore.
[11] Dimitrova N A and Dimitrov G V 2003 *J. Electromyogr. Kinesiol.* **13** 13–36.
[12] Ma L, Chablat D, Bennis F and Zhang W 2009 *Int. J. Ind. Ergon.* **39** 211–20.
[13] Bosch T, Mathiassen S E, Visser B, de Looze M P and van Dieën J H 2011 *Ergonom.* **54** 154–68.
[14] Baron R M and Kenny D A 1986 *J. Pers. Soc. Psychol.* **51** 1173–82.
[15] Hayes A F 2009 *Commun. Monogr.* **76** 408–20.
[16] Hayes A F 2013 *Introduction to mediation, moderation, and conditional process analysis a regression-based approach* New York Guilford Press
[17] Konrad P 2005 *The ABC of EMG. A practical introduction to kinesiological electromyography.* Noraxon Inc. USA.
[18] Takala E 2002 *Scand. J. Work. Environ. Heal.* **28** 211–3.
[19] Xu Z Ko J Cochran D J and Jung M 2012 *Comput. Ind. Eng.* **62** 431–41.

Laser clad NiCrBSi alloy wear-resistance coating with RE addition on heavy duty spur gear flank

N Zhao[1, 2], L Tao[1, 2*], H Guo[1, 2] and M Q Zhang[1, 2]

1 School of Mechatronics, Northwestern Polytechnical University, Xi'an 710072, China;
2 Shaanxi Engineering Laboratory for Transmissions and Controls, Xi'an 710072, China

Abstract. In this research the wear-resistance composite coating successfully produced on heavy duty gear work surface by laser was reported. The coating containing 99 wt.% NiCrBSi alloy and 1 wt.% RE (rare earth element) oxidation powder. The RE addition coupled with laser operating parameters optimization caused elimination of both cracks and pores meanwhile further enhanced comprehensive properties of the laser layer. The coating microhardness, microstructure, phase construction and wear behaviors were tested by hardness tester, SEM equipped with EDS, XRD and tribometer, respectively. The results reflected the fact that the RE addition enhanced the coating ability of wear resistance and laser clad layer properly bonded with the gear flank. The wear volume loss rate of coating was half of that of the gear flank metal the COF curve of coating kept bellow that of the gear flank steel.

1. Introduction

The heavy duty gear have undergone sever abrasive wear and this cost huge amount of money and time in the replacement of failure gears. Laser cladding have been widely utilized in the production of protective wear resistance layers [1-14]. Laser layer applications range from aviation engine, unclear industry to navy defensive marine. The fine microstructure and small crystal size caused by quick heating and solidification cycle which are unique in laser processing and the metallurgical bond which allowed the cladded gear flank qualified the power transmission underground, based on the above reasons laser clad wear-resistance coating has become a hot research topic recently. Bearing these merits of laser coatings in mind, the laser produced layer is competent for solving the problem of gear abrasive and fatigue failure.

RE has been proved to be able to refine the grain size of laser clad alloy layer [7-17]. MMC (metal matrix composite) coatings reinforced by RE has attracted lots of research interests. The laser clad alloy layer has finer grain than other conventional approaches like thermal spray, TIG welding and so on, however, the added RE can further refine grain size. Laser clad RE reinforced wear resistance layer exhibited excellent properties if the amount of RE addition is controlled properly in permissible content otherwise the RE particle aggregation would damage the mechanical properties of MMC coating.

2. Experiment details

The gear flanks were grind by sand wheel and the fatigue layer were completely removed before cleaned and fixed on fixture for laser processing. The mixed composite alloy powders were commercially

Content from this work may be used under the terms of the Creative Commons Attribution 3.0 licence. Any further distribution of this work must maintain attribution to the author(s) and the title of the work, journal citation and DOI.
Published under licence by IOP Publishing Ltd

available NiCrBSi self-fluxing alloy blended with 1 wt.% RE oxidation particles. All the data used in cladding procedure were listed in table 1.

Table 1. Laser cladding parameters.

Parameter	Laser Parameter Value
Power	1.8 kW
Laser Track Wide	3 mm
Laser beam diameter	2 mm
Overlap Rate	30 %
Cladding Mode	Single layer, straight line
Max Angle of Inclination	30°
Feed Powder	6.0 rpm
Shield Argon	600 ml/min
Laser Scan Speed	8 mm/s

In addition, the nominal compositions of base metal gear steel and NiCrBSi alloy were shown in table 2.

Table 2. Nominal compositions of NiCrBSi alloy powder and substrate steel 17CrMoNi6 in wt.%.

	Cr	S	P	Si	B	C	Fe	Ni	Mn	Mo
NiCrBSi Alloy	17	-	-	4.2	3.7	0.85	7.1	Bal.	-	-
17CrMoNi6	1.50-1.80	0.035	0.035	<0.40	-	0.15-0.20	Bal.	1.40-1.70	0.40-0.60	0.25-0.35

2.1. Gear flank processing

The gear steel used in this study is 17CrMoNi6. The work surfaces were grind and polished until 0.2mm depth was removed from original involute profile before laser cladding. This process was designed to ensure the cladded layer surface can match the original gear flank involute shape with proper post processing.

2.2. Powder blending and drying

The NiCrBSi alloy mixed with 1 wt.% La_2O_3 powder and fully blended in yttrium stabled zircon balls in 750 rpm for 12h. Then mixed powders were heated to 150℃ in a vacuum furnace and kept for 3h for dehumidification.

2.3. The laser cladding on gear flank

The cladding procedure was shown in figure 1. The coating tracks were cladded from gear root to gear top.

In this part it is important to note that there must be a certain angle of inclination (from the normal direction of the cladding involute surface) to make sure the laser beam could reach the roots of gear flank. The maximum deviation angle could be 30° in this case.

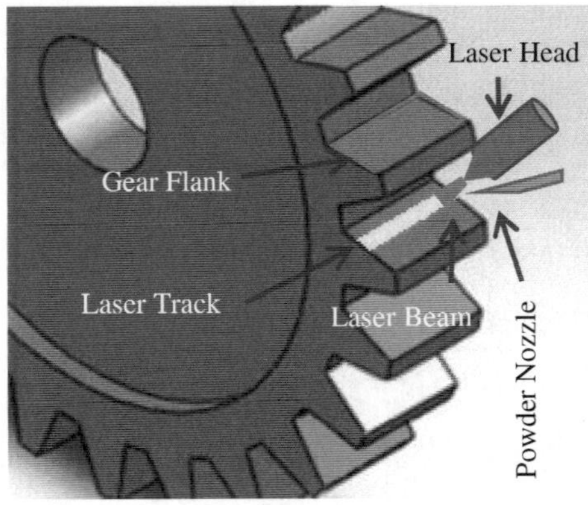

Figure 1. Sketch map for laser cladding on heavy duty spur gear flank.

3. Results and Discussion

3.1. Microstructure analyze

The SEM of overlap area of cladded layer with RE addition should be seen as follows:

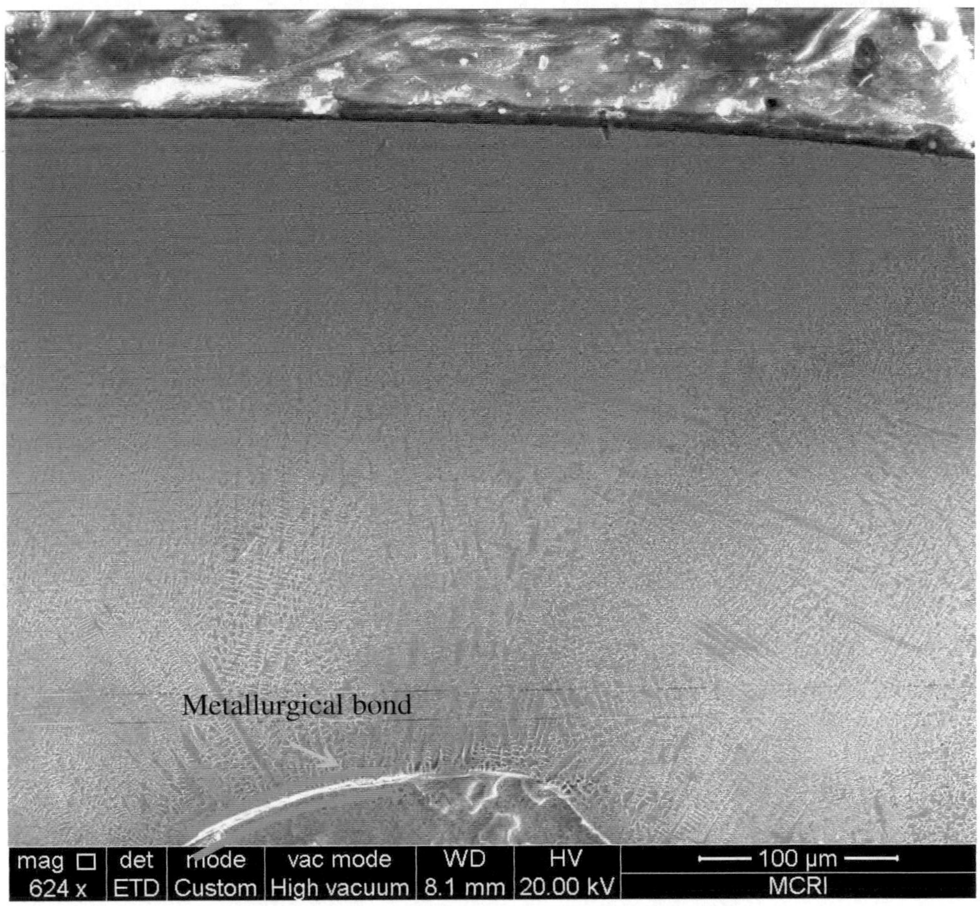

Figure 2. SEM of overlap area of laser clad layer.

It is universally known that the overlap area is the weakest part in laser produced layer. Therefore, this study focus on the microstructure of this overlapped region. It can be seen the top of the layer is flat and neat. The dendrites distribution were homogeneous and compact. This can be explained by the nature of La element: the La atoms is active and are inclined to chemical react with sulfur and phosphorous elements forming low-density chemical reaction product on laser clad layer top. Such compound of lanthanum increased liquid density in solidification cycle and restricted grain growth [7-12]. The dendrites in top area were tinier than that in bottom area. The arrow marked white belt curve was the sign of metallurgically bonding laser coating [7, 8].

3.2. Lanthanum distribution

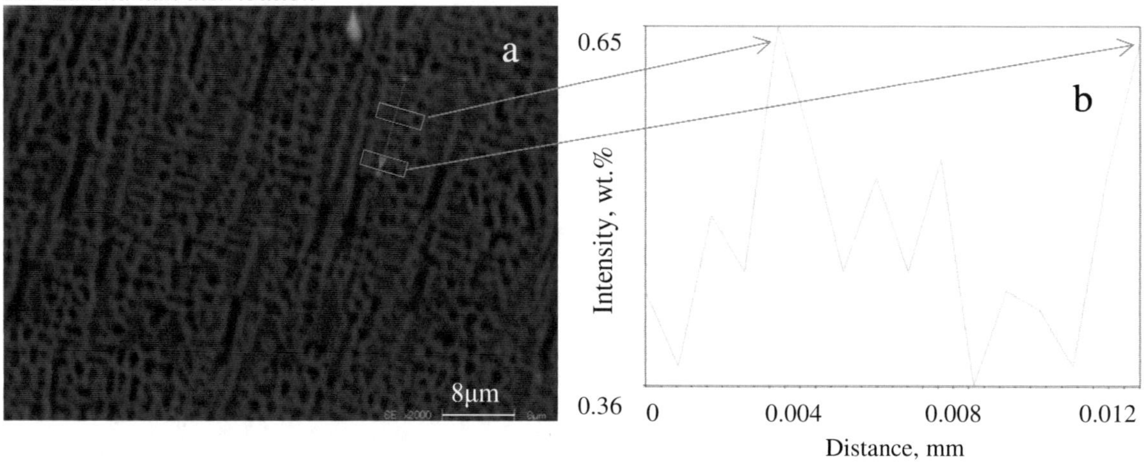

Figure 3. EPMA of La atoms distribution : (a) scan trace location and (b) La content.

In figure 3, the EPMA line scan results proved the La is more likely stay inter-dendrite. Then the drag of grain size increase in molten pool solidification is thus possible. And the top surface dendrites are in fact smaller than that in middle and bottom part, seen in figure 2.

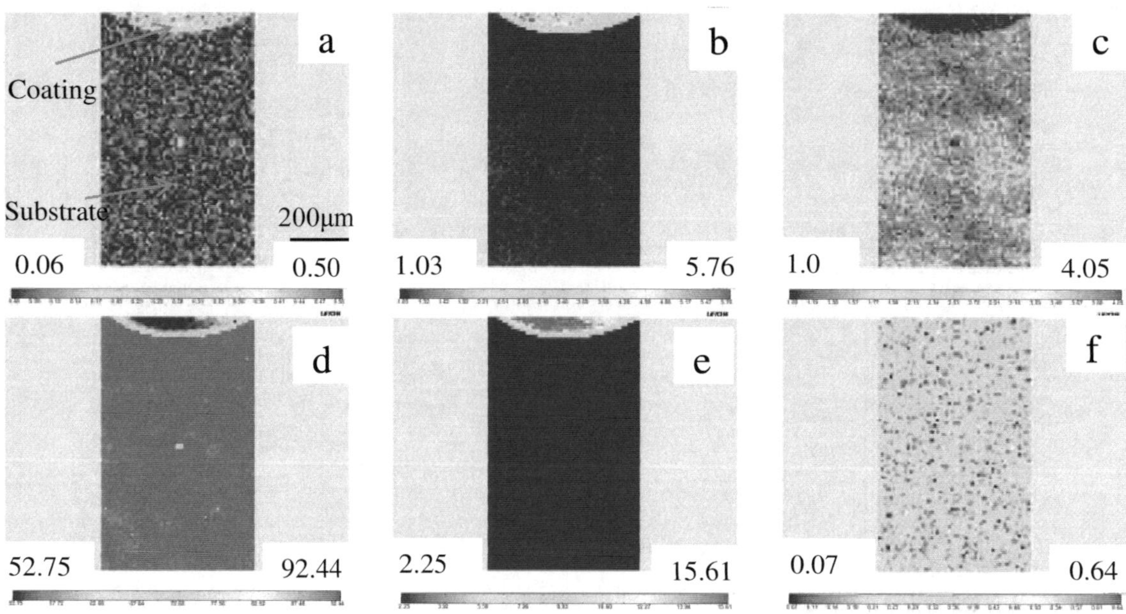

Figure 4. EPMA of different elements distribution:
(a) B, (b) Cr, (c) Mo, (d) Fe, (e) Ni and (f) La. (in wt.%).

From figure 4 (f), a conclusion can be reached that the lanthanum almost evenly distributed in the heat affected zone of substrate and the content density has no obvious differ with that in laser layer zone. This phenomena justified the homogeneously dispersion strengthen of RE to gear steel 17CrMoNi6.

3.3. Wear test

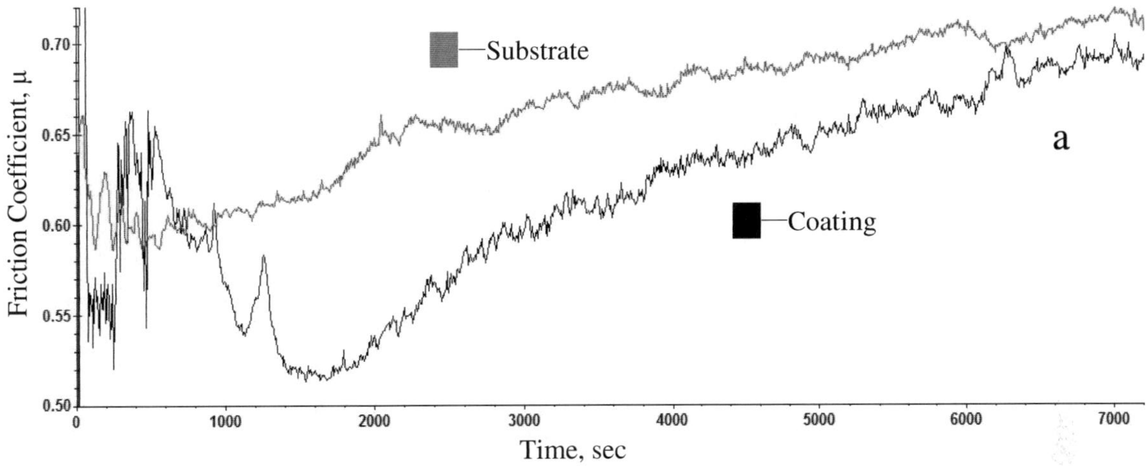

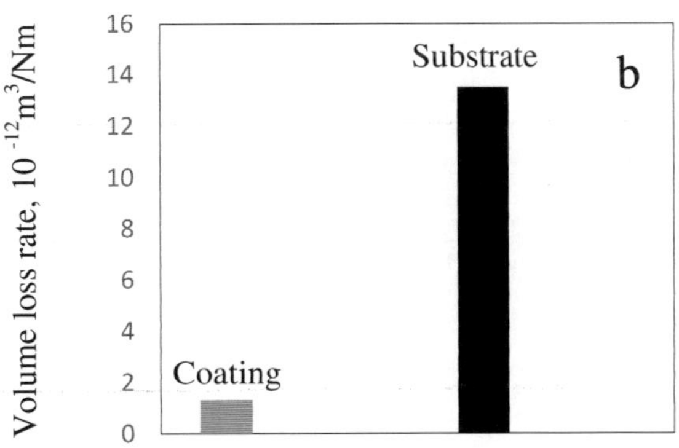

Figure 5. Wear resistance of composite coating: (a) COF and (b) wear volume loss rate.

The COF of composite coating keep below the COF of base steel 17CrMoNi6 in figure 5 (a). As shown in figure 5 (b) the wear volume loss rate average value of 3 tested parallel specimens were 1.3 and 13.5 for composite coating and substrate gear steel 17CrMoNi6, respectively.

4. Conclusions

- The RE element La can be traced evenly dispersed in laser clad layer and deep in the base gear steel 17CrMoNi6.
- The La element mainly segregated in inter-dendrite space and effectively controlled grain size growth.
- Even the overlapped region has both flat and neat top surface and tiny, homogeneous microstructure and dendrites. The laser clad wear-resistance layer has no crack no pore and metallurgically bond with base metal 17CrMoNi6 gear steel.

5. References

[1] Paul C P et al. 2013 *J. Opt. Laser Tech.* **50** 155.
[2] Ocelik V et al. 2015 *J. Surf. Coat. Tech.* **197** 303.

[3] Desale, G R, Paul C P and Gandh B K et al. 2009 *J. Wear* **266** 975.
[4] St Georges L 2007 *J. Wear* **263** 562
[5] Amado J M, Tobar M J and Alvarez J C et al. 2009 *J. Appl. Surf. Sci.* **255** 5553.
[6] Amado J M, Montero J and Tobar M J et al. 2012 *J. Phys. Proc.* **39** 362.
[7] Sharma S P, Dwivedi D K and Jain P K 2009 *J. Wear* **267** 853.
[8] Farahmand P et al. 2014 *J. Ceram. Int.* **40** 15421.
[9] Zhang S H et al. 2008 *J. Mater. Chem. Phys.* **112** 668.
[10] Zhou X W and Shen Y F 2014 *J. Surf. Coat. Tech.* **249** 6.
[11] Sen R J, Das S and Das K et al. 2011 *J. Surf. Coat. Tech.* **205** 3847.
[12] Cai B et al. 2013 *J. Trans. Nonferr. Met. Soc. China* **23** 2002.
[13] Zhang K M et al. 2015 *J. Trans. Nonferr. Met. Soc. China* **22** 1817.
[14] Li X L, He S M and Zhou X T et al. 2014 *J. Mater. Charact.* **95** 171.
[15] Xiao C B and Han Y F 2001 *J. Mater. Sci.* **36** 1027.
[16] Song S H, Sun H J and Wang M 2015 *J. Acta. J. Rare Earth.* **33** 1204.
[17] Serres N, Hlawka F and Costil S et al. 2011 *J. Wear* **270** 640.

Acknowledgments
This work was financially supported by the National Natural Science Foundation of China (51075328, 51675424) and Fundamental Research Funds for the Central Universities (3102015BJ (II) MYZ28).

IOP Conf. Series: Materials Science and Engineering **248** (2017) 012017 doi:10.1088/1757-899X/248/1/012017

Laser re-manufacturing of failure 18Cr2Ni4WA gear in low-speed heavy-load mining machine transmission

X F Chi[1]

[1]Ordos Vocational College, the Inner Mongolia Autonomous Region, China

Abstract. This article investigated laser re-manufacturing technology application in mining industry. The research focused on green re-manufacturing of failure spur. Leave the main gear body stay intact after the dirty, rust, fatigue and injured part were removed completely before the green re-manufacturing procedure begin. The optimized laser operating parameters paved the road for excellent mechanical properties and comparatively neat shape which often means less post processing. The laser re-manufactured gear surface was systematically examined, including microstructure observation, and dry wear test at room temperature. The test results were compared with new gear surface and used but not broken gear surface. Finally, it proved that the green re-manufactured gear surface displayed best comprehensive mechanical properties, followed the new gear surface. The resistance of dry wear properties of used but not broken gear surface was the worst.

1. Introduction

The laser technology, rapidly developed in the recent decades have boost and give birth to a huge amount of new laser research branches and facility applications. The merits of finer microstructure, metallurgical bond with substrate and rapid heating and cooling procedure which are unrivalled and unique in laser technology made it the almost the best fitted technology in green re-manufacture industry for the re-manufacturing of components made of expensive and rare metal. Nickel base alloy have been widely researched and used in corrosive, abrasive and high temperature work conditions [1-4]. There is no doubt that nickel base alloy is more compatible than ferrous alloy in the study of protective re-manufacturing strengthen coating for mining entry-driving machine transmission components.

This article systematically researched the laser re-manufacture of 18Cr2Ni4WA steel low-speed, heavy-load spur gear in entry-driving machine transmission system. And the wear test without lubrication proved the good properties of re-manufactured strengthen coating on gear surface.

2. Experiment

The failure gears were cleaned and got rid of rust and fatigue layer. This part of layer has a deepest depth of 0.25mm on gear surface therefore the removal depth for each gear top surface should be 0.3mm indeed to make sure the entire fatigue layer can be ground clearly.

2.1. The alloy powder and substrate

The chemical compositions of both the nickel base alloy powder for laser treatment and the substrate steel were listed in table 1. The average diameter of alloy powder is 80μm and particles diameter range from 25-125μm. The original substrate surface was carburized and the micro-hardness reached 60 HRC on top surface.

Content from this work may be used under the terms of the Creative Commons Attribution 3.0 licence. Any further distribution of this work must maintain attribution to the author(s) and the title of the work, journal citation and DOI.
Published under licence by IOP Publishing Ltd

Table 1. Chemical compositions of Nickel base alloy powder and substrate steel 18Cr2Ni4WA in wt.%.

	Cr	S	P	Si	B	C	Fe	Ni	Mn	Mo	W
Niickel base Alloy	18	-	-	4.1	3.7	0.85	7.0	Bal.	-	-	-
18Cr2Ni4WA	1.35-1.65	< 0.025	< 0.025	0.17-0.37	-	0.13-0.19	Bal.	4.00-4.50	0.30-0.60	0.25-0.35	0.80-1.20

2.2. Laser operating parameters

The alloy powder should be dry in vacuum drying oven at 200℃ for 1h. The gear surface were grind and polished and cleaned before laser operating. Both carrying and shielding gas were argon. The laser parameters were listed in table 2 in detail.

Table 2. The parameters in laser cladding.

Parameter	Value
Power	2.0 kW
Laser Track Wide	3 mm
Laser beam diameter	2 mm
Overlap Rate	25 %
Cladding Mode	Single layer straight track
Feed Powder	8.0 rpm
Shield Argon	600 ml/min
Laser Move Speed	8 mm/s

2.3. Gear surface cladding

The figure 1 is the gear cross sectional sketch drawing of laser re-manufacture. The orange semi-spherical shape showed the laser tracks which filled the flank of low-speed heavy-load 18Cr2Ni4WA gear.

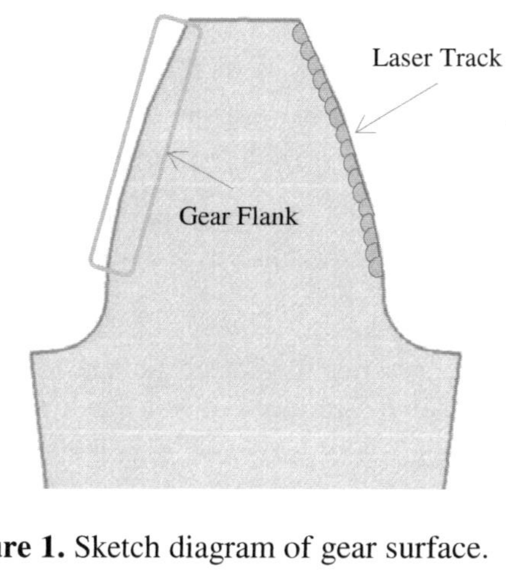

Figure 1. Sketch diagram of gear surface.

3. Wear tests & Discussion

3.1. Microstructure

The cross sectional of laser re-manufactured gear sample was shown in figure 2:

IOP Conf. Series: Materials Science and Engineering **248** (2017) 012017 doi:10.1088/1757-899X/248/1/012017

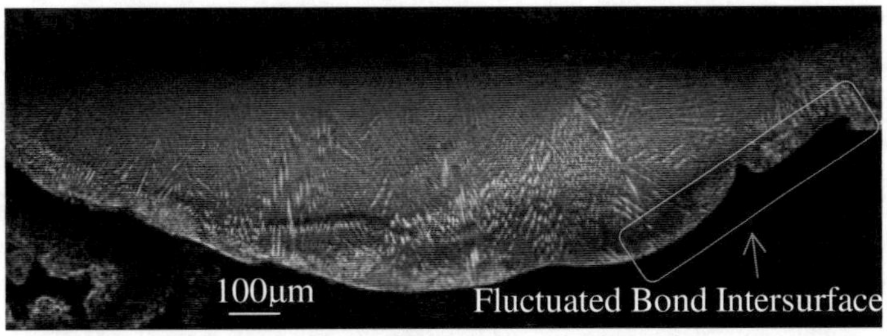

Figure 2. Optical microscope of laser re-manufactured gear.

The OM photograph of figure 2 showed the microstructure distribution of laser clad re-manufactured gear flank. In the middle and bottom of coating there appeared some coarse acicular structure. The bond interface displayed fluctuated curve especially in overlapped area. This may be caused by previous carburization hardening of gear flank. Generally the microstructure of coating is compact and small.

3.2. Wear test

The wear tests were conducted on UMT-2 tribometer and the load was 25N, test time 2h, ball-on-disk strait reciprocating mode was adopted.

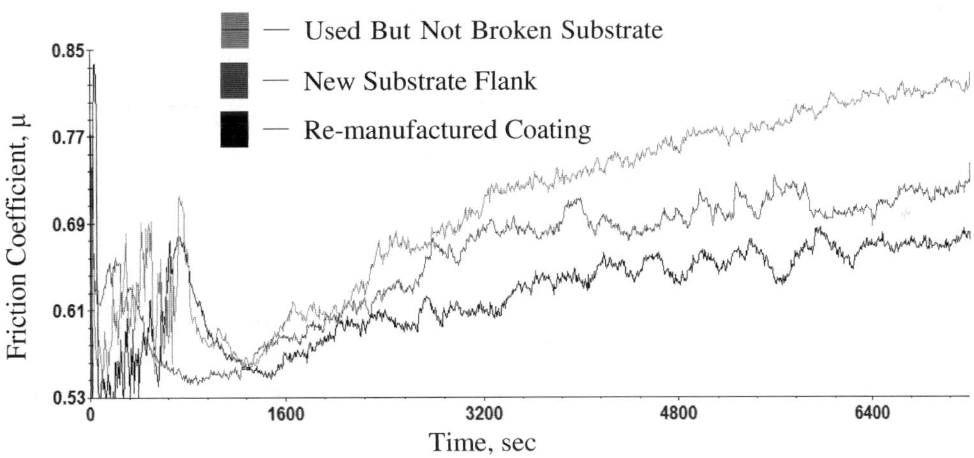

Figure 3. COF of the laser re-manufactured gear coating, old gear surface and new gear surface.

In figure 3, the coefficient curve of re-manufactured coating dropped below the coefficient curves of both new and used substrate steel flank specimens science 1600s when the stable wear procedure began.

By comparing figure 4 (a, b, c) a conclusion can be reached that the scratches in the worn surface of laser clad re-manufactured coating was the slightest and lightest. The figure 4 (c, f) showed the micro groove which may stem from fatigue caused by heavy-load transmission in the worn surface of old, but usable gear surface sample. The latent fatigue spotting and peeling region may not visible but can be shown in wear test. Such fatigue damage might cause deep scratches in figure 4(c) and micro plough in figure 4(f). The laser clad re-manufactured coating and newly produced gear flank samples did not form micro groove in worn surface, the friction pair GCr15 ball didn't appear micro plough neither.

IOP Conf. Series: Materials Science and Engineering **248** (2017) 012017 doi:10.1088/1757-899X/248/1/012017

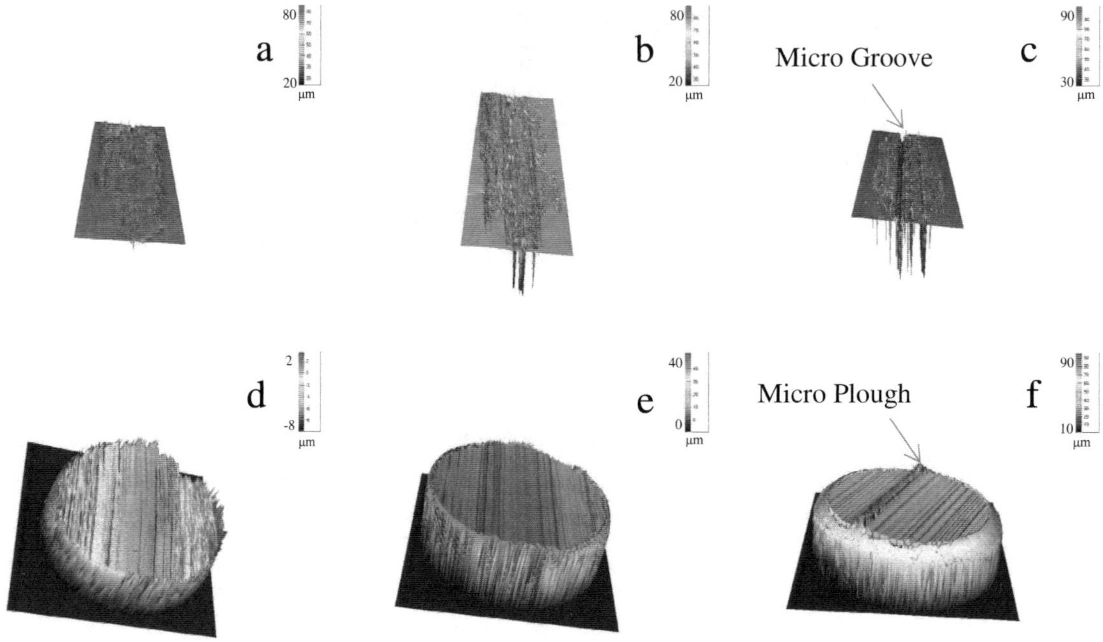

Figure 4. Worn surfaces of (a) laser re-manufactured gear coating, (b) newly produced gear flank, (c) used but still usable gear flank, (d) the friction pair of GCr15 ball for laser re-manufactured gear coating, (e) the friction pair of GCr15 ball for newly produced gear flank, (f) the friction pair of GCr15 ball for used but still usable gear flank.

4. Conclusions

The main worth while to note points in this research are:

- The microstructure in re-manufactured gear flank coating was compact and the bottom interface appeared fluctuated curve which probably may be caused by carburization hardening of gear surface.
- The COF curve of laser re-manufactured coating kept lower than that of both newly produced and used, but still usable.
- The scratches in the worn surface of laser re-manufactured coating sample was the lightest compared with new and old gear flank samples. The micro groove appeared in the worn surface of used, but still usable gear flank specimen which could be stem from the top surface fatigue. Both of laser re-manufactured coating and new gear flank samples did not have such problem.

5. References

[1] Raj D, Tyagi R, Chaubey S K, Souza A D, Palani I A and Jain N K 2015 *J. Mater. Today: Proc.* **2** 1755.

[2] Bobzin K, Brögelmann T, Stahl K, Stemplinger J P, Mayer J and Hinterstoißer M 2015 *J. Surf. Coat. Tech.* **284** 290.

[3] Savaria V, Bridier F and Bocher P 2016 *J. Int. J. Fatig.* **85** 70.

[4] Lv Y, Lei L Q and Sun L N, 2017 J. *Int. J. Fatig.* **98** 121.

[5] Farahmand P and Kovacevic R 2015 *J. Surf. Coat. Tech.* **276** 121

IOP Conf. Series: Materials Science and Engineering **248** (2017) 012018 doi:10.1088/1757-899X/248/1/012018

Effect of Strength Matching on Mechanical Properties of WELDOX 960 Steel Welded Joint

Zhiling Wang[1]

[1]Jiangsu province xuzhou technician institute, Xuzhou, Jiangsu 221151, China

Abstract. In this paper, WELDOX960 high strength steel sheet was used as the research object, and the welding method was adopted with Ar-CO2 mixed gas welding method. The change of microstructure and properties under different strength matching was studied by changing the strength matching method of welded joint. The main research contents are as follows: The comprehensive mechanical properties of the weld are tested, including the tensile, impact and microhardness tests. The changes of the comprehensive mechanical properties of the welds under different strength matching are obtained, and the selection of the best strength matching under different application conditions is put forward.

1. Introduction

Along with the automobile crane lifting capacity and reliability improvement, as far as possible, reduce the boom weight increase the lifting capacity of crane, so with the strength of steel is also more and more high. Welding in the strength is higher than that of high strength steel of 700 MPa, the stress concentration is serious backing weld position will often appear fracture [1]. So far, the high strength steel crane production, in order to get good toughness, often at the expense of the weld metal strength, selection of welding material with low matching, this method makes the weld is guaranteed in flexibility, but has greatly reduced the weight of crane.

The tensile strength of the automobile crane used for WELDOX960 steel 985 MPa as the research object, using Ar ten C02 mixed gas welding and tensile strength were 840MPa (80), 880 MPa (85), 950MPa (90) of 3 kinds of wire, for multi-layer welding. In the stress concentration layer is serious, considering the influence of backing welding on the properties of welding joint, design of each layer all adopt high strength wire 3 and the strength of 525MPa (50) of 3 groups of low strength welding backing, a total of 6 groups of strength matching form, i.e. 50-80, 80-80, 50-85, 85-85, 50-90, 90-90, matching of welding strength affect the microstructure and properties of welded joint of WELDOX960 steel.

2. The experimental materials and specimen preparation

In this study, WELDOX960 low alloy high strength steel sheet was used as the research object, and the effects of different strength matching welds on the microstructure and properties of low alloy high strength steel were studied by adopting different intensity matching methods. In the six groups of 50-80, 80-80, 50-85, 85-85, 50-90, 90-90, one sample was tested in each group. The test load was 4.9035N and the packing time was 10s.In this experiment, four kinds of welding wires with different tensile strength are used. Their grades are: ER50-6, T union GM100, T union GM110, T union GM120, and their chemical composition and mechanical properties are shown in table 1 and table 2.

Content from this work may be used under the terms of the Creative Commons Attribution 3.0 licence. Any further distribution of this work must maintain attribution to the author(s) and the title of the work, journal citation and DOI.
Published under licence by IOP Publishing Ltd

Table 1. The Chemical composition of four kinds of welding wire.

Wire grade	C	Si	Mn	Cr	Mo	Ni
ER50-6	0.1	0.9	1.54	-	-	-
Tunion GM100	0.08	0.6	1.7	0.2	0.5	1.5
Tunion GM110	0.09	0.7	1.7	0.3	0.6	1.65
Tunion GM120	0.1	0.8	1.8	0.35	0.6	2.3

Table 2. The Mechanical properties of four kinds of welding wire.

Wire grade	Yield strength/MPa	Tensile strength/MPa
ER50-6	440	525
T union GM100	780	840
T union GM110	790	880
T union GM120	890	950

It can be seen from table 2 that the tensile strength of the ER50-6 is 525 MPa, which is lower than that of the base metal, but has good toughness and is used as a primer. Because in the low-alloy high-strength steel multi-layer welding process, the backing welding shall bear the welding stress of the rear weld bead, and if the plastic toughness of the bottom is not good, especially in the tip of the groove, it is easier to cause the stress concentration of the backing layer, resulting in cracking of the weld [2]. The welding groove form is shown in figure 1. Especially in the larger thickness conditions, the relative residual stress produced by welding is larger, the occurrence of cracks is more serious, the use of less strength ER50-6 can effectively reduce the occurrence of this situation.

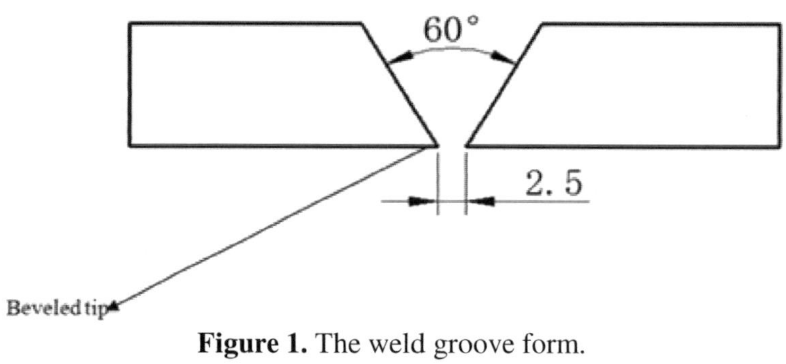

Figure 1. The weld groove form.

The tensile strength of Tunion GM100, T union GM110 and T union GM120 is 840MPa, 880MPa and 950MPa respectively, and these three kinds of welding wire are incremented step by step and are used as main welding materials to achieve different strength matching with the base metal and reach the expected intensity match effect.

The experiment was divided into 6 groups, which were welded by 6 pairs of 300mm*150mm*12mm WELDOX960 steel plate (yield strength of 960MPa) into 6 steel plates with specifications of about 300mm*302.5mm*12mm. Each group was subjected to different intensity matching for welding, and the specific strength matching form can be seen in table 3.

Table 3. Welding Material Tensile Strength Matching Form.

Groups		50-80	80-80	50-85	88-85	50-90	90-90
Backing welding	σb/MPa	525	840	525	880	525	950
Fill and cover welding	σb/MPa	840	840	880	880	950	950

As shown in Table 3, in the group 50-80, the backing welding uses a wire with a tensile strength of 525MPa (hereinafter referred to as 50 wire), and a welding wire with a tensile strength of 840MPa is used to fill and cover the surface (hereinafter referred to as 80 wire); n the group 80-80, the bottom and cover welding both use 80 wire. In group 50-85, the backing welding uses a 50 wire with the filler and cover wire using a tensile strength of 880MPa (hereinafter referred to as 85 wire). And as such, the last set of welding wires use a yield strength of 950MPa (hereinafter referred to as 90 wire). In this way, 6 sets of welds with increasing strength are obtained, namely, six sets including 50-80, 80-80, 50-85, 85-85, 50-90 and 90-90. The following experiments are classified according to this method.

This experiment uses the active gas protection welding (MAG); the protection gas is 86%Ar+14%CO2; the joint form is the butt joint, the plate thickness 12mm; the weld form is 60 degrees V shaped groove and the groove gap is 2.5mm. In order to reduce the influence of welding parameters on different groups, the current, voltage and so on should be kept in a smaller range as much as possible during welding. Therefore, the welding parameters of the different groups are similar and change within the same range. Table 4 shows the welding process parameters used in the experiment. The bead level is 3 layers, 6 passes welding, preheating 100℃ before welding.

Table 4. Welding parameter.

Program	Welding current(A)	Voltage(V)	Welding speed mm/min	Gas flow L/min
Cover welding	110-130	16.7~17.0	61	15
Filling welding	110-130	16.7~17.0	67	15
Backing welding	110-110	16.5~16.8	80	15

3. Experimental results analysis

3.1. The tensile properties of the welded joint

The tensile test results of welded joints are shown in table 1. In the 6 sets of experiments, there are obvious yield outside, and the fracture condition is ductile fracture. As can be seen from the table, the tensile strength of these three groups of 50-80, 50-85, 50-90 is gradually increasing. Although the backing welding material is the same, its influence is small, because the main part of the welding force is filling welding. Under normal circumstances, the mechanical properties of thick plate welds are determined by filling welding, and the role of backing welding as the connection and prevention of cracks is much smaller than that of filling welding [3]. For the plate used in the paper, the 12mm thickness welding proportion is not too large, generally not more than 2mm, but the cover welding and filling welding material is the same, which can play the same role as filling welding, and can withstand greater tension to ensure welding performance. In table 5, the tensile strength of 50-80 is 860.54MPa while the tensile strength of 50-85 is 872.08MPa. The difference between these 2 groups is small, only 11.54MPa, indicating that in terms of the tensile strength, the group of 50-85 did not highlight the corresponding advantages while in the group of 50-90, the tensile strength advantage was most obvious, reaching 946.3MPa. At the same time, the tensile strength of 90-90 reached 973.40MPa, 27.1MPa higher than that of 50-90. This might be due to the higher tensile force, when the 50 welding wire was broken first, the crack source was formed, and then it expanded to the filling and cover surface welding parts, which leads to fracture when the tensile force is relatively low.

Table 5. The tensile strength and elongation of welded joints.

Number	Tensile strength /MPa	Elongation /%
50-80	860.5	12.75
80-80	922.0	10.70
50-85	872.1	12.50
85-85	960. 0	9.30
50-90	946.3	12.33
90-90	973.4	9.20

It can be seen from table 5, the arrangement from large to small of the elongation is just in adverse to that of the tensile strength, that is, the higher the strength of the weld, the lower the elongation, and the lower the plasticity. As can be seen from the table, the elongation in 50-80 was 12.75% while in 50-85, it was 12.50%. There is little difference between the two groups, indicating that the low matching primer layer has little effect on the strength of the plasticity. Considering that the tensile strength of 50-85 is only 11.54MPa higher than that of 50-80, the conclusion is that the overall mechanical performance of 50-80 is better than that of 50-85. The tensile strength of 90-90 is the highest, but its elongation is only 9.20%, nearly half of that of 50-90, which is 12.33%. This shows that the bottom of the 50 welding wire can improve the overall plasticity of the weld, and the difference between the two elongation reached 3.13%. Therefore, the weld strength of 50-90 and 90-90 is relatively higher, especially in the group of 50-90, the plasticity is close to the base metal. The strength of the 50-80 is relatively low, and the plasticity is better. In the group of 50-85, it appears moderate, the strength and plasticity of which are between 50-90 and 50-80, so its comprehensive performance is better.

3.2. The impact testing results and analysis

The impact test is carried out on the JBN-300 pendulum impact test machine, and the results of the absorbing energy of weld center impact are shown in table 6.

Table 6. The absorbing energy of weld center impact.

Groups	Area	Average value AKV/J
50-80	Welds	129.39
80-80	Welds	130.35
50-85	Welds	142.46
85-85	Welds	147.83
50-90	Welds	120.02
90-90	Welds	104.79

Table 6 is the absorbing energy of weld center impact. As can be seen from the table, the weld center impact absorption energy (Akv) in 50-80, 80-80 are 129.39J and 130.35J. The value is basically the same, but there is no difference due to different welding materials. In the same case, there are 50-85 and 85-85. The impact absorption work (Akv) of the two groups are 142.46J and 147.38J respectively. The difference between the two is less than 5J, but the two groups of 50-90 and 90-90 are completely different. The Akv value of 50-90 is 120.02J, and the Akv value of 90-90 is 104.78J, and the difference between the two is 15.24J, which shows that the impact property of the 90 welding seam is improved due to the existence of the 50 welding wire backing layer. We note that in the 50-90 and 90-90, these two groups of weld have the highest tensile strength and lowest elongation, which shows that for high strength, low plasticity of weld, the soft layer will improve the impact property of weld seam, while for weld bead of low strength and better plasticity, such as 80-80 and 85-85, low strength rendering welding has less impact on impact properties, and can even be neglected. Therefore, the Akv value of the two groups of 50-85, 85-85 is the highest, which is to say, its impact toughness is the best, followed by 50-80 and 80-80, and 50-90, 90-90 is the worst.

3.3. Analysis of micro hardness test results

The experimental results of microhardness are shown in figure 2, 3 and 4.

3.3.1 Cover and filling welding hardness. As shown in figure 2, all of the cosmetic welding hardness curves follow the same law: In the weld zone, all the curves are basically paralleled. After entering the fusion zone, the hardness rises suddenly and varies in amplitude, then the curve enters the quenching zone, the hardness decreases greatly, and finally enters the base metal zone, and the hardness rises or falls to reach the hardness of the base metal. As shown in the weld zone, the hardness of the weld is different due to the different welding materials. The change range of 50-80 cosmetic welding hardness is about 300-310HV; the hardness of 80-80 is about 320HV -330HV, and although 50-80 is using the same welding material of cosmetic welding (80 wire), there is still a difference of 10HV-20HV, which is due to small differences in the welding parameters. 50-85, 85-85 welding hardness of the cosmetic welding can be considered the same, and their range of change is smaller than the first two groups, and are at around 270HV -280HV, but it's about 30HV smaller than 50-80 and 80-80. From the previous tensile test data, we can see that the tensile strength of 50-85 is 872MPa, 860MPa greater than 50-80, that is to say, although the micro-hardness of 50-85 is lower by nearly 30HV, it does not mean that its tensile strength is less than that of 50-80. The cosmetic welding hardness of 50-90 is relatively high, reaching about 380HV, and the cosmetic welding hardness of 90-90 is relatively high too, between 360HV-370HV. The difference between the two groups is about 10HV, and the reason is the same as that of the first and second groups, which is the slight difference in welding process.

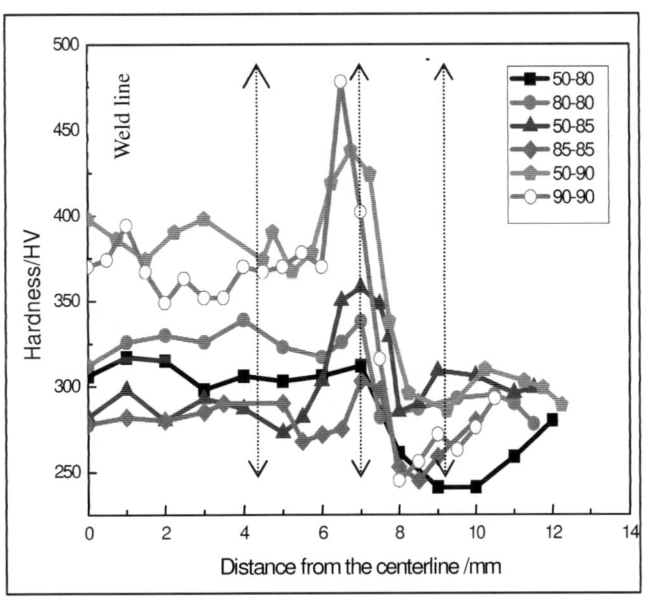

Figure 2. The distribution curve of Cover weld hardness.

In the fusion zone, the hardness of all the basic groups has an upward trend, but the rates of increase are different, in which 90-90 curve is the most sharp, and the highest hardness reaches more than 450HV, followed by 50-90, which is about 440HV. The hardness of the remaining group here is much smaller than that of the two groups. Following 50-90, the group of 50-85 is 350HV; 80-80 is about 340HV; 50-80 is about 320; and 85-85 is the minimum, at around 300HV. The fusion zone is the junction of the weld and the base metal. The part is an incomplete mixing zone, the micro-structure and properties being uneven, and there are a lot of dislocations, so the hardness is relatively higher. However, higher hardness results in embrittlement of heat affected zone, which results in the increase of cold cracking tendency of welded joint and the reduction of impact toughness of heat affected zone. Therefore, the cold cracking resistance of 90-90 and 50-90 is the worst, and the cold cracking resistance of 85-85 and 50-80 is the best, and 50-85 and 80-80 are between them.

In the quenching zone, the hardness values of each group are decreased, and the hardness of the quenching zone will decrease to about 280HV regardless of the hardness of the fusion zone. This is because the quenching zone in the heat affected zone of each group is basically the same, This is because the quenching zone in the heat affected zone of each group is basically the same, which are carbide, ferrite and austenite decomposition products, and this hybrid tissue has lower hardness and less resistance to plastic deformation [4]. In the base metal zone, the hardness of each group has not changed much. The hardness of the base metal is gradually transited from the quenching zone to the base metal, and the hardness of the base metal is about 300HV.

3.3.2 Hardness of filling welding. Figure 3 shows the hardness distribution of the filling welding. Its change rule is basically similar to that of cosmetic welding hardness. From the welding center to the fusion line, the hardness value remains basically unchanged, and the hardness decreases sharply after the curve enters the heat affected zone, and then slowly rises to the base metal region.

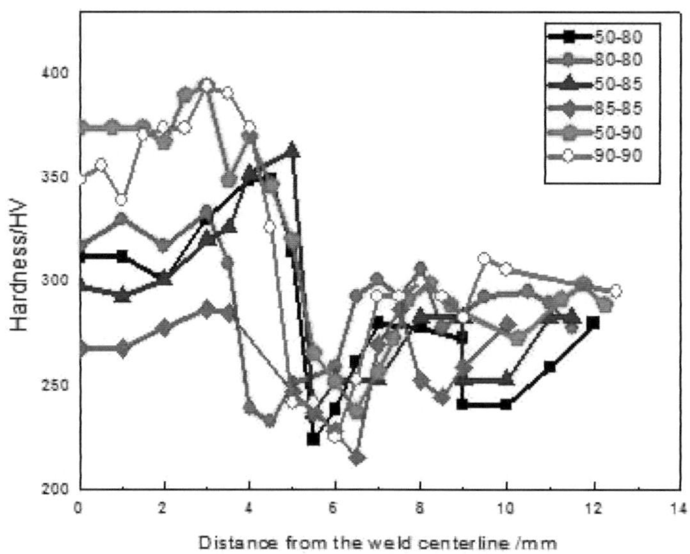

Figure 3. The distribution curve of filling welding hardness.

Unlike cosmetic welding, the hardness of the filling welding does not rise significantly at the fusion line, or let's say, at a very small increase. This is because when the cover welding is carried out, the welding heat input carries out an annealing treatment of the filling welding, so that the non-uniformity of the micro-structure and the performance of the fuse line is greatly improved. Thus, the hardness value is reduced to approximately equal to that of the weld zone. Also, it reduces the hardness of the fusion line and improves the cold cracking resistance of the heat affected zone. In figure 3, the hardness strength of weld zone status are filled and cover welding. 50-90 and 90-90 are the highest, at about 370HV-390HV; 80-80, 50-80 in the middle, the hardness is about 310HV-320HV; and the lowest hardness is the two groups of 50-85, and 85-85, with the hardness respectively 300HV and 280HV. In the heat affected zone, the lowest hardness decreased to less than 250HV, 30HV less than the cosmetic welding, which is 280HV. This is also because the amount of heat on the cosmetic welding softens the annealing process of the filling welding. The hardness curve passes to the base metal zone, and the hardness is about 300HV, which is the hardness of the base metal.

3.3.3 Backing welding hardness of the wire priming in 50-90 and 90-90. The hardness of the backing weld in the group of 50-90 in figure 4 is about 260HV, and the hardness value of the deposited metal is almost 100HV higher than that of the ER50-6 wire deposited metal, which is 159HV. This is because the softer 50 wire metal is attached to the harder 90 wire metal, and the higher the strength of the latter, the higher the restraint on the former, resulting in higher hardness. It can also be explained that the higher the strength of filling and cosmetic welding, the worse the plasticity and toughness. The smaller the stress released by the plastic deformation after welding, the higher the residual stress and the higher the hardness of the backing welding. At the junction of the weld and the heat affected zone, the hardness starts to rise substantially, and the high-point of hardness is higher than the hardness of the base metal because of the uneven number of dislocations in the fusion zone.

The hardness of the backing welding in the 90-90 groups is about 415HV, much higher than that of the 90-90 filling welding (360-370HV) and the hardness of the cover welding (about 380 HV). The main reason is the influence of the backing bead on the welding bead, which leads to the increase of hardness. But in the fusion zone and the heat affected zone, the hardness change law of the 90-90 backing welding and filling welding is similar. In the heat affected zone, the hardness does not rise significantly, then decreases directly, and then smoothly transfers to the base metal.

The higher hardness will lead to embrittlement of the weld, which results in the increase of cold cracking tendency of the welded joint and the decrease of the shock toughness of the weld metal. Thus, the cold cracking resistance of 90-90 and 50-90 is the worst, and the cold cracking resistance of 85-85 and 50-80 is the best, and 50-85 and 80-80 are between them.

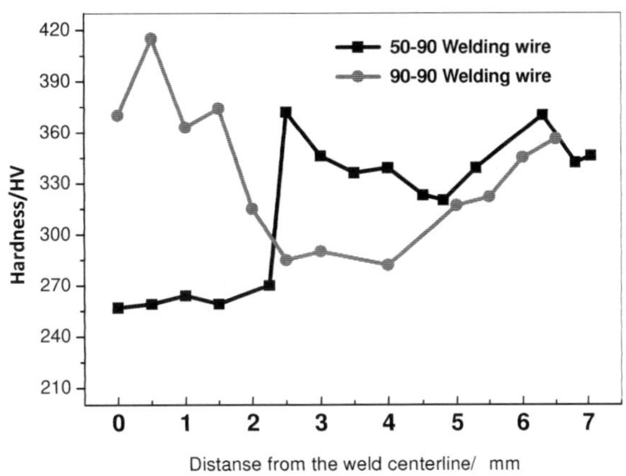

Figure 4. The hardness of backing welding substrate of 50-90 and 90-90 groups.

4. Summary

Except for 50-80, the tensile specimens of each group have obvious deformation after breaking, and there is a certain yield in tensile process, which belongs to ductile fracture.

The tensile strength of welds in the two groups of 50-90, 90-90 is the highest, but the plasticity and weld impact toughness are the worst; the weld strength in 50-80 is the lowest, but the plasticity and weld impact toughness are better; the strength and plasticity of 50-85 are between that of 50-90 and 50-80; the toughness of welding seam is the best, and the comprehensive mechanical property is better.

IOP Conf. Series: Materials Science and Engineering **248** (2017) 012018 doi:10.1088/1757-899X/248/1/012018

For high strength and low plasticity welds, low strength backing layers can significantly improve the impact properties of welds; but for low strength and better plasticity welding bead; low strength backing welding has little impact on impact properties, which could even be ignored.

The micro-hardness of all welds in the two groups of 85-85, 50-85 is the lowest; that of 50-80 and 80-80 are slightly higher, and that of 50-90 and 90-90 are the highest hardness.

Reference

[1] Yan K 2007 *Mech. Eng. Autom.* **(1)** 171-2.

[2] Wang J 2007 *Study on Microstructure and properties of welded joint of bogie frame.* Dalian Jiaotong University, [42] Tor-Bjorn Laraaon, Tomas Berlund, Handbook on WELDING of Oxelosund steels

[3] Guo A M, Li S R, Guo J et al. 2008 *Effect of zirconium addition on the impact toughness of the heat affected zone in a high strength low alloy pipeline steel.* Materials Characterization, 134-9.

[4] Han L & Chrysanthou A 2008 *Mater. Eng.* **29(2)** 458-68.

Research on Kinematic Trajectory Simulation System of KUKA Arc Welding Robot System

Min Hu[1]

[1]Jiangsu province xuzhou technician institute, Xuzhou, Jiangsu 221151, China

Abstract. In this paper, the simulation trajectory simulation of KUKA arc welding robot system is realized by means of VC platform. It is used to realize the teaching of professional training of welding robot in middle school. It provides teaching resources for the combination of work and study and integration teaching, which enriches the content of course teaching.

1. Arc welding KUKA robot system and kinematic modelling

1.1. Arc welding KUKA16 robot system

In this paper, KUK16 model of arc welding robot as the background of the robot movement trajectory simulation.

Figure 1. The physical robot.

Robot is composed of the robot body, robot control cabinet, teaching box composition. KR16 arc welding robot mainly by the welding system mainly includes the robot system, welding system, the surrounding equipment, safety devices and other accessories and other components. Welding system mainly by the welding power supply, robot wire feeder, cable assembly, robot welding gun, as shown in figure 2. Other accessories include anti-collision, clear gun shear mechanism, dust removal equipment.

Content from this work may be used under the terms of the Creative Commons Attribution 3.0 licence. Any further distribution of this work must maintain attribution to the author(s) and the title of the work, journal citation and DOI.
Published under licence by IOP Publishing Ltd

IOP Conf. Series: Materials Science and Engineering **248** (2017) 012019 doi:10.1088/1757-899X/248/1/012019

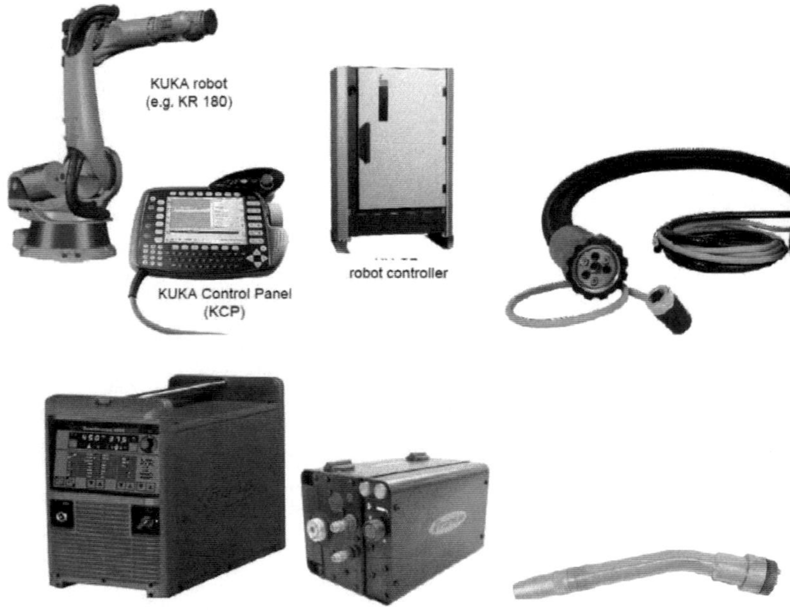

Figure 2. The composition of the arc welding robot.

1.1.1 Robot body. KUKA16 model body weight 45kg, the shape of the structure from the 6 joints, each rotation joint and joint movement has a degree of freedom [1], that is the number of robot joints is its degree of freedom, so the robot has six free degree. Each joint is driven by a servo motor. A4, A5, A6 three joints to determine the location of the hand, the robot can be configured according to the function of the corresponding end of the corresponding device. In this paper, the end of the robot is equipped with arc welding torch; the other three joints A1, A2, A3 control robot pose. As shown in figure 3.

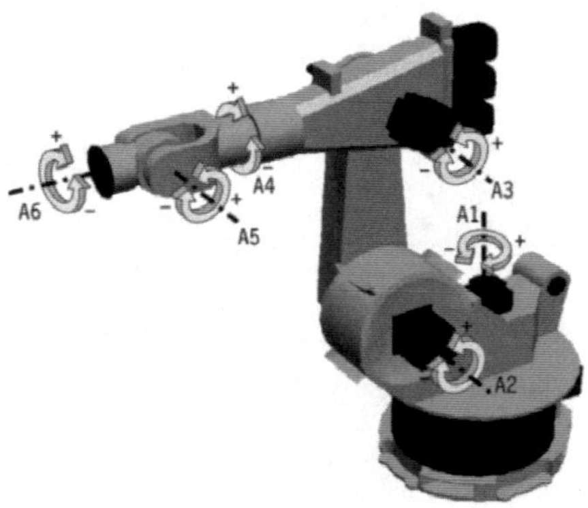

Figure 3. The main body of the robot

Robot mechanism performance indicators are measured by multiple aspects of the main consideration of two points: First, the economy, the second is the use of performance. The standard is: 1) the degree of freedom of the robot, that is, the number of independent coordinates required for the joint movement of the object space; 2) the working range, that is, the range of the end of the torch, the working range of the robot is 1600mm; 3) ; 4) repeat positioning accuracy.

IOP Conf. Series: Materials Science and Engineering **248** (2017) 012019 doi:10.1088/1757-899X/248/1/012019

Table 1. The partments of KUKA16 robot.

Structure		Vertical multi-articulation (6 degrees of freedom)
Maximum Single axis operating range	Axis 1 (turn around)	±185°
	Axis 2 (Upper arm)	+35°, -155°
	Axis 3 (Lower arm)	+154°, -130°
	Axis 4 (Wrist turn)	±350°
	Axis 5 (Wrist swing)	±130°
	Axis 6 (Wrist turn)	±350°
	Axis 1	156°/s
	Axis 2	156°/s
	Axis 3	156°/s
	Axis 4	330°/s
	Axis 5	330°/s
	Axis 6	615° / *s*

1.1.2 Control cabinet. The teaching robot uses KRC2 KUKA robot second generation control system. Use familiar Windows system interface, standard industrial computer, hard disk, optical drive, floppy drive, print interface, I/O signal, a variety of bus interface, remote diagnosis and other configurations. The exterior of the control cabinet and the internal control components are shown in figure 4 and figure 5.

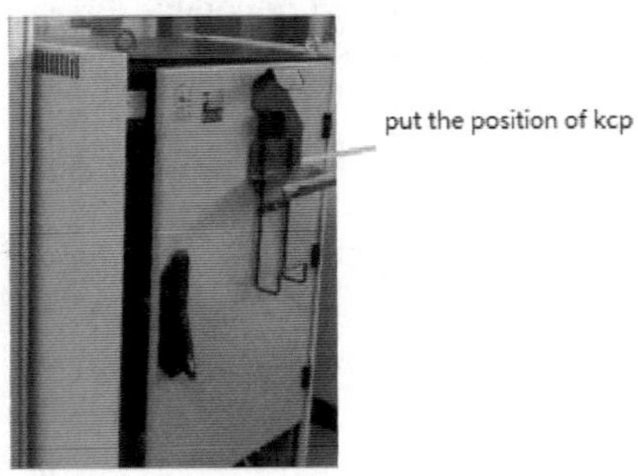

Figure 4. External components of the control cabinet.

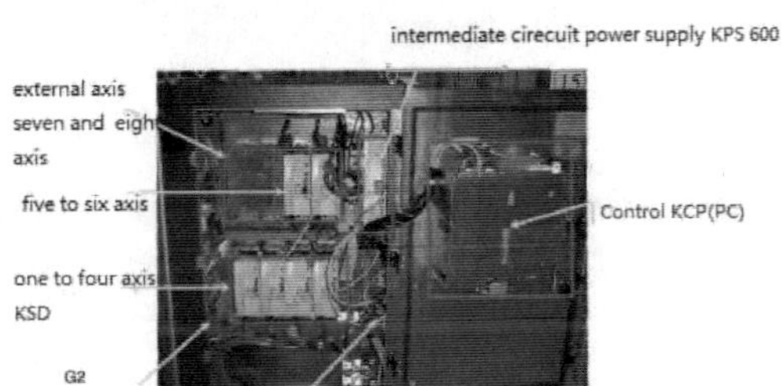

Figure 5. Internal components of the control cabinet.

IOP Conf. Series: Materials Science and Engineering **248** (2017) 012019 doi:10.1088/1757-899X/248/1/012019

As shown in figure 5, the internal components of the control cabinet are controlled by the computer section. The robot body shaft, the external shaft drive control circuit, the middle control power supply G1 KPS600, the supply power supply G2 KPS27 composition. The robot body has 6 axes, can add 6 external shaft, more than 2 external shaft control cabinet need to add the top cabinet, set the top cabinet can be equipped with four external shaft drive. Computer control PC part shown in figure 6. In the computer control part of the COM interface, you can store the file.

Figure 6. Composition of the computer PC.

1.1.3 Demonstration system. Demonstration system is the machine man-machine interaction interface, through the teaching device can manipulate the robot movement, to achieve different forms of welding trajectory, complete teaching programming [2]; to set the movement coordinates, to achieve TCP and external axis calibration, through the display program Status code information on the machine fault diagnosis and maintenance; but also to achieve the safety of the role of the robot, the appearance of teaching parcels shown in figure 7.

The display of the LCD screen is a VGA color image can be intuitive to show the operation of the robot and programming action. The system is used in our familiar Windows operating system, many features in the user interface is similar to the PC, so the operator is easy to get started. On the right side of the display there is a flange on the right side of the wheel, which is used in combination with a permitted switch on the back of the monitor to move the robot, and the operator is free to use it. The appearance of the entire KCP in accordance with the principles of ergonomic design, user-friendly design, light and smart, easy to operate.

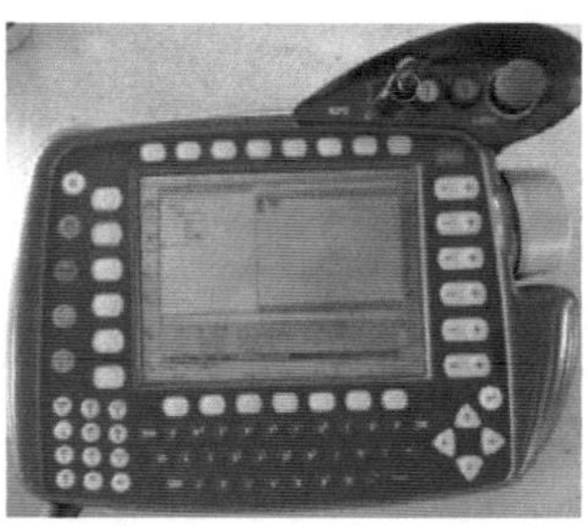

Figure 7. Appearance of the demonstrator.

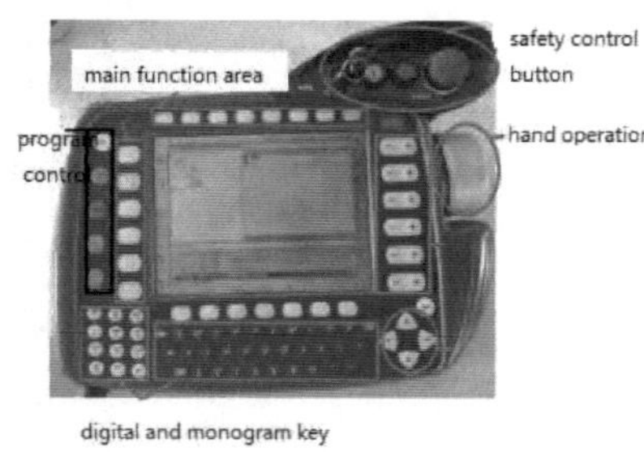

Figure 8. The function of the demonstrator.

The demonstrator is mainly composed of several parts shown in figure 8, and the operation safety control button area can perform the welding mode adjustment, the driving setting, the emergency stop button and so on, and plays the role of safety protection during the operation. The main function of the operating menu has files, edit, configure, display, prepare to run, instructions, craft and help. The file editor is mainly used to manage the file saved by the welding program. The configuration and display are for the system setting of the robot body. The function of the running menu is mainly to measure, check the zero point, set the robot parameters and install the additional software. Instructions and processes used in the programming of welding robots programming statement writing and welding procedures of the switch, the role of help is the operator encountered difficulties in the operation of the guide.

1.1.4 Welding system. The welding system is an important part of the arc welding robot. The welding system directly affects the weld quality. The welding system in this paper mainly includes the German EWM welder, the welding wire system and the clear gun program, as shown in figure 9 and figure 10.

Figure 9. The welding machine of PHOENIX 522.

Figure 10. The wire feeding mechanism.

2. Programming languages

KUKA robot programming instructions according to the functional division can be divided into: motion control, environment definition settings, computing functions, program control input and output functions. Motion control has two control methods, PTP control (Point to Point) trajectory control and CP Continuous path control [3]. The point-to-point control method is to control the location of the torch through the position of the robot, not taking into account the trajectory path between the point and the point; the continuous trajectory control is to control the trajectory between the two points, interpolate a lot of short points to complete the continuous Of the curve trajectory. Trajectory interpolation methods are: linear interpolation method, circular interpolation method, free curve interpolation, joint interpolation.

KUKA robot movement has three main ways, namely, point to point PTP, linear LIN, and circular CIRC.

The form of PTP is to determine the route from point to point, and the trajectory between points and points is not precisely determined, so there are many possible paths.

Linear motion LIN, Robot axis installation TCP or workpiece reference point along the line to move to the end of the way to control the robot torch to the specified speed along the exact path to reach the specified end.

Circular CIRC interpolation motion, TCP or workpiece reference point along an arc to the target point. Use the start point, the auxiliary point, and the end point to define the path. The direction of the entire path has been changed, CIRC movement in the case of accurate positioning of the CIRC movement, the robot accurately stopped at the end of each.

The current robot programming has teaching programming, offline programming and robot language programming. At present, most of the operation of the KUKA arc welding robot is taught by teaching. Teaching programming is simple, reliable, fast, for the simple part of the majority of enterprises using teaching programming. However, teaching programming has a drawback, mainly for motion control and sequential action as the center, if the teaching programming is not the result of our imagination, to re-teach. The off-line programming can be far away from the complex welding environment, in the simulation environment to achieve the welding path; therefore, off-line programming is generally used in large quantities to repeat the production of complex shapes of the workpiece.

3. Robot coordinate system setting

Arc welding robot commonly used coordinate system with the base coordinates, joint coordinates, tool coordinates, workpiece coordinates (also known as the target coordinate system) and table coordinate system. And the earth directly in contact with the base coordinate system, also known as the earth coordinates, often used Cartesian coordinate system, is to control the overall coordinate system of the robot; robot at work, the direct impact of the operation of the tool coordinate system and workpiece coordinates, the two The study of coordinates is also a long-term problem of arc machining robot kinematics. The coordinated movement of the two plays an important role in the application of motion simulation software in the next section.

IOP Conf. Series: Materials Science and Engineering **248** (2017) 012019 doi:10.1088/1757-899X/248/1/012019

The tool coordinate system is called the hand coordinate system, which is the robot end operator. For the arc welding robot is the welding torch, in order to study his movement, the fixed coordinate system origin set in the folder or the end of the torch or the middle of the middle of the folder position [4]. In order to describe the robot end position we usually use a matrix T to represent. The tool coordinate system is called the hand coordinate system, which is the robot end operator. For the arc welding robot is the welding torch, in order to study his movement, the fixed coordinate system origin set in the folder or the end of the torch or the middle of the middle of the folder position. In order to describe the robot end position we usually use a matrix T to represent.

$$T=[n \ \ o \ \ a \ \ p]=\begin{bmatrix} n_x \ o_x \ a_x \ p_x \\ n_y \ o_y \ a_y \ p_y \\ n_z \ o_z \ a_z \ p_z \\ 0 \ \ 0 \ \ 0 \ \ 1 \end{bmatrix} \tag{1}$$

Referring to figure 11 where the z-axis is represented by the vector a, the y-axis is represented by the azimuth vector o, and the x-axis is determined by the right-hand rule: n = o × a is denoted by the vector n. Where vector P describes the position of the robot in the base coordinate and is used to determine the robot end position. The vectors a, n, and o are used to describe their gestures.

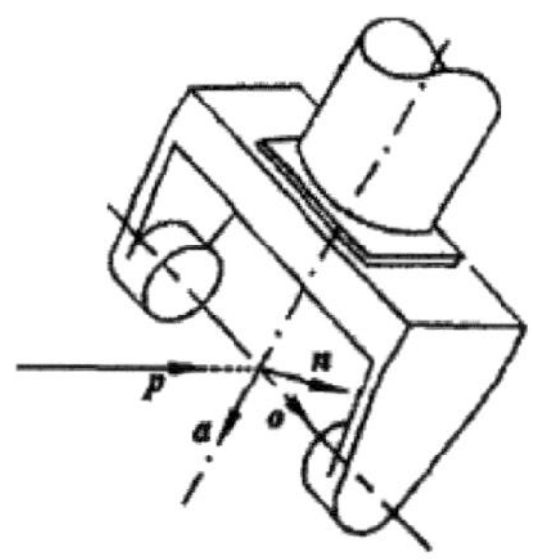

Figure 11. The tip position of the robots.

For joint robots, the motion of each joint will have an effect on the end pose. Their movements are independent of each other, and each joint move around the respective coordinate system. Therefore, it is necessary to use the homogeneous coordinate transformation of the previous section to complete the coordinate transformation between the two coordinate system.

In this paper, the coordinates of the KUKA16 arc welding robot coordinate, the Word coordinates (using Cartesian coordinates, the origin on the robot body base), TOOL coordinates, the origin of the robot 6-axis flange center point, Base coordinates (workpiece coordinates), The origin on the workpiece. In the joint coordinate system, each axis of the robot can operate in both positive and negative directions. The rotation of each axis in Cartesian coordinates is shown in figure 12

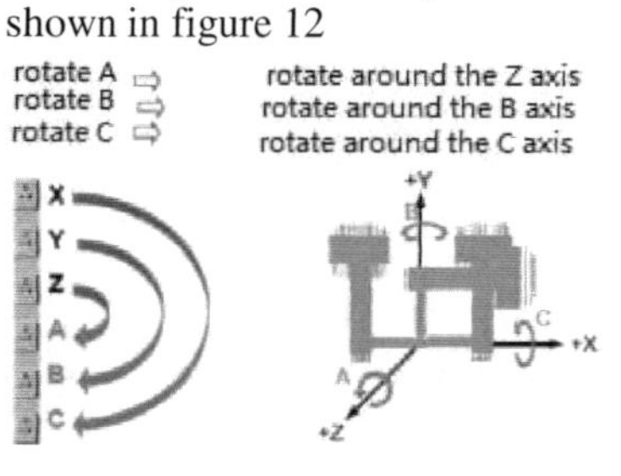

Figure 12. The rotation of the descartes coordinate

4. Kinematics modelling of the robot

IOP Conf. Series: Materials Science and Engineering **248** (2017) 012019 doi:10.1088/1757-899X/248/1/012019

Robot kinematics is the theoretical basis for the study of the simulation of arc welding robot. There are two main problems in the research of robot kinematics, one is kinematics and the other is inverse kinematics.

The so-called positive kinematics is known as geometric parameters of the robot bar and the joint variables, the end of the actuator relative to a given coordinate system position and attitude, the given coordinate system is the robot system in the earth coordinate system, the world coordinate system (World Coordinate).

Inverse kinematics and positive kinematics are antisplot modes, known geometric parameters of the bar, given the position and attitude of the end effector relative to the overall sitting association, and determining the size of the joint variable. In general, the joints of the robot arm are independent variables, the end of the actuator position changes are usually reflected in the overall coordinate system, therefore, according to the end of the actuator in the overall coordinate system to determine the robot joint variables to be carried out The movement is the inverse of the kinematics solution. The solution of this problem is one of the important contents of the robot motion control system.

KUKA16 arc welding robot kinematics modeling First of all, the kinematics analysis of the robot, the establishment of the corresponding robot model is to achieve the robot control and motion trajectory simulation of the prerequisite. In order to obtain the KUKA robot motion model, the direct method is to study the KUKA robot motion algorithm. As the manufacturers of robot control system internal kinematics algorithms are confidential, so many of the robot to study. The D-H parameter design method is used to study the D-H parameter design method. In this paper, the KUKA16 arc welding robot is used to design the D-H rules. The specific rules of the D-H parameter method are not specified in this paper.

In the analysis, the joints of the robot are simplified as rods, and the connection between the joints is not affected. The simplified result will not affect the analysis results. Figure 13 is to simplify the KUKA robot coordinate system, combined with the geometric dimensions of the robot parameters DH kinematic parameters are shown in table 2.

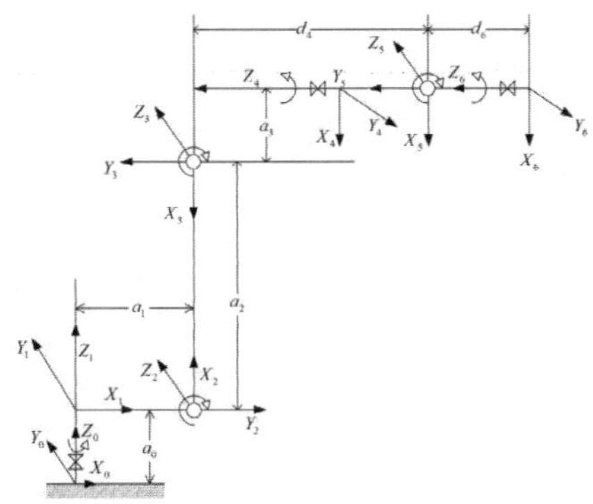

Figure 13. The coordinate system of the robot.

Table 2. D-H parameters.

Joints	θ_i(°)	α_{i-1}(°)	a (mm)	d (mm)	Joint angle range
1	$-\theta_1$	-90	260	0	-185~185
2	θ_2	0	680	0	-155~35
3	θ_3+90	90	35	0	-130~154
4	$-\theta_4+180$	90	0	670	-350~350
5	θ_5	-90	0	0	-130~130
6	$-\theta_6$	0	0	0	-350~350

In the table, θ_i indicates the rotation angle around the z-axis joint, d indicates the vertical distance between the common lines of the two links. α_{i-1} refers to the twist angle of the connecting rod, and a represents the joint offset, which refers to the distance between the two joints.

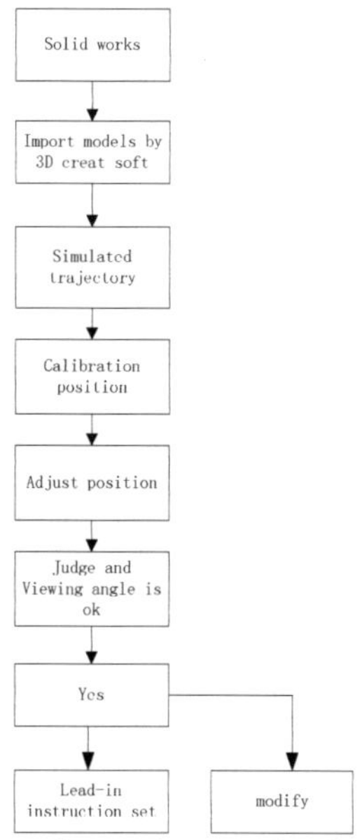

Figure 14. The robot trajectory simulation flowchart.

5. Application of trajectory simulation of the arc welding robot

The software in the arc welding robot simulation is in Visual Component Suit 3D Creat simulation software to achieve. Arc welding robot teaching system is based on the interactive computer graphics technology and robot kinematics theory to support the basis of geometric modeling is generated in the three-dimensional drawing software and display. In the process of teaching practice, the use of the software simulation function, help to establish the concept of robot theory, improve the practical operation of the robot programming capabilities. With the software, students in the practice process can simulate the welding process, and experience the application of positioning function, better grasp the welding process of welding gun posture settings and changes, improve welding quality. In the simulation of the use of the process [5], the software has a strong real-time and authenticity of the characteristics, to simulate the image and the actual work of production status is very similar. But the important problem that exists in the simulation is the deviation between the simulation and the actual product, which is what we are going to solve.

The following is mainly for this problem made the following study to complete the simulation trajectory process shown in Fig. 14.

5.1 Solid works software

SolidWorks is a professional three-dimensional mechanical modeling software, can be features, database and parametric design. Has a strong design product features, the use of stretching, rotating, shell, feature array and other functions can be achieved complex geometric model. Using SolidWorks, you can also dynamically observe the movement of parts, modeling can be generated after the workshop general drawing drawings. In the production of weldment model, you can call the model library to create tubes, rods, angle iron, etc. to model. The software is one of the most powerful software in 3D drawing software.

In the simulation, SolidWorks draws a 1: 1 three-dimensional model according to the drawing size, and saves the model as .STL format. Load the .STL file through the third-party simulation software to open the model.

5.2 Visual component suit 3D creat software

Visual Component Suit 3D Creat is developed in the Windows environment, with a reusable 3D intelligent database, a powerful graphics editing authoring environment, can be quickly created and released 3D set up to achieve automation control, is a powerful Simulation software. The software can provide special programs to add interface ADD-ON to extend the software function of the application. You can also embed the 3D creat window in other programs. In the third-party software or production line controller can be based on COM-based interface for production verification and optimization design. The software is mainly used in process and mechanical simulation, storage and logistics simulation, off-line programming and robot unit development and flexible manufacturing system simulation and so on [6].

In this paper, the application of ADD-ON embedded in the development of the welding process, can be combined with KUKA robot to achieve offline programming function, the software embedded in the VB development of the main window of the teaching software as a function module call, through visual teaching The method enables students to have a clearer understanding of the movement of each axis of the robot during the welding process.

6. Summary

With the Chinese manufacturing 2025 and industrial 4.0 put forward, the industrial intelligence and automation should meet higher requirements. As an important part of the industrial robot family, the welding robot is widely used in the ship, automobile, home appliance and other industries. In order to meet the requirements of automation development, the robot has strengthened the professional basis and general education of the robot, so that the students can adapt to the job and brand application, The simulation system based on VC platform is developed to help the professional teaching of welding robot.

7. Reference

[1] Ni N 2005. *Neural network based welding robot CO2, welding process parameter optimization,* HeFei University of Technology.

[2] Tian Z Q 2004 *Research and development of a robot mechanism with five degrees of freedom and kinematic analysis*. Xi'an University of technology.

[3] Zhang X 2006 *Research and implementation of five DOF serial robot control system.* Northeastern University.

[4] Li X X 2014 *Space operation trajectory and welding torch attitude design of arc welding robot.* Shenyang University.

[5] Zhao Y Y. *Inverse kinematics analysis and Simulation of robot arm chess game.* Qiingdao University.

[6] Kang D 2013 *Design and implementation of curriculum management system based on project teaching*. Zhongshan University, 10.

IOP Conf. Series: Materials Science and Engineering **248** (2017) 012020 doi:10.1088/1757-899X/248/1/012020

Waveform distortion by 2-step modeling ground vibration from trains

F Wang[1], W Chen[1], J Zhang[1], F Li[1], H Liu[1], X Chen[2], Y Pan[1], G Li[1], F Xiao[1]

[1] School of civil engineering, Heilongjiang University, Harbin, Heilongjiang 150080, China
[2] School of Civil Engineering, Central South University, Changsha, Hunan 410075, China

Abstract. The 2-step procedure is widely used in numerical research on ground vibrations from trains. The ground is inconsistently represented by a simplified model in the first step and by a refined model in the second step, which may lead to distortions in the simulation results. In order to reveal this modeling error, time histories of ground-borne vibrations were computed based on the 2-step procedure and then compared with the results from a benchmark procedure of the whole system. All parameters involved were intentionally set as equal for the 2 methods, which ensures that differences in the results originated from the inconsistencies of the ground model. Excited by wheel loads of low speeds such as 60 km/h and low frequencies less than 8 Hz, the computed responses of the subgrade were quite close to the benchmarks. However, notable distortions were found in all loading cases at higher frequencies. Moreover, significant underestimation of intensity occurred when load frequencies equaled 16 Hz. This occurred not only at the subgrade but also at the points 10 m and 20 m away from the track. When the load speed was increased to 350 km/h, all computed waveforms were distorted, including the responses to the loads at very low frequencies. The modeling error found herein suggests that the ground models in the 2 steps should be calibrated in terms of frequency bands to be investigated, and the speed of train should be taken into account at the same time.

1. Introduction

As a train moves along a track, uneven contact occurs between the wheels and rails, which excite coupled vibrations of the train-track-subgrade system. This impacts the environment of the surrounding area in a negative way. Research on this vibration and its propagation in terms of mathematical modeling and numerical simulation has been progressively conducted in the past 2 decades. These models can be categorized into 2 groups according to the integrality of the vibration system involved. Typical works in the first group consist of mechanical models of the 3 fundamental parts, i.e., the vehicle, the track, and the ground. The vibrations of these 3 parts are then simulated by the complete whole-system model [1-5]. In the second group, the modeling procedure consists of 2 steps, each with a separate mechanical model, as shown in figure 1 [6-9]. In the first step, the train and the track are carefully modeled to simulate their dynamic interactions, while the ground is simplified to a manageable damped-springs system. These damped springs were not able to describe wave propagation in the subgrade. The ground vibration needs to be simulated by means of another refined model in the second step, such as a multi-layered half-space, finite element model, or finite difference model.

Theoretically, the whole-system procedure is more practical than the separated 2-step procedure. However, the former has to be formulated in a Cartesian system of reference moving with the train,

Content from this work may be used under the terms of the Creative Commons Attribution 3.0 licence. Any further distribution of this work must maintain attribution to the author(s) and the title of the work, journal citation and DOI.
Published under licence by IOP Publishing Ltd

which requires a time-invariant mechanical system in a moving coordinates system. It is common that engineering practices do not fulfill this rigorous requirement of a moving coordinate system, since any structure with a limited length in the track direction, a railway station for example, could make the model time-variant in the moving reference system. Therefore, in many cases, researchers utilize the separated modeling procedure.

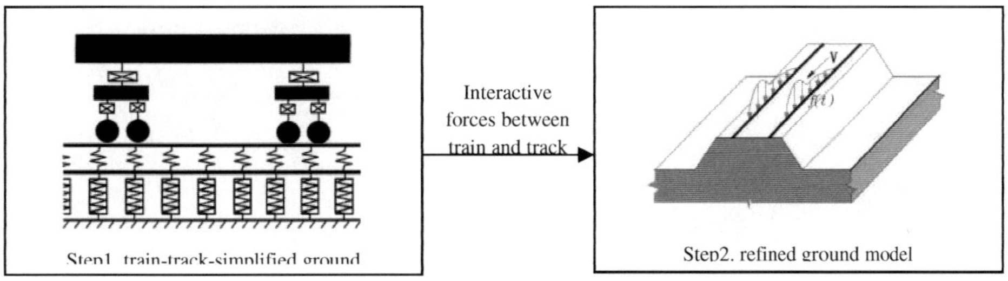

Figure 1. Computational scheme of the 2-step procedure.

One of the key conditions of the 2-step procedure is that the springs in the first step need to be equivalent to the ground model in the second step, since both of them represent the same physical ground. The equivalent stiffness of the springs is commonly computed by comparison of the static responses of both the ground models with an identical static load, which likely maintains equivalence in a static state. However, the train-induced vibrations are typically dynamic processes. Thus, it is uncertain whether the springs are still equivalent to the refined ground model in the dynamic case of vibration. If it is no longer equivalent, the calculated ground vibration is distorted and inaccurate. Numerical results may no longer be reasonable and can be interpreted incorrectly. This inaccuracy can be overcome if the influence of the modeling error is well understood.

The objective of this research is to reveal the modeling error of the 2-step procedure in the computed vibration of the ground and subgrade. A whole-system model is also formulated as a benchmark. Their results are compared, and some features of the vibrations are subsequently discussed.

2. Whole-system model of vehicle-track-ground

Figure 2 illustrates a well-developed whole-system model of the vibration for the vehicle-track-ground [4]. The vehicle is simplified to be a mass-spring-dashpot system, while the matrix $\boldsymbol{H}_{\mathrm{w}}$ represents its receptance at the M axles. The track is infinitely long in the x direction, and has a contact width of $2b$ with the ground. The rails are represented by an Euler beam that has a mass per unit length of m_{R}, and the flexural rigidity of EI. The sleepers are modeled by a distributed mass under the rail beam with a mass per unit length of m_{s}. The rail pads are approximated by a continuous spring distributed along the x direction with a stiffness of k_{p}. Consistent mass approximation is leveraged to simplify the embankment with a mass per unit length m_{B}, while only the vertical stiffness k_{B} is taken into account. Material damping effects are represented by a loss factor, and their imaginary parts are incorporated into the complex stiffness and modulus. Only the vertical dynamics of the vehicle and track are accounted for, and the interactive stress between the embankment and the ground is assumed to be vertical and uniformly distributed along the y axis

Only the harmonic wheel/rail unevenness with respect to the x-coordinate is used in this paper. Any random unevenness is composed of such fundamental ingredients according to Fourier analysis. The spatial wavelength of the unevenness can be written as $\lambda=2\pi c/\Omega$, thus the axles oscillate up and down, with respect to the z axis, in a time-harmonic manner with the frequency of $\Omega/2\pi$ when the train is moving at the speed of c. The motions of the axles result in a series of time-harmonic forces acting on the wheel-rail contacts moving along the track. Regarding the equilibrium of forces and compatibility of displacement, the amplitudes of the interactive forces at the contacts can be obtained and written in a vector form as

$$\boldsymbol{P} = -\left[\boldsymbol{H}_{\mathrm{R}} + \boldsymbol{H}_{\mathrm{w}} + \boldsymbol{K}_{\mathrm{h}}^{-1}\right]^{-1} \boldsymbol{Z} \tag{1}$$

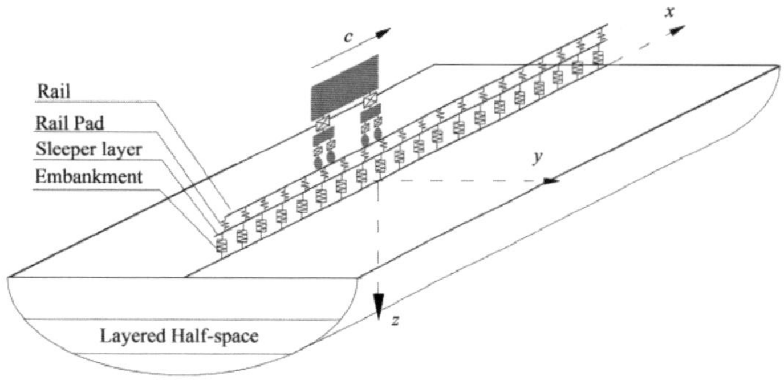

Figure 2 The Whole-system model of the coupled vibrations of vehicle-track-ground

The vector $\boldsymbol{Z}=\{e^{i\Omega a_1/c}, e^{i\Omega a_2/c}, \ldots, e^{i\Omega a_M/c}\}^{\mathrm{T}}$ collects the wheel/rail unevenness perceived by all the wheels, and $a_1, a_2, \ldots, a_M$ are the initial x-coordinates of the M axles at time t=0. $\boldsymbol{K}_{\mathrm{h}}$ is a matrix representing the stiffness of the linearized Hertz spring between the rails and the wheels. The matrix $\boldsymbol{H}_{\mathrm{R}}$ represents the transfer function of the track-ground subsystem. The elements at the lth row and mth column are

$$H_{lm}^{\mathrm{R}} = w(a_l - a_m),\ l,\ m=1,M, \tag{2}$$

where (a_l-a_m) is the horizontal distance of the 2 wheel/rail contacts with the initial position $x=a_l$ and $x=a_m$. The function $w(x_*)$ in the moving system of reference $(x_*=x-ct,\ y)$ can be obtained by the inverse Fourier transform of,

$$\tilde{w}(\beta) = \frac{\Delta_1\Delta_2 - \Delta_4^2 H}{\Delta_1\Delta_2\Delta_3 - \Delta_3\Delta_4^2 H - \Delta_1 k_{\mathrm{p}}^2} \tag{3}$$

in the wave number domain (β, γ). In equation (3), $\Delta_1=1+k_{\mathrm{B}}H-\omega^2 m_{\mathrm{B}}H/3$, $\Delta_2=k_{\mathrm{p}}+k_{\mathrm{B}}-\omega^2(m_{\mathrm{s}}+m_{\mathrm{B}}/3)$, $\Delta_3=EI\beta^4-\omega^2 m_{\mathrm{R}}+k_{\mathrm{p}}$, $\Delta_4=k_{\mathrm{B}}+\omega^2 m_{\mathrm{B}}/6$. ω denotes the circular frequency of the particles in the fixed frame of reference

$$\omega=\Omega-\beta c. \tag{4}$$

The function H is the receptance of the ground

$$H = -\frac{1}{2\pi}\int_{-\infty}^{+\infty} \delta(\beta,\gamma)\frac{\sin\gamma b}{\gamma b}d\gamma \tag{5}$$

in which $\delta(\beta, \gamma)$ is the dynamic Green's function of the layered half-space in the wave number domain (β, γ) [10].

Excited by the wheel/rail forces $\boldsymbol{P}=(P_1, P_2, \ldots, P_M)$, the vibration pattern of displacement in the wave number domain (β, γ) can be computed in terms of

$$\tilde{U}_{\mathrm{P}}(\beta,\gamma) = \tilde{U}(\beta,\gamma)\sum_{l=1}^{M} P_l e^{-i\beta a_l} \tag{6}$$

where $\tilde{U}(\beta,\gamma)$ is the displacement pattern exited by the time-harmonic load moving along the track [11]. For a particle located at (x, y, z) of the ground, the displacement time history is

$$U_{\mathrm{P}}(x, y, t) = e^{i\Omega t} U_{\mathrm{p}}(x - ct,\ y) \tag{7}$$

where the vibration pattern $U_{\mathrm{p}}(x_*, y)$ in the moving reference frame $(x_*=x-ct,\ y)$ can be obtained by the inverse transform of $\tilde{U}(\beta,\gamma)$ with respect to β and γ.

3. 2-step model of train induced vibration

For the purpose of comparison, the 2-step model in this paper has been intentionally set to be the same as the above whole-system, except for the ground representation in the first step. As shown in figure 3a, the ground is replaced by the damped Winkler spring.

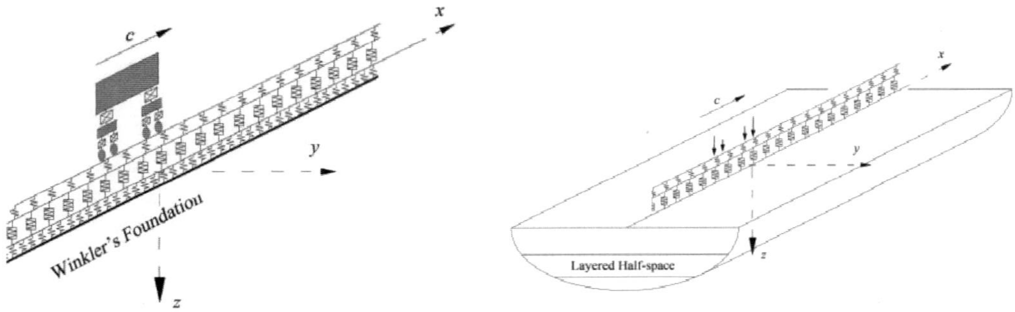

a) b)

Figure 3. The 2-step model with a) for the Tain-track-simplified ground model of the first step, and b) for the Track-ground model of the second step.

The receptance of the ground is now

$$H = \frac{1}{2bk_s} \tag{8}$$

in which k_s denotes the foundation modulus and is NOT frequency-dependent. Regarding the receptances in equation (5) and equation (8) yields,

$$k_s = -\frac{\pi}{b\int_{-\infty}^{+\infty} \delta(\beta,\gamma)\frac{\sin\gamma b}{\gamma b}d\gamma} \tag{9}$$

Equation (9) is used to maintain the static equivalence of the 2-step model and the whole-system model. Note that the Green's function in equation (9) is of the static state with frequency ω=0 and is simply a special case of the one in equation (5). The wheel/rail interaction forces $\boldsymbol{P}$ are also obtained by the equation (1), but the function H in equation (3) should be calculated in terms of equation (8) instead of equation (5).

The model of the second step, shown in figure 3b, is identical to the whole-system model without the vehicle. The wheel/rail forces $\boldsymbol{P}$ generated in the first step are inputted as a series of moving excitations on the rail beam. Responses in time domain are still calculated according to equation (7).

All other parameters involved are the same as those in the whole-system model, which helps ensure that every difference of the response originates from separate modeling.

4. Numerical results and discussion

An example was computed according to the above 2 models respectively. The model's ground had 2 upper sand layers overlaid on a harder half-space of Berea sandstone. The properties of the ground, the 2-suspension vehicle, and the track structure are detailed in references [10] and [11].

It was assumed that the amplitude of harmonic unevenness was 1 mm, thus the frequency perceived by the wheels, $\Omega/2\pi$, equaled 0.1, 4, 8, 16, 31.5, 63 Hz. Then the vibration of a particle in the ground was calculated in terms of equation (7). The computed time histories of the vertical displacement at x=0, y=0, and z=0.2m, located directly below the midline of the track within the sandy subgrade, are presented in figure 4. t=0 corresponds to the moment that the first wheelset passed over the particle, while the speed of the vehicle was c = 60 km/h.

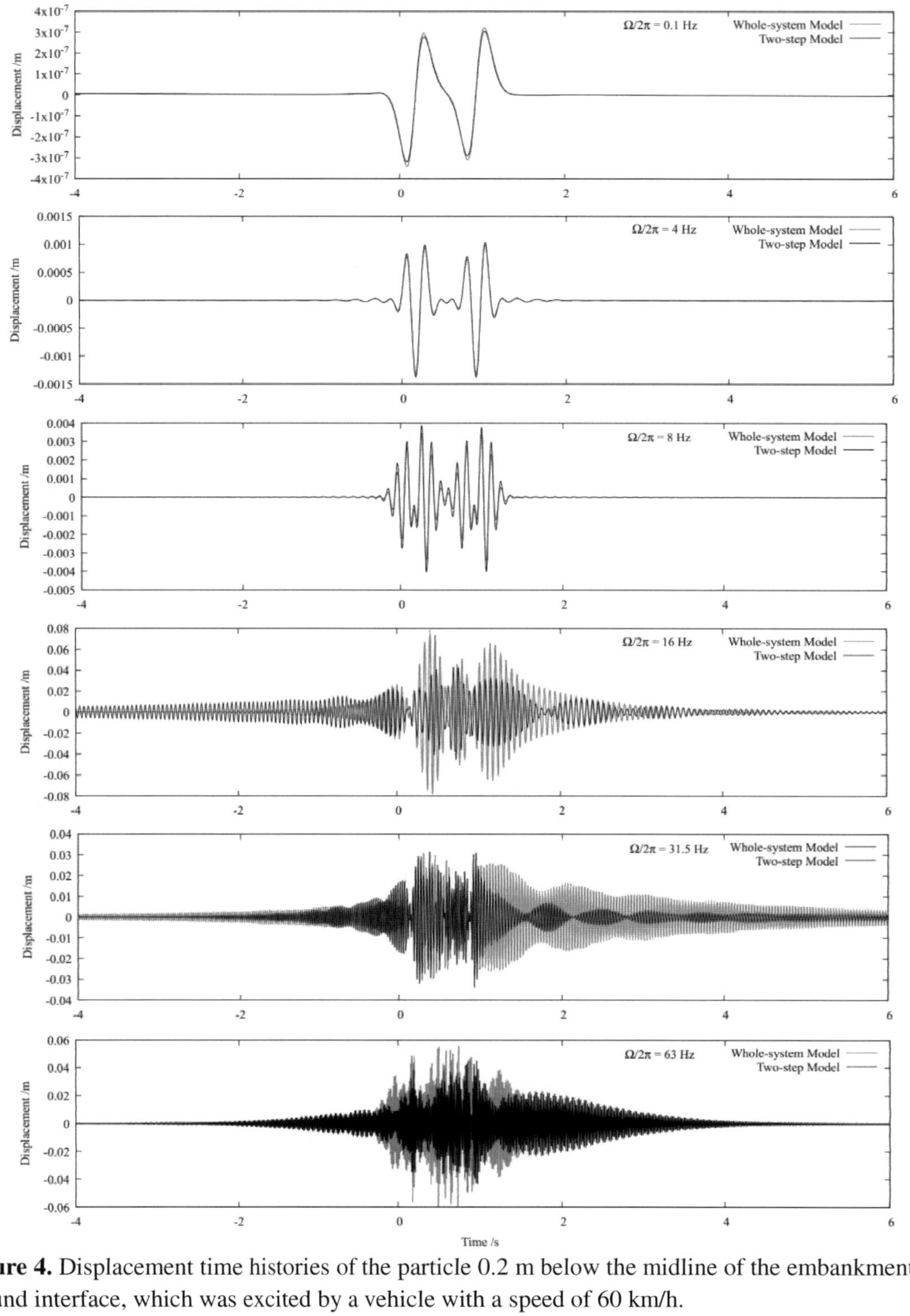

Figure 4. Displacement time histories of the particle 0.2 m below the midline of the embankment-ground interface, which was excited by a vehicle with a speed of 60 km/h.

In the cases of low excitation frequencies such as 0.1 Hz and 4 Hz, the curve obtained by the 2-step model coincides with that of the whole-system model. As the frequency increased to 8 Hz, subtle differences were found between the 2 results. When the frequency approached 16 Hz, remarkable distortion of the 2-step modeling waveform occurred. For higher frequencies of 31.5 Hz and 63 Hz, this discrepancy was still evident enough that it should not be ignored. The good agreement between models at very low excitation frequencies was expected, since the 2 models are equivalent under the static conditions where the vibration frequency equals zero. The distortion of the waveforms of 16 Hz, 31.5 Hz, and 63 Hz indicates that separate modeling may bring significant error to the simulation, since the main frequency range of the ground vibration approaches 80 Hz according to field measurements [12-14].

The computed results also showed that the distortion can be found not only in the immediate vicinity of the track but also in the far field of the ground, as shown in figure 5. This is where the time histories at the 2 points with horizontal distances 10 m and 20 m to the track were presented. Compared with the magnitude of curves from the whole-system model, the maximum displacement computed by the 2-step model was much lower, which indicates that the error of separate modeling may lead to an unfavorable underestimation of the vibration in certain circumstances.

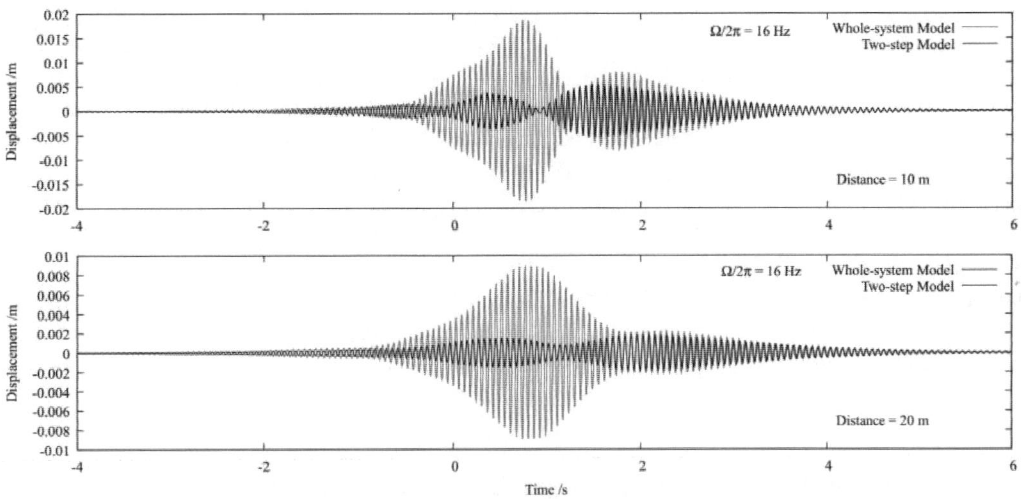

Figure 5. Displacement time histories of the 2 particles 0.2 m below the ground surface with horizontal distances to the midline of the track of 10 m and 20m, which were excited by a vehicle with a speed of 60 km/h.

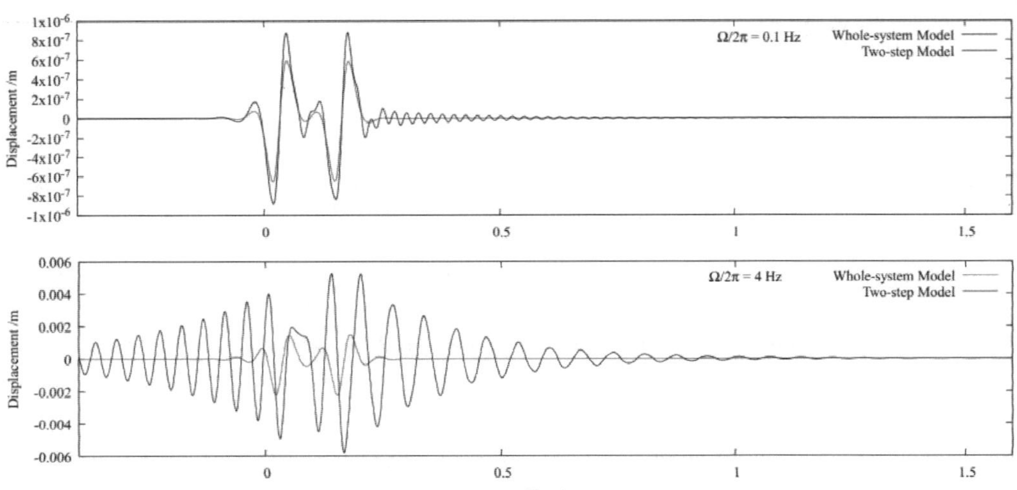

Figure 6. Displacement time histories of the particle, 0.2 m below the midline of the embankment-ground interface, excited by a vehicle with the speed of 350 km/h.

When the speed of the vehicle was increased to 350 km/h, the routine speed of CRH in China, the modeling error was more severe than the lower speed cases. Distortions were clearly found in the curves of vibration responses to all excitations of unevenness, even in the low frequencies of 0.1 Hz and 4 Hz, as shown in figure 6. The 2-step model failed to retain good performance under very low frequencies in the low-speed cases, which might be attributed to the high frequency content of ground vibrations generated by the high speed of the wheel/rail loads.

Conclusion

As the whole-system model of the coupled train-track-ground vibrations is unable to represent some complicated engineering conditions, the separated 2-step procedure has been commonly employed to numerically investigate ground vibration. The major concern of this method is the inconsistency of the ground model in the 2 steps, which may lead to modeling errors in the simulated results.

It was found that the modeling error depends on both the train's speed and the frequency of the wheel/rail excitation. In the case of low-speed trains and very low frequencies, the 2-step model generated satisfactory simulation results. However, for higher frequencies of excitation, the distortion of the waveforms was remarkable. This modeling error was not only found in the immediate vicinity of the track but also in the far field of the ground. Moreover, underestimation of the vibration intensity might take place because of the separate modeling. For high-speed trains, the distortion was found in the waveforms of vibration excited by every frequency of unevenness, which was more common than in the low-speed cases.

Regarding the modeling error of the 2-step method, it is strongly recommended that the stiffness of ground model in the first step be calibrated by the frequency response of the ground model in the second step, thus accounting for the influence of the train's speed.

References

[1] Galvín P, François S, Schevenels M, Bongini E, Degrande G and Lombaert G 2010 *Soil. Dyn. Earthq. Eng.* **30** 1500-12.
[2] Gupta S, Liu W F, Degrande G, Lombaert G and Liu W N 2008 *J. Sound Vib.* **310** 608-30.
[3] Lombaert G, Degrande G, Kogut J and Ois S F 2006 *J. Sound Vib.* **297** 512-35.
[4] Sheng X, Jones C J C and Thompson D J 2004 *J. Sound. Vib.* **272** 937-65.
[5] Sheng X, Jones C and Thompson D J 2003 *J. Sound. Vib.* **267** 621-35.
[6] Fernández Ruiz J, Alves Costa P, Calçada R, Medina Rodríguez L E and Colaço A 2017 *Struct. Infrastruct. Eng.* **13** 652-64.
[7] Kouroussis G, Van Parys L, Conti C and Verlinden O 2014 *Adv. Eng. Softw.* **70** 63-76.
[8] Lu F, Gao Q, Lin J H and Williams F W 2006 *J. Sound Vib.* **298** 30-42.
[9] Paolucci R, Maffeis A, Scandella L, Stupazzini M and Vanini M 2003 *Soil. Dyn. Earthq. Eng.* **23** 425-33.
[10] Wang F, Tao X X, Xie L L and Siddharthan R 2017 *Earthq. Eng. Eng. Vib.* **16** 311-28.
[11] Sheng X, Jones C and Petyt M 1999 *J. Sound Vib.* **228** 129-56.
[12] Xia H, Zhang N and Cao Y M 2005 *J. Sound Vib.* **280** 1017-29.
[13] Chen J, Xia H, Chen S and Su M 2010 *Eng. Mech.* **27** 98-103.
[14] Gao G, Li Z, Feng S and Sun Y 2007 *Rock Soil Mech.* **28** 1817-22.

Acknowledgments

This research is supported by the National Natural Science Foundation of China (Grant No. 51478482, 51678221), Natural Science Foundation of Heilongjiang Province of China (Grant No. E201221), Heilongjiang Undergraduate Training Programs for Innovation and Entrepreneurship (Grant No. 201710212034), and the Fundamental Research Fund for Heilongjiang Provincial Universities,

Heilongjiang University Grants (Grant No. HDJCCX-201626). We thank LetPub (www.letpub.com) for its linguistic assistance during the preparation of this manuscript

IOP Conf. Series: Materials Science and Engineering **248** (2017) 012021 doi:10.1088/1757-899X/248/1/012021

Transverse vibration analysis of stiffened plates with elastic support

Z Li[1, 2], N J Ma[3], M Chen[3]

[1]China Construction Steel Structure Corp. Ltd, Shenzheng, 3331 center road, China
[2]Hunan University, Hunan, south road of Lushan, China
[3]South China University of Technology, Guangdong, 381 Wushan raod, China

Abstract. The transverse vibration of stiffened plates with elastic support boundary conditions is studied in this paper. First, strain energy and kinetic equations of the deck, girders and ribs are respectively established. Second, governing equations for the structure are obtained according to the theorem of conservation of energy and solved using Ritz method. According to the relationship between the motherboard and the main girder deformation coordination, the displacement function of the deck can be got based on the displacement function of the main girder. At last, finite element model is built to check the accuracy of theoretical analysis. It can found that the theoretical solution and the finite element solution have the same variation trend, and the error between the two methods is small. Theoretical analysis method is reliable, and it has certain significance for further research and engineering design.

1. Introduction

In practical engineering, the orthotropic bridge deck is used more and more widely. The deck slab is the most important bearing structure in the bridge, and the load on it is complex. So we need to conduct the thorough research to the characteristics of the bridge deck. There are several kinds of methods, such as grillage method [1], Raileigh-Ritz method [2], finite element method [3], differential quadrature method [4] and no element method [5]. However, the work has been done confined to the boundary conditions of four edges simply supported or clamped, and the research to the boundary conditions for elastic supports is very rare. Zhong W.H et al. [6] considered the material nonlinearity and static load in their paper. Han Q. [7] studied the free vibration problem of rectangular plates with four edges simply supported. Ma N.J. et al. [8, 9] studied the local vibration of orthotropic steel bridge deck and the free vibration problem of rectangular plates with four boundary fixed.

In this paper, the transverse vibration of stiffened plates with elastic support is analyzed. The law of conservation of energy is used to establish the control equations, and Ritz method is used to solved it. In order to check the accuracy of theoretical analysis, finite element model is built by Midas Civil.

2. Building energy formula alone

The energy contains the energy in the orthotropic deck and in the main beam. The strain energy in the deck is

$$V_p = \frac{D}{2}\int_0^a\int_0^b \left\{ \left(\nabla^2 w_p\right)^2 + 2(1-\mu)\left[\left(\frac{\partial^2 w_p}{\partial x \partial y}\right)^2 - \frac{\partial^2 w_p}{\partial x^2}\frac{\partial^2 w_p}{\partial y^2}\right]\right\} dxdy \tag{1}$$

Content from this work may be used under the terms of the Creative Commons Attribution 3.0 licence. Any further distribution of this work must maintain attribution to the author(s) and the title of the work, journal citation and DOI.
Published under licence by IOP Publishing Ltd

Where
$$D=\frac{Eh^3}{12\left(1-\mu^2\right)} \tag{2}$$

where, a and b are length and width, E is the Young modulus of the material, h is the thickness of the deck, μ is the Poisson ratio of the material, w_p is the deflection of the deck.

The strain energy in the section i stiffening rib in the x-direction and y-direction is

$$\left.\begin{aligned} V_{sxi} &= \frac{1}{2}EI_x\int_0^a\left(\frac{\partial^2 w_p}{\partial x^2}\right)^2 dx\Bigg|_{y=y_i} \\ V_{syj} &= \frac{1}{2}EI_y\int_0^b\left(\frac{\partial^2 w_p}{\partial y^2}\right)^2 dy\Bigg|_{x=x_j} \end{aligned}\right\} \tag{3}$$

The strain energy in the main beam is

$$\left.\begin{aligned} V_{gx} &= \frac{1}{2}EI_{gx}\int_0^a\left(\frac{\partial^2 w_x}{\partial x^2}\right)^2 dx \\ V_{gy} &= \frac{1}{2}EI_{gy}\int_0^b\left(\frac{\partial^2 w_y}{\partial y^2}\right)^2 dy \end{aligned}\right\} \tag{4}$$

From Equations (1) and (4), the strain energy in the whole structure is obtained

$$V=V_p+2\square V_{gx}\left(x=0,a\right)+2\square V_{gy}\left(y=0,b\right)+V_s \tag{5}$$

where,

$$V_s=\sum_{i=1}^{N_x}V_{sxi}+\sum_{j=1}^{N_y}V_{syj} \tag{6}$$

N_x and N_y are the number of stiffening ribs in the x-direction and y-direction.

The kinetic energy in the deck is

$$T_p=\frac{1}{2}\rho h\int_0^a\int_0^b\left[\left(\frac{\partial w_p}{\partial t}\right)^2\right]dxdy \tag{7}$$

The kinetic energy in the section i stiffening rib in the x-direction is

$$T_{sxi}=\frac{1}{2}\rho A_x\int_0^a\left(\frac{\partial w_p}{\partial t}\right)^2 dx\Bigg|_{y=y_i} \tag{8}$$

where, ρ is density of the material, A_x is sectional area of the stiffening ribs.

The kinetic energy in the section j stiffening rib in the y-direction is

$$T_{syj}=\frac{1}{2}\rho A_y\int_0^b\left(\frac{\partial w_p}{\partial t}\right)^2 dy\Bigg|_{x=x_j} \tag{9}$$

The kinetic energy in the main beam is

$$\left.\begin{aligned} T_{gx} &= \frac{1}{2}\rho A_{gx}\int_0^a \left(\frac{\partial w_x}{\partial t}\right)^2 dx \\ T_{gy} &= \frac{1}{2}\rho A_{gy}\int_0^b \left(\frac{\partial w_y}{\partial t}\right)^2 dy \end{aligned}\right\} \tag{10}$$

From equations (7) and (10), we have the kinetic energy in the whole structure, as following

$$T = T_p + 2 \cdot T_{gx}\left(x = 0, a\right) + 2 \cdot T_{gy}\left(y = 0, b\right) + T_s \tag{11}$$

where,

$$T_s = \sum_{i=1}^{N_x} T_{sxi} + \sum_{j=1}^{N_y} T_{syj}$$

3. Displacement function

To apply Ritz method, we need first select a Displacement function for the orthotropic deck. It has been mentioned before that the deck is laid on four girders simply supported at their ends.

When a thin plate free vibrate at a certain frequency ω and vibration mode *W(x,y)*, its instantaneous deflection can be expressed as

$$w = A\cos\left(\omega t\right)W\left(x, y\right) \tag{12}$$

where,

$$W\left(x, y\right) = A_1\left(-\frac{1}{2}x^4 + ax^3 - \frac{a^3}{2}x\right) + B_1\left(-\frac{1}{2}y^4 + by^3 - \frac{b^3}{2}y\right) + C_1\sin\left(\frac{m\pi x}{a}\right)\sin\left(\frac{n\pi y}{b}\right) \tag{13}$$

4. Approximate solutions from energy method

According to the principle of conservation of energy, the maximum value of the strain energy is equal to the maximum of the kinetic energy. We have

$$V_{\max} = T_{\max} \quad \text{that is} \quad Q = V_{\max} - T_{\max} = 0 \tag{14}$$

Based on the assumption of the vibration type function and different constants, *Q* obtain the minimum value, as follow

$$\frac{\partial Q}{\partial A} = \frac{\partial Q}{\partial B} = \frac{\partial Q}{\partial C} = 0 \tag{15}$$

From equations (15), we have

$$\frac{5}{32}A + \frac{5}{32}B + C = W \tag{16}$$

From equations (16), constants A B C and lowest order natural frequency *ω* can be obtained. Due to limited space, the results of the formula will not be listed.

5. Numerical calculations and discussions

The geometry of the finite element model of the orthotropic bridge deck is shown in figure 1 with $b_r = 0.05m$ and $h_r = 0.1m$. The size of the main girder is also shown in figure 1 with $b_g = 0.1m$ and $h_g = 0.2m$. The thickness of the deck is $0.05m$. The model is simply supported at the four corners illustrated in figure 2 (b). The main girders and ribs are simulated by beam element. The bridge deck is simulated by plate element. Figures 2 (a) illustrates the geometry of this bridge system established by MIDAS Civil. The width of the plate is $b=2m$. The aspect ratio a/b is a parameter. Three longitudinal ribs located in $3b/4$, $b/2$ and $b/4$ and three transverse ribs located in $3a/4$, $a/2$ and $a/4$ are selected .

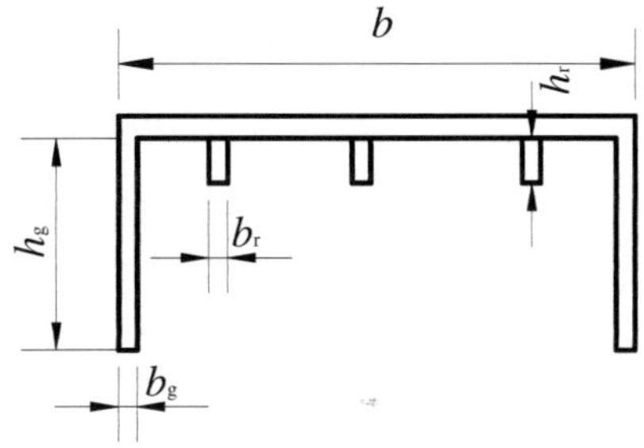

Figure 1. Geometry of the cross section in breadth direction.

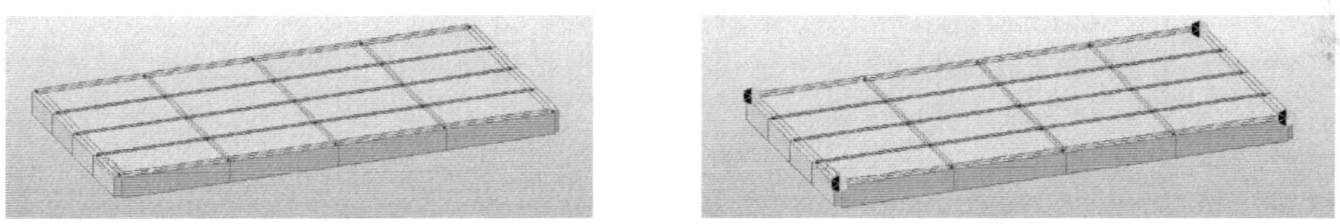

(a) (b)

Figure 2. Model established by MIDAS Civil.
(a) bridge system (b) boundary conditions

All parameters of the material and structural were brought into theoretical analysis and MIDAS Civil analysis. The natural frequency "ω" of two methods and MIDAS Civil and errors between them are listed in table 1. The aspect ratio a/b of the bridge deck is usually in the range of 1 to 10 in application.

Table 1. Natural frequency of two methods and errors between them.

	The natural frequency "ω" (*hz*)							
a/b	1	1.5	2	2.5	3	4	5	10
Theoretical solution	71.83	41.24	25.36	17.02	12.19	7.12	4.66	1.22
MIDAS Civil	72.2	41.36	25.23	16.7	11.8	6.75	4.36	1.11
The error (%)	-0.5	-0.3	0.5	1.9	3.3	5.5	6.9	10.3

From table 1, we can get some conclusions as following:

(1) The trend of two methods that the natural frequency decrease with the increase of the aspect ratio a/b is the same. The main cause for such trend is that the stiffness of the bridge deck decrease with the increase of the aspect ratio a/b and is proportional to the natural frequency, such as the hypothesis of the small deflection of the sheet.

(2) The error between the theoretical solution and finite element solutions increases with the increase of the aspect ratio a/b. First reason is some assumptions or simplified are taken in the process of theoretical analysis. Second, the nature of the orthotropic plate is more significant with the increase of aspect ratio. But energy method did not reflect this characteristics. Third, the finite element model also has some errors itself. For example, multiple Ritz vector method is taken to calculate the natural frequency, and it will be affected by the number of units.

(3) The errors between two methods are relatively small when the aspect ratio is from 1 to 5. It reflects the theoretical analysis method has higher precision and the displacement function satisfies the boundary conditions and meet the accuracy requirements.

6. Conclusions

Transverse vibration of stiffened plates with elastic support is analyzed in this paper. Theoretical model and finite element model are built and compare with each other. From section 6 in this paper, we can get that the errors between two methods are relatively small. Displacement function that is selected to simulate the elastic boundary conditions is reliable, and it can be provide reference for further research.

References

[1] Balendra T and Shanmugam N E 1985 *J. Sound Vib.* **99** 333-50.
[2] Bedair O K and Troitsky M S 1997 *Int. J. Mech. Sci.* **39** 1257-72.
[3] Voros G M 2009 *Thin-Walled Str.* **47** 382-90.
[4] Zeng H and Bert C W 2001 *J. Sound Vib.* **241** 247-52.
[5] Peng L X, Liew K M and Kitipornchai S 2006 *J. Sound Vib.* **289** 421-49.
[6] Zhong W H, Hao J P, Lei L and Yan X L 2008 *J. Vib. Shock,* **27** 46-8.
[7] Han Q 2008 *J. South China Univ. Tech. (Nat. Sci. Ed.)* **28** 78-82.
[8] Ma N J, Wang R H and Han Q 2012 *J. of Vib. Shock,* **31** 60-4.
[9] Ma N J and Wang R H 2011 *Chinese J. Theoret. Appl. Mech.* **3** 922-30.

Study on Network Error Analysis and Locating based on Integrated Information Decision System

F Yang[1], Z H Dong[1]

[1]Astronatics Engineering University, Beijing, 101416, China

Abstract. Integrated information decision system (IIDS) integrates multiple sub-system developed by many facilities, including almost hundred kinds of software, which provides with various services, such as email, short messages, drawing and sharing. Because the under-layer protocols are different, user standards are not unified, many errors are occurred during the stages of setup, configuration, and operation, which seriously affect the usage. Because the errors are various, which may be happened in different operation phases, stages, TCP/IP communication protocol layers, sub-system software, it is necessary to design a network error analysis and locating tool for IIDS to solve the above problems. This paper studies on network error analysis and locating based on IIDS, which provides strong theory and technology supports for the running and communicating of IIDS.

1. Introduction

Modern high techniques rely on high efficient and real time information services working in order, which can stably run on mobile terminal end. Integrated information decision system (IIDS) integrates multiple sub-system developed by many facilities, including almost hundred kinds of software, which provides with various services, such as email, short messages, drawing and sharing. This requires IIDS to be running stably and efficiently. However, IIDS has the following problems:

First, because of the complexity of actual tasks and complexity of information communication, IIDS always has the problem of information communication blocked or communication error, and leads to that tasks cannot be assigned or information cannot be shared. The current system lacks of effective error locating of system information errors which affect the operation of major tasks.

Second, the design of task planning requires for restrict timing and has uncertainty. Major task planning also requires the support of IIDS. Thus, operators have to make full use of IIDS, and in the meantime, to quickly diagnosis and solve the errors of IIDS.

Current work to solve the problem of error diagnosis and locating has considered many kinds of errors that may happened in the network. For example, if a cable gets problems, computers that connected to the cable cannot connect to the network. Some other errors may be caused by server, or some processes are not working, which may lead to information lost. In order to locate the error, and improve the quality of experience for users, [1] proposed an error locating model based on K-neighbor machine learning method, which can solve the errors caused by network bandwidth is too narrow and signal strength is too low. [2] mainly considered the exception of network message, and proposed an error analysis methods, which can be used to multi-network message analysis systems. Current work only focused on network errors that under specific situations, but cannot be applied to the communication problems in IIDS.

In order to solve the above problems, we analyze the system requirement to find out in which way we can analyze and locate the network error information. Figure 1 shows system requirement analysis. From the viewpoints of assignment operation, the information flow is transmitted from assignment

Content from this work may be used under the terms of the Creative Commons Attribution 3.0 licence. Any further distribution of this work must maintain attribution to the author(s) and the title of the work, journal citation and DOI.
Published under licence by IOP Publishing Ltd

release to assignment planning. From the viewpoints of system of system, we can analyze network health status of network information system and information system. From the viewpoint of information, information transmission status is unclear. And from the viewpoint of physics, IIDs also lacks of feedbacks of evaluation results. As the path of information transmission is not clear, it is hard to evaluate assignment results. According to the above discussion, this paper proposed a network information analysis and locating tool based on PHM technique, which uses distributed error analysis and locating tool to collect information transmitted in the IIDS network, to determine network health status, to obtain information of the errors, and to find out where is the error, in order to solve the communication problems of IIDS.

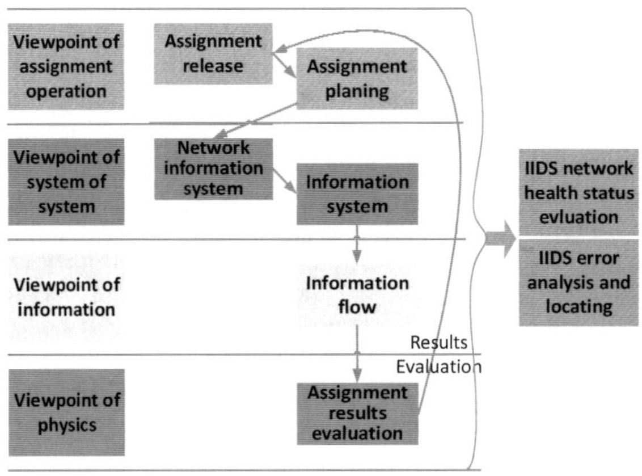

Figure 1. System requirement analysis.

2. Related work of error analysis and locating

PHM (Prognosties and Health Management) is a whole new resolution for management of health state, which integrates the latest research results of modern information technology, and artificial intelligent [3]. PHM monitors the exchanging service operation state of external information for the whole IIDS, and prejudges and diagnoses the probably errors, by monitoring the data in data transmission network in real-time.

Currently, the PHM technology in abroad countries has extended from program design to project verification, which has been applied to huge complicate system, including aerospace, Industrial manufacturing, water conservancy, and nuclear power, etc. [4]. In our country, many trades is developing studies on PHM technologies and related theories, which has made Industry standards and development specifications of PHM[5].

In our country, PHM has been mainly applied to the domain of aerospace, which is mainly used to monitor the performance state of airplane and their motors. In the meantime, some work also took research on the theory of PHM, the construction of PHM system architecture, and the verification of the effect of PHM technology on maintenance mode [6].

Reference [7] has made comparison analysis on various program of Airborne PHM system architecture, and proposed a 4-layer architecture structure which contains unit layer PHM, sub-system layer PHM, domain layer PHM, and platform layer PHM. Reference [8] builds up topology of system equipment according to the error characteristics of UVA system, and constructs the logical system structure of UVA PHM system. Reference [9] proposed an architecture of integrated modular aerospace electronic system and PHM, which is suitable for the online upgrade and maintenance of aerospace electronic system.

Currently, the maintenance of devices mainly contains periodic inspection and after-maintenance, which costs huge human and material resources, and the efficiency is very low. PHM has the following functions of error detection and isolation, error diagnosis, error prediction, health management, and component lift tracking.

IOP Conf. Series: Materials Science and Engineering **248** (2017) 012022 doi:10.1088/1757-899X/248/1/012022

PHM system has been applied to different domains including aerospace and industry, which has the same basic theory, and the difference is that the technologies and methods in specific applications in different domains are different. The development of PHM has been gone through 3 stages including error diagnosis, error predication, and system integration. Currently, the study on PHM aims to improve the error diagnosis and prediction accuracy, and extend the application domains. PHM has been successfully applied to military and civil aviation, which has broad prospects in various domains.

3. Network Error Analysis and Locating Program Design

3.1 Study Objects

This paper has developed technique study on network error analysis and locating for IIDS. The proposed technique adopts distributed architecture, which is setup to each node in the network, to obtain data transmitted in the network, analyze the obtained data, and also find out the position of errors. Actually, to design such a network error analysis and locating tool is required to solve the following problems: First, to effectively recognize and locating the errors of system information status. Second, to carry out effective assistant decision and technique maintenance for existing errors.

And the general study framework of network error analysis and locating method of IIDS is shown as figure 2. The main research objective is to solve the above problem and construct an error analysis and locating system to achieve the transparency of assignment operation in each unit, and accuracy of technique maintenance.

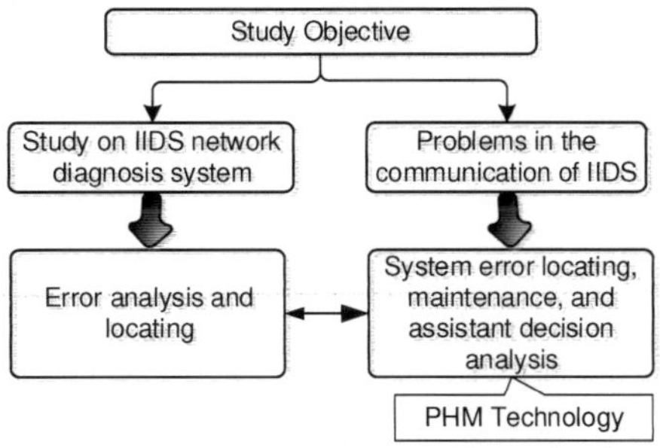

Figure 2. General study framework.

In order to improve the ability of assignment planning, decision, and operation based on information system, and provide strong theory and technology supports for the running and communicating of IIDS, we has applied PHM technology to analyze data and locate errors follow the following steps [10].

(1) Data collection. This is using different sensors to detect and collect related parameters of the detected system, which makes valid information transfer and information transmission using collected data.

(2) Information Induction Processing. This part of outputs includes sensor data that is after filter, compress and simplify, and also includes spectral data and other characteristic data.

(3) State detection. Its function is mainly to compare this data with predetermined failure criterion, in order to monitor current state, and also to provide error alert ability according to those predetermined threshold of various parameters.

(4) Health evaluation. It mainly evaluates the health state of detected system (or subsystem, component), if they have the phenomenon of parameter degradation, and generates error diagnosis records and makes sure the possibility of errors.

(5) Error prediction and decision. This part can use the above mentioned data to evaluate and predict the future health state of the monitored system, and gives decision, suggestion and solution.

(6) Maintenance and decision. It mainly includes human-machine interface and machine-machine interface.

3.2 System Design

Error analysis and locating system is to analyze the errors of IIDS and its device operation status, and makes sure the reason and location of errors, as well as the supporting services of the errors, in order to make the system continuously and stably running, and solve the problems in time. The main contents of error analysis and locating includes, to analysis the reliability of network node links, to analysis information transmission and flow, to analysis the effect of operation environment and external environment on system operation, and to determine error information and error location.

According to the characteristics of IIDS information dynamically changing in real time, and the data transmission network open in full time, it has to control the system network in fulltime. First, real time performance analysis and dynamic display, to measure in real time, analyze, and evaluate the standard of network and device, and form a series reports of device exception report, data capacity analysis, and network performance characteristics, etc. By comparing with different types of data, to analysis the operation status of network and devices, configuration, analysis and integration of the task information, external information exchanging information of IIDS, and information flow states. Second, the real-time error prejudgment and diagnosis based on PHM, has the function of error automatically answer, locating the position of errors in time, and precisely determination of the error nature and degree, which can find out the reasons of errors quickly, and provide with assistant decision for error maintenance.

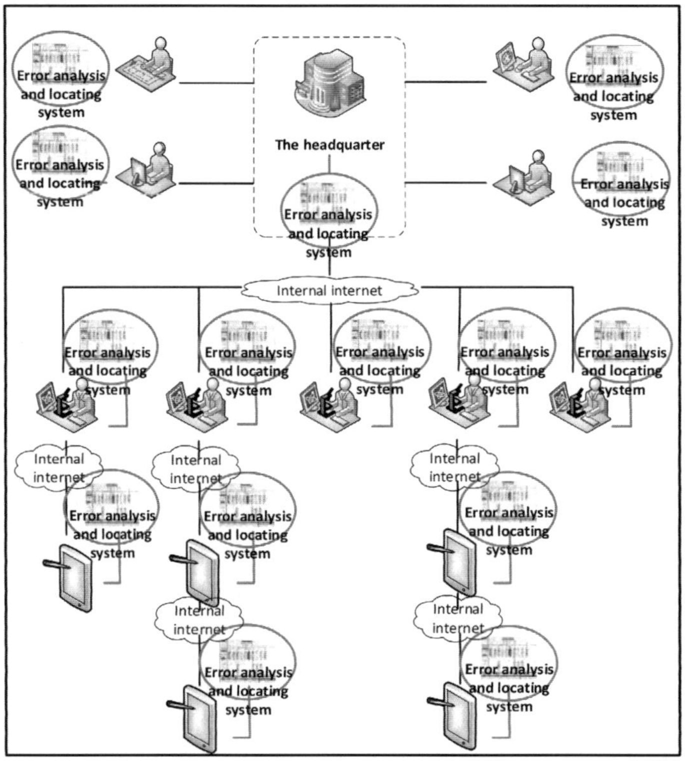

Figure 3. System general deployment structure based on IIDS

3.2.1 System General Structure. In order to make full use of the functions of IIDS, and solve the problems, such as communication, information sharing, information security, and information monitoring, etc. in the IIDS, this paper proposes to construct an online monitoring tool to monitor the operation states and system failures of IIDS. The main function is precise error locating, error maintenance, and assistant decision. The external exchanging information operation status and data transmission status of the devices in the error analysis and locating system. When a supporting service in the IIDS has problems, data links have errors or information flow blocked the system, the error analysis and locating system can alarm in time, and quickly and precisely locating the errors, and provides with assistant decision function for error maintenance.

This paper studies on an error analysis and locating system based on IIDS, which deployment is based on IIDS. When the IP resources of the whole IIDS are limited and other factors that make the system hardware cannot be used, the deployment that is based on IIDS can be adopted, which is setting up the error analysis and locating system software to each computer and sharing the hardware platform with IIDS. The advantage is not taking IP resources, but it requires to have high coupling degree with IIDS. The general deployment is shown as figure 3.

The proposed system carries out error prejudgment and diagnosis in real-time for IIDS based on PHM, which has the functions of automatically answering, locating the positions of errors in time, and precisely obtaining the nature and extend of the errors.

3.2.2 System Software Design. The proposed system adopts PHM to prejudge and diagnosis the IIDS. PHM has the functions of error testing and diagnosis, performance testing, error prejudgment, health management. The proposed analysis and locating system, which has been deployed to all nodes in the IIDS, can complete tasks like performance testing, error diagnosis of IIDS, and achieve the error predicting of key system and components.

The work flow of error analysis and locating system is shown as figure 4, which includes data monitoring in real time, feature extraction, state monitoring, error diagnosis, error predicting, error locating, and maintenance decision, etc.

Network error analysis and locating system has been embedded into IIDS, which can get all system operation states of IIDS by the information monitoring system. The proposed error analysis and locating system can extract features of monitored information by error analysis and decision making function, and get the operation state information of key parts and key links, and also can make diagnosis and prediction when compared with history monitoring data and find the location of errors. The proposed system can also summary and analyze the error diagnosis results and location information. Given the error problems and solutions in the error decision making repository, the proposed system can provide maintenance and decision making information.

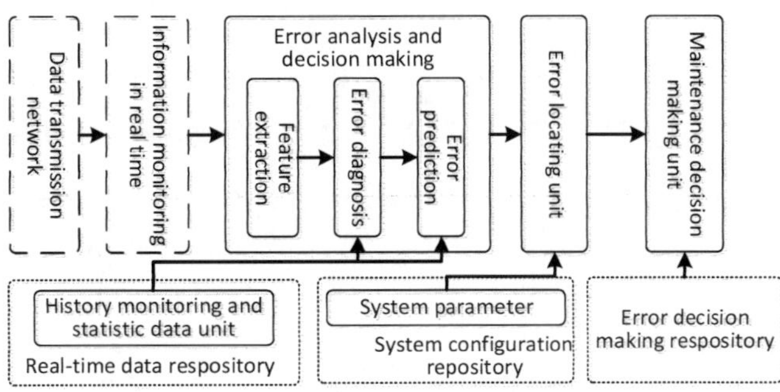

Figure 4. Error analysis and locating system.

The network error analysis and locating system can get all data including the operation states of starting services in each server and client, and data links communication states in real time for IIDS. The proposed system can diagnosis, locate and alarm for the software service error and data link communication error for IIDS. In the meantime, the error information can report to higher unit in real time, until sending to the server of headquarter. The proposed system can make operator understand the error states of IIDS and data transmission network, and change the working plan in time, so that they can adapt to the various assignment demanding.

4. System Prototype Implementation

In order to guarantee the adaptability and reliability of online monitoring system in practice, this paper also implements a prototype of this system, which analyzes and locates IIDS system errors. System prototype can display the analysis and locating results for monitored information, and provide with the results of error diagnosis. For each devices in the network, this prototype can display exchanging service operation states of external information, and data transmission states among devices, which is shown as figure 5.

The device states of each position in IIDS use different color to present the states of clients and servers, according to the monitored states of starting services of clients and servers in IIDS. The green box represents that the device or server works normally. The red box represents that the system is running, but the IIDS software is turned off. The yellow box represents that some important services in the server or devices have gone wrong. The gray box represents that the data links gone wrong, and cannot get the operation state of this device. Clicking any box can show all external information exchanging states of the server or clients in a table.

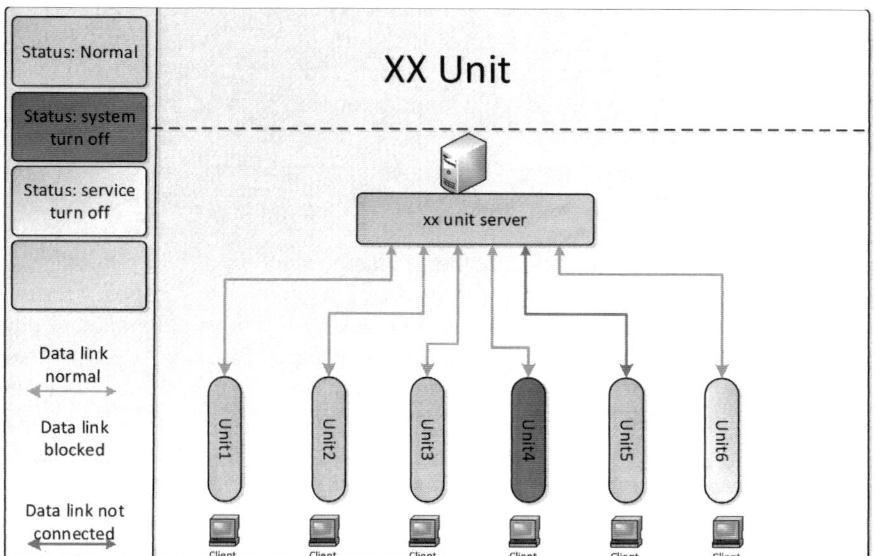

Position name	Service type	State	Suggestion
Position6	Email	Normal	
	Short message processing service	Normal	
	Long message sending	Abnormal	Please check or restart Indor MDS service process
	Long message transmission	Abnormal	Please check or restart Indor RTS service process
	Alarm service	Normal	
	Contacts service	Normal	
	Time unified service	Normal	

Figure 5. The states of each position in xx Unit.

5. Conclusion

The proposed network error analysis and diagnosis tool can determine and locate the errors of IIDS in 3 minutes, and can give alarm and change the position states for the whole system at the same time. The proposed tool can provide solutions of many major problems happened in the joint tests and interconnections of IIDS, including the position of errors which is hard to precisely and effectively located, and providing error assistant decision and technique maintenance in real time, which is also very hard to carry out in practice.

6. References

[1] Wang Z H, Liu Y L, Hu Y J, Zhao Z J, and Ci S 2014 *Interact. Ind. Univ. Res. Tech. Innovat.* **5** 83-90.
[2] Xu C B, Feng Y, and Xiong L, 2014 *East China Electr. Pow.* **42(12)** 2543-7.
[3] Luo R Z, Sun B, Zhang L and Liu H 2013 *Spacecraft Eng.* **22(4)** 95-102.
[4] Zhang B Z 2008 *Comput. Meas. Contr.* **16(5)** 591-4.
[5] Jing B, Tang W, Huang Y F and Yang Z 2014 *J. Electr. Meas. Instr.* **28(12)** 1301-7.
[6] Wu M Q, Fang H Z, Wen B W and Yi D W 2011 *Comput. Meas. Contr.* **19(1)** 98-101.

IOP Conf. Series: Materials Science and Engineering **248** (2017) 012022 doi:10.1088/1757-899X/248/1/012022

[7] Zhang L, Zhang F M, Li J T and Li Y B 2008 *J. Air Force Eng. Univ. (Nat. Sci. Ed.),* **9(2)** 6-9.
[8] Luo H, Rong H and Peng L L, *Aircraft design.* **29(4)** 54-7.
[9] Zhao N S, Zha Z J and Wang G Q 2011 *Meas. Contr. Tech.* **30(1)** 1-6.
[10] Lu Y C, Jiang L and Zhao H L 2014 *Acta ArmamentarII,* **35(1)** 68-75.

Dynamic response analysis of surrounding rock under the continuous blasting seismic wave

P F Gao[1, 2*], Q Zong[1], Y Xu[1] and J Fu[1]

[1]School of Civil Engineering and Architecture, Anhui University of Science and Technology, 232001, Huainan, Anhui, China
[2]Anhui Jiangnan Blasting Engineering Co., Ltd. 242300 Xucheng, Anhui, China

Abstract. The blasting vibration that is caused by blasting excavation will generate a certain degree of negative effect on the stability of surrounding rock in underground engineering. A dynamic response analysis of surrounding rock under the continuous blasting seismic wave is carried out to optimize blasting parameters and guide underground engineering construction. Based on the theory of wavelet analysis, the reconstructed signals of each layer of different frequency bands are obtained by db8 wavelet decomposition. The difference of dynamic response of the continuous blasting seismic wave at a certain point caused by different blasting sources is discussed. The signal in the frequency band of natural frequency of the surrounding rock shows a certain degree of amplification effect deduced from the dynamic response characteristics of the surrounding rock under the influence of continuous blasting seismic wave. Continuous blasting operations in a fixed space will lead to the change of internal structure of the surrounding rock. It may result in the decline of natural frequency of the whole surrounding rock and it is also harmful for the stability of the surrounding rock.

1. Introduction

The continuous harmful effect of blasting vibration may cause the instability of the surrounding rock of underground carven, then resulting in huge losses of lives and properties [1, 2]. The safety criterion of blasting earthquake effect has attached the attention of a large number of blasting researchers because of it is directly related to the safety of buildings and residents around the blasting area. The particle vibration velocity has been used as the building blasting seismic effect safety criteria for a long time. With the progress of a large number of monitoring and experimental and theoretical research works, the frequency is found to be a non-negligible factor for the safety criterion of blasting seismic effect [3, 4]. Research on safety criterion of the blasting seismic effect has become an urgent problem in the field of engineering blasting [5, 6].

Wavelet transform theory, which with a good time and frequency domain analysis characteristic, has been introduced into the analysis of typical non-stationary stochastic signals such as blasting vibration signals for a long time. The application of the wavelet transform theory has already achieved fruitful results so far. Ling et al. [7] used the time-energy distribution of wavelet transform to determine the actual delay time of the micro-blasting. The energy distribution characteristic that is based on the time-frequency characteristic and hierarchical decomposition of wavelet transform was analyzed [8]. Based on the time-frequency characteristic of wavelet transform, the characteristics of vibration signals of different measuring points, inside and outside the building, were analyzed. In the meanwhile, the dynamic response characteristics of building under blasting earthquake were extracted [9]. Based on the wavelet transform theory, Song et al. [10] established a time-energy density function

Content from this work may be used under the terms of the Creative Commons Attribution 3.0 licence. Any further distribution of this work must maintain attribution to the author(s) and the title of the work, journal citation and DOI.
Published under licence by IOP Publishing Ltd

to analyze the time-frequency domain characters of response signals for the surrounding rock in underground tunnel.

In this paper, the variation characteristics of the vibration signals at the same monitoring point are analyzed for continuous blasting. The time-frequency characteristics of the blasting vibration signals are obtained by using the decomposition and hierarchical reconstruction technique according to the wavelet transform theory. In the meanwhile, the dynamic effect of continuous blasting for surrounding rock structure is analyzed.

2. Test site

The main and deputy plant of Jixi pumped storage power station is located in Anhui province. The upstream is connected with the upstream reservoir through the diversion system. The downstream is connected with the main change hole through the bus hole. The surrounding rock structure of the underground powerhouse is mainly composed of block and sub-block structure. Parts of the underground powerhouse are composed of the mosaic structure. The carven wall of the underground plant is basically stable.

The stability of the plant wall is very important for engineering construction. The plant wall is the key research area for blasting damage control. The side wall of the plant is blasted by using the method of prescribing glossy layer, while 6 rows of holes are drilled. There is a row of peripheral holes with the spacing of 40-50mm and the charge line density of 120g/m. A row of slow punching holes with the distance of 70mm and the spacing of 80mm and the charge quantity of 1.4kg per hole are arranged outside. The spacing of the remaining 4 rows of holes is 100mm. The row spacing is 95mm and the charge quantity of every single hole is 2.4kg, respectively.

A vibration meter is placed in the 3# bus hole as a blasting monitoring point. The distances of the No.1 and No.2 blasting points from the monitoring point are all 34.5m. The total charge quantity of the first blasting is 72kg, while the maximum single-shot charge quantity is 14.4kg. The total charge quantity of the second blasting is 48kg. The number of blasting holes for the second blasting is less than 10 of the first blasting. The maximum single-shot charge quantity is 14.4kg. The segmented blasting technique is used, while the delaying time is 460ms.

Because of the main vibration frequency of the blasting vibration signal is generally below 200Hz, the sampling rate of the signal is set to 5000 Hz according to the sampling theorem. As a result, the Nyquist frequency is 2500 Hz. The error of signal reconstruction by using db8 wavelet is relatively small, so, the db8 wavelet is used to deal with the blasting vibration signal in this paper. The monitoring signal is decomposed to 8 layers. According to the principle of wavelet decomposition, 9 frequency bands can be got. Where a8 is the low frequency band, its bandwidth is 0~9.77Hz. And d8~d1 are the high frequency bands, their frequency distribution (Hz) as described below. There are d8: 9.77-19.53 Hz; d7: 19.53-39.06 Hz; d6: 39.06-78.13 Hz; d5: 78.13-156.25 Hz; d4: 156.25-312.5 Hz; d3: 312.5-625 Hz; d2: 625-1250 Hz; d1: 1250-2500 Hz.

3. Wavelet analysis of blasting vibration signal

3.1. Principle of wavelet transform

Given $\psi(t) \in L^2(R)$ (where $L^2(R)$ is the signal space with a limited energy), its Fourier transform is $\psi(t)$. Then when $\psi(t)$ meets the allowable condition: $C_\psi = \int_R \frac{|\psi(t)|^2}{|t|} dt < \infty$, herein, $\psi(t)$ is a basic wavelet. After the basic wavelet is scaled and translated, a wavelet sequence can be obtained.

The continuous wavelet transform for any function of $f(t) \in L^2(R)$ can be described as below.

$$W(a,b) = \langle f, \psi_{a,b} \rangle = |a|^{-\frac{1}{2}} \int_R f(t)\overline{\psi(\frac{t-b}{a})}dt \qquad (1)$$

Where $\langle f, \psi_{a,b} \rangle$ is the inner product of f . $\psi_{a,b}$. $\overline{\psi(\frac{t-b}{a})}$ is the conjugate function of $\psi(\frac{t-b}{a})$, where a is the scaling factor and b is the translation factor.

The inverse transformation of continuous wavelet is:

$$f(t) = \frac{1}{C_\psi} \int_R \int_R \frac{1}{a^2} W_f(a,b) \psi(\frac{t-\mathrm{b}}{\mathrm{a}}) dadb \tag{2}$$

While the discrete sequence of wavelet sequences is:

$$\psi_{j,k}(t) = 2^{\frac{-j}{2}} \psi(2^{-j} t - k) \tag{3}$$

Where $a = 2^j$, $b = 2^j k$, $(j,k \in z)$.

The discrete wavelet transform is:

$$C_{j,k} = \int_R f(t) \psi_{j,k}(t) dt = \langle f, \psi(j,k) \rangle \tag{4}$$

The wavelet reconstruction formula is:

$$f(t) = C \sum_R \sum_R C_{j,k} \psi_{j,k}(t) \tag{5}$$

3.2. Wavelet signal analysis results

According to the wavelet analysis theory, the db8 wavelet is used as the fundamental wave. The blasting vibration signal is decomposed into 8 layers, after that, the wavelet decomposition coefficient for each layer is reconstructed. The calculation and analysis procedures are carried out by using the Matlab software. The original signal and the reconstructed signal of each frequency band for different blasting points are shown in figure 1 and figure 2.

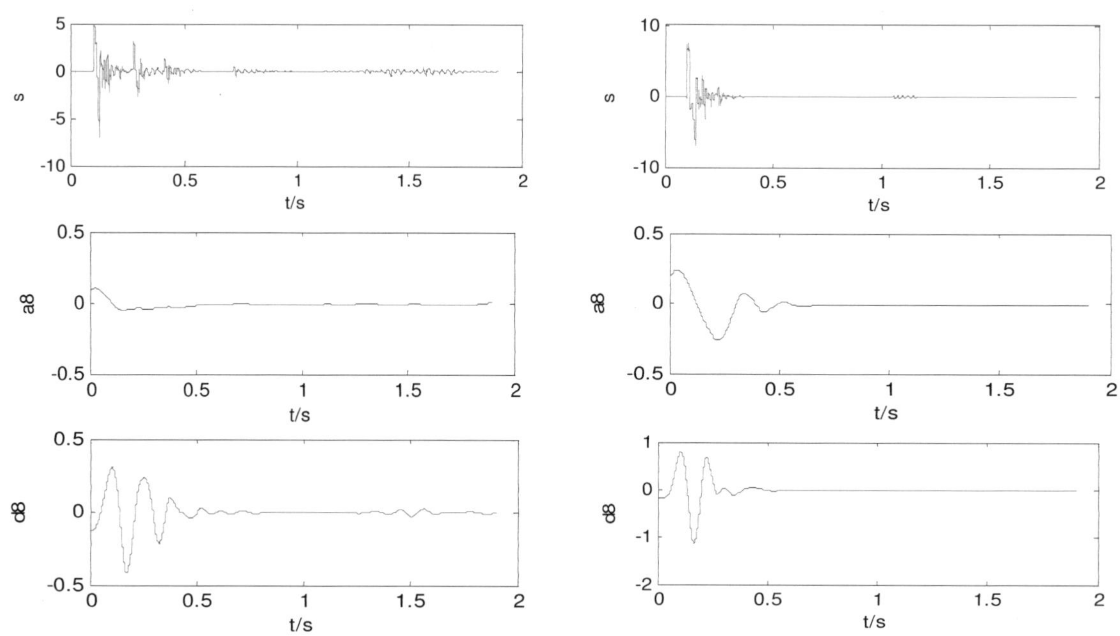

IOP Conf. Series: Materials Science and Engineering **248** (2017) 012023 doi:10.1088/1757-899X/248/1/012023

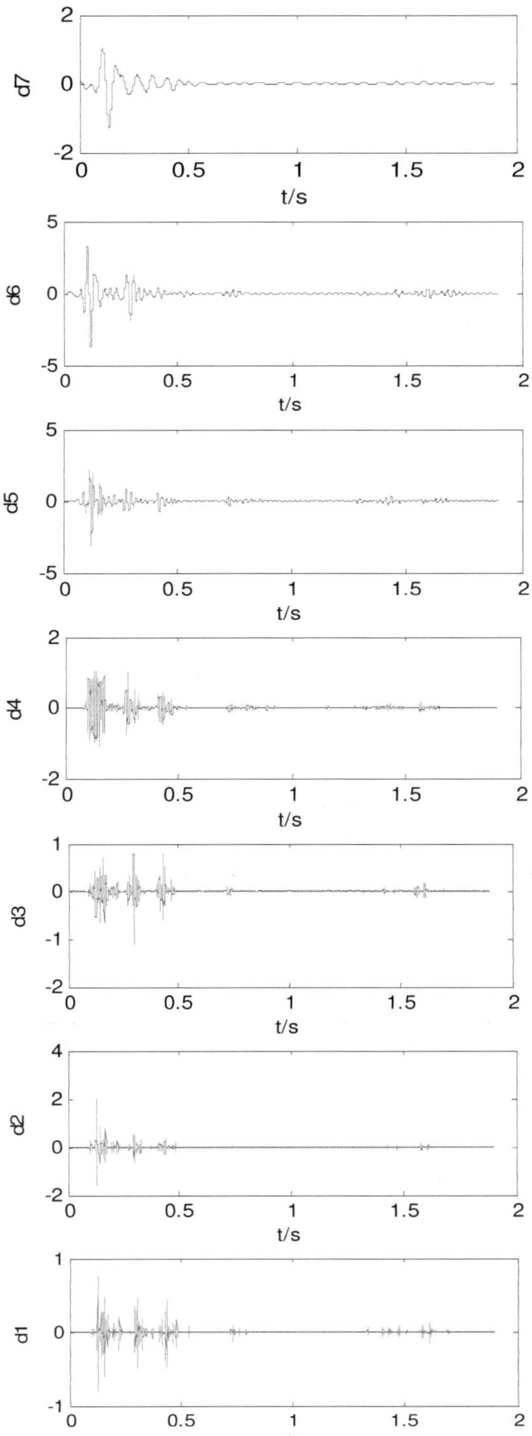

Figure 1. The original signal and reconstructed signal of the No.1 blasting point.

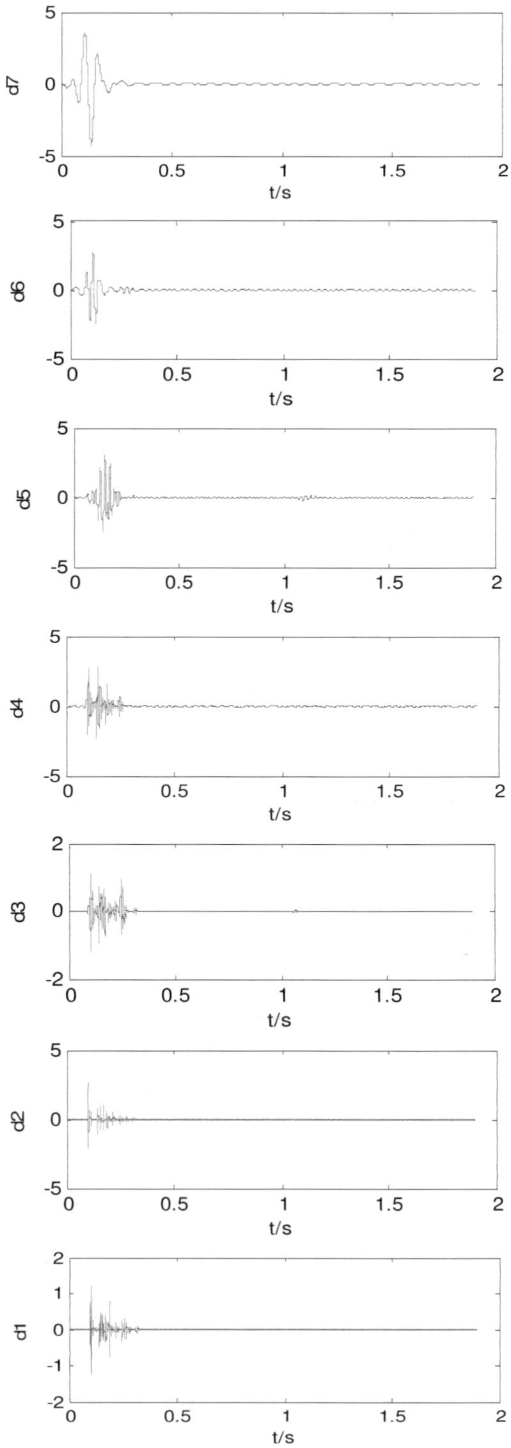

Figure 2. The original signal and reconstructed signal of the No.2 blasting point.

It can be seen from the original signal vibration waveforms that the peak value of the first blasting vibration velocity is smaller than the peak value of the second blasting vibration velocity. Even though the total charge quantity of the first blasting is higher than the second blasting and the maximum single-shot charge quantity is similar to each other.

The reconstructed waveforms of the a8 low frequency band (0-9.77Hz) of the blasting vibration signal show a certain degree of amplitude in the first half stage, but fluctuate at zero value in the second half of the signal. It is indicated that when the blasting seismic wave reaches the measuring point, the energy of the frequency band signal generates some impacts for the side wall of bus hole in the first half section.

For the reconstructed waveforms of the d8 frequency band (9.77-19.53Hz), it can be seen that the side wall show a large amplitude vibration during the first 0-500ms, while the amplitude of vibration decreases to 0 after 500ms. It is indicated that when the blasting seismic energy reaches to the side wall of the bus hole, due to the dynamic response characteristics of the surrounding rock, some frequency bands of the blasting vibration signal will maintain a certain degree of amplitude for a relatively long time, showing a slowly energy attenuation characteristic.

It is found that the peak velocity of d7 (19.53-39.06Hz) and d6 (39.06-78.13Hz) are relatively larger compared with the peak velocities of other frequency bands of the two blasting. Therefore, it can be concluded that the natural frequency of the surrounding rock is located in the range of these two bands, in which shows a significant amplitude increasing phenomenon.

The vibration signals show a process that continuously increase to the peak value and then decrease to zero within 500ms by comparing with the vibration signals of frequency band d3 (312.5-625Hz), d4(156.25-312.5Hz) and d5 (62.5-125Hz) of the first blasting, and the vibration signals of frequency band d4 (156.25-312.5Hz), d5 (62.5-125Hz) and d6 (39.06-78.13Hz) of the second blasting. The similarity of the signal amplitude of these two bands shows that the blasting vibration signal does not affected by surrounding rock structure. The similarity of the blasting vibration waveform attenuation of the two blasting shows that the vibration response characteristics of rock structure at the same measuring point are similar for different blasting sources. And it also shows that the natural frequency of the surrounding rock is lower before the first blasting.

4. Conclusion

The time-frequency characteristics of the blasting vibration signal obtained from the wavelet analysis can clearly reflect the difference of the blasting vibration signal caused by the slight difference of the single-shot charge quantity. It can be used to guide engineering blasting operation and optimize blasting parameters. Blasting signals may last for a long time and accompanying with a certain degree of increment when blasting signals input into the surrounding rock for the low frequency band. The continuous blasting at the same point of the surrounding rock will result in some structure changes, then reducing the natural frequency of the surrounding rock. When the next blasting seismic wave enters the surrounding rock with a frequency band near the natural frequency of the surrounding rock, the blasting vibration signal will persist for a long time and accompanying with higher amplitude. It is not conducive to the stability of the surrounding rock. The peak vibration velocity does not increase on the increase of the delayed blasting segments. Even the rational use of delayed blasting, the increase of the amount of detonation will not necessarily lead to the increase of the peak vibration velocity for the protected engineering.

5. Reference

[1] Duan Y S 2007 *Water Resour. Hydropow. Eng.* **38** 24.
[2] Ye H W, Zhou J M, Yu H B, Wang C and Lei T 2016 *Blasting* **33** 12.
[3] Jiao Y B 1995 *Blasting* **12** 45.
[4] Yang S Q, Liao X K and LIU B C 2001 *Explos. Shock Wav.* **21** 223.
[5] Wu L, Chen J P and Shu J H 1999 *Explos. Mater.* **28** 24.
[6] Chen S H, Wei H X, Zhang A K and Mao X H 2010 *Explos. Shock Wav.* **30** 377.
[7] Lin T H and Li X B 2004 *Chinese J. Rock Mech. Eng.* **23** 2260.
[8] Yan J W, Long Y and Fan Xiang 2007 *Explos. Shock* **27** 405.
[9] Hu K Y, Wu L Q and Cao Y 2015 *Blasting* **32** 194.
[10] Song B and Cao Y 2013 *Rock Soil Mech.* **34** 234.

IOP Conf. Series: Materials Science and Engineering **248** (2017) 012024 doi:10.1088/1757-899X/248/1/012024

Compaction algorithm for orthogonal packing problems

Vladislav A Chekanin[1] and Alexander V Chekanin[1]

[1]Department of Theoretical Mechanics and Strength of Materials, Moscow State University of Technology «STANKIN», 3a Vadkovsky lane, Moscow, 127055, Russian Federation

Abstract. The article is devoted to an algorithm for increasing the density of placement schemes in orthogonal packing problems. The proposed heuristic packing compaction algorithm is based on the principle of local replacement of placed objects located as far from the origin of a container as well as located near the local free spaces inside it. The algorithm uses a list of six rules which select objects from a container for deleting and subsequent reallocation of them into freed spaces of the container. The article contains results of computational experiments carried out on the standard strip packing problems.

1. Introduction

Solution of many different practical optimization problems, including resources saving problem, cutting of materials, problems in logistics and manufacturing, scheduling and planning comes down to the orthogonal packing problem [1–3]. This problem is a particular case from a wide class of resource allocation problems. All these problems are classical problems of the mathematical theory of operations research. The solution of these problems is to find the most rational ways of allocating resources (materials) of one type (called objects) among resources of another type (called containers). A detailed classification of resource allocation problems was proposed by G. Wascher, H. Haubner and H. Schumann in 2007 [4], in preparation of which 445 scientific articles were considered. The wide area of practical applications of resource allocation problems in industry and the economy makes the problem of developing effective algorithms and methods for solving these problems urgent [5], as evidenced by the presence of a large number of scientific publications [1, 4].

The orthogonal packing problem is the NP-complete problem in the strong sense [6]. To obtain exact solutions of such problems is necessary to use the resource-intensive optimization algorithms, which are ineffective on practice due to heavy spending of time resources. Therefore, to solve NP-complete packing problems approximate heuristic optimization algorithms are usually used. They provide obtaining of suboptimal solutions for an acceptable time [7–10].

In this article is proposed a packing compaction algorithm based on the idea of reallocation of placed into a container objects. Since the problem of developing of an iterative algorithm intended for compacting of the placement schemes was first introduced here, there is no practical possibility of comparing its effectiveness with other similar algorithms. The described algorithm is designed to improve the quality of suboptimal solutions (i.e. to increase the density of the placement schemes), which were previously found using various optimization heuristic and metaheuristic algorithms, the development of which is the goal of most researchers in the field of the cutting and packing.

2. Statement of the orthogonal packing problem

Consider the statement of D-dimensional orthogonal packing problem of n objects into one container with dimensions $\{W^1, W^2, \ldots, W^D\}$. All objects and the container are presented in the form of D-

Content from this work may be used under the terms of the Creative Commons Attribution 3.0 licence. Any further distribution of this work must maintain attribution to the author(s) and the title of the work, journal citation and DOI.
Published under licence by IOP Publishing Ltd

dimensional parallelepipeds with dimensions $\{w_i^1, w_i^2, \ldots, w_i^D\}, i \in \{1, \ldots, n\}$. We denote the position of an object i located inside the container by $(x_i^1; x_i^2; \ldots; x_i^D)$. The goal is to find the most density placement of all the objects into the container under the following conditions of correct placement:

- all edges of the objects are parallel to edges of the container;
- all packed objects do not overlap with each other:
$$\left(x_i^d \geq x_k^d + w_k^d\right) \vee \left(x_k^d \geq x_i^d + w_i^d\right) \quad \forall d \in \{1, \ldots, D\}, \quad i, k \in \{1, \ldots, n\}, \quad i \neq k\,;$$
- all packed objects are within the bounds of the container:
$$\left(x_i^d \geq 0\right) \wedge \left(x_i^d + w_i^d \leq W^d\right) \quad \forall d \in \{1, \ldots, D\}, \quad \forall i \in \{1, \ldots, n\}.$$

When objects are placed into the container any rotation of them is not allowed.

A solution of any dimensional packing problem is represented by a so-called placement string (also known as a chromosome) which contains a sequence of objects selected for placing. To obtain a packing is necessary to decode the placement string for this is solved the problem of selecting the objects to be packed and free spaces of the container. The statement of D-dimensional orthogonal packing problem includes specifying of a load direction of the container as the priority selection list $\{L\} = \{L^1; L^2; \ldots; L^D\}$ of its coordinate axes, where $L^d \in [1; D]\, \forall d \in \{1, \ldots, D\}$, $L^i \neq L^j\ \forall i \neq j$. This list defines the order of selecting of free spaces of the container for placing the objects into them.

The quality of the placement is estimated by the criteria of length minimization of filled with the objects container's part which is measured along the coordinate axis $l = L^D$, i.e. the minimized fitness function is calculated as $S = \max\left(x_i^l + w_i^l\right),\ i = 1 \ldots n$ and it defines the position of the most remote object in the container.

To describe placement schemes of orthogonal objects in containers was chosen the previously developed packing representation model – the model of potential containers. The efficiency of this model for orthogonal packing problems is shown in paper [11]. This model describes the current state of the container with a set of so-named potential containers in form of orthogonal objects with the maximal dimensions which potentially can be placed in the container. A potential container k is described by a vector $\{p_k^1; p_k^2; \ldots; p_k^D\}$ of its dimensions as well by a vector $\{x_k^1; x_k^2; \ldots; x_k^D\}$ which defines the position of the origin of the coordinate system of this potential container relative the origin of the coordinate system of the containing it container. The fast access to potential containers is provided by usage of the effective multilevel linked data structure described in papers [12, 13].

3. Compacting packing problem

In the basis of the proposed packing compaction algorithm lays usage of the rules for selecting the placed objects for deleting of them with subsequent more rational filling of the freed spaces in the container. The implementation of an algorithm for deleting objects designed especially for the model of potential container model is described in detail in [14]. The rules of selecting objects for reallocation of them are based on the methods given below.

Method AREA selects from the container all objects contacting with the most remote edges of an orthogonal object with the dimensions $V^d = W^d\ \forall d \neq l$, $V^l = \max\left(x_i^l + w_i^l\right),\ i = 1 \ldots n$ which encloses the placement (figure 1).

Method LAST selects from the container all objects contacting with potential containers the edges of which along the coordinate axis l match to the border of the container, i.e. the potential containers for which $x_k^l + p_k^l = W^l$ (figure 2).

Method MAX selects from the container all objects contacting with a potential container that has the maximal volume (area) and located inside the placement, i.e. the potential container for which $x_k^l + p_k^l < W^l$ (on figure 3 this potential container is represented by a rectangle without color filling).

Method DOUBLE selects from the container all objects contacting with two potential containers which have maximal volumes (areas) and located inside the placement (figure 4).

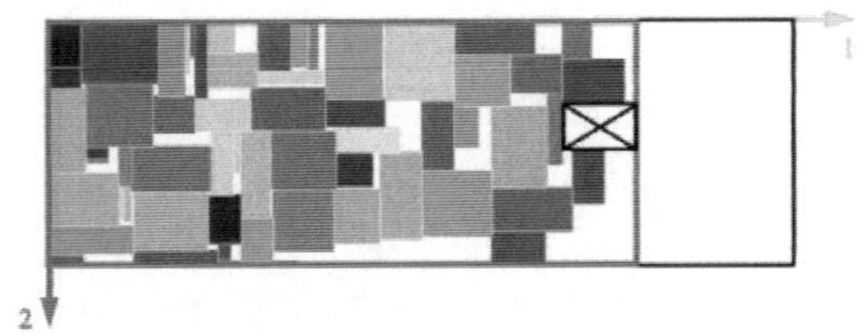

Figure 1. Objects selected by method AREA.

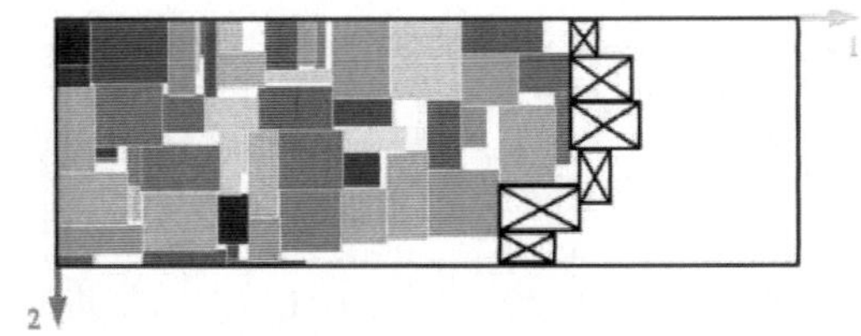

Figure 2. Objects selected by method LAST.

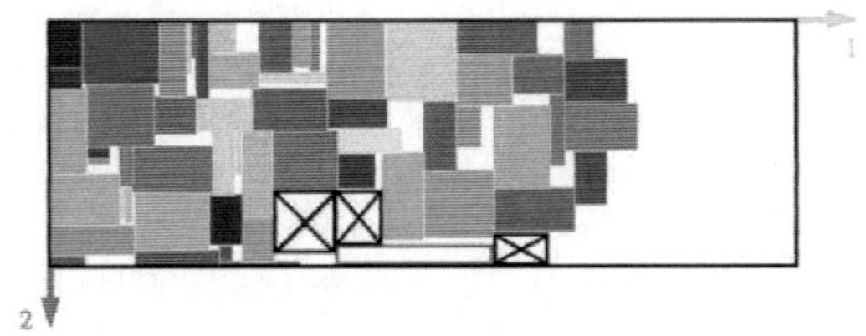

Figure 3. Objects selected by method MAX.

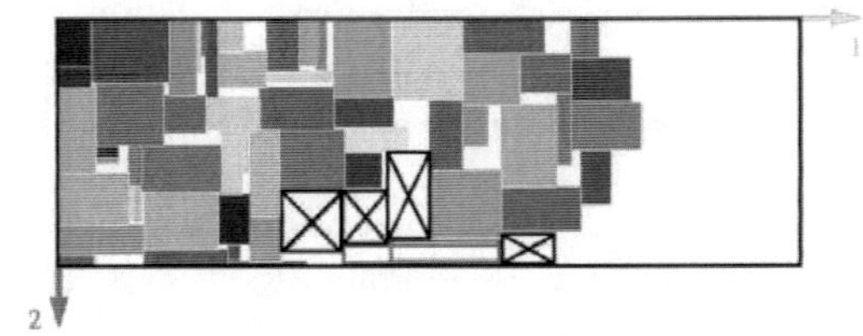

Figure 4. Objects selected by method DOUBLE.

The methods of the first group (AREA and LAST) are intended to select objects located as far from the origin of the container. The methods of the second group (MAX and DOUBLE) are intended to select the objects located close to the largest free spaces in the container. Based on these methods, six object selection rules for local reallocation of placed objects were implemented (table 1).

Table 1. Object selection rules.

No.	Rule	Method 1	Method 2	Description
1	MAX_AREA	AREA	MAX	Cooperative application of two methods.
2	MAX_LAST	LAST	MAX	Cooperative application of two methods.
3	DOUBLE_AREA	AREA	DOUBLE	Cooperative application of two methods.
4	DOUBLE_LAST	LAST	DOUBLE	Cooperative application of two methods.
5	MAXMAX_AREA	AREA	MAX	Application of the method AREA as well as application of the method MAX to the placement scheme obtained after deleting of all objects previously found by the method MAX.
6	MAXMAX_LAST	LAST	MAX	Application of the method LAST as well as application of the method MAX to the placement scheme obtained after deleting of all objects previously found by the method MAX.

The proposed packing compaction algorithm is a heuristic algorithm which sequentially applies each of the described rules from the special ordered list (this order is given in the section of computational experiments of the article). This algorithm works with a population of 10 chromosomes (the placement strings), composed as follows:

- chromosome 1 contains sequence of objects in the order as they recorded in the problem;
- chromosome 2 contains sequence of objects ordered by descending of their volumes (areas);
- chromosome 3 contains sequence of objects ordered by ascending of their volumes (areas);
- chromosome 4 contains sequence of objects ordered by descending of their dimensions measured along the coordinate axis l;
- chromosome 5 contains sequence of objects ordered by ascending of their dimensions measured along the coordinate axis l;
- chromosomes 6–10 contain pseudorandom sequences of objects.

If any chromosome provides obtaining of a placement scheme with the better density (i.e. with the smaller value of the fitness function), then the best found solution is saved, after which is performed an attempt to compact this placement again and for that are applied all the rules from the ordered list starting with the first rule. This process is repeated iteratively while the algorithm provides a better density of the placement or until the stop criteria will be performed, which is the achievement of a predefined number of iterations or the use of all rules without any improvement of the density.

4. Computational experiments

The effectiveness of the compaction algorithm is investigated on the standard two-dimensional strip packing problems (rectangular cutting problems) from the paper of O. Berkey and P. Wang [15] (classes of instances C1–C6) and from the paper of S. Martello and D. Vigo [16] (classes of instances C7–C10). Were solved 500 problem instances (10 classes of 50 instances each). Every class contains from 20 to 100 rectangular objects. The maximal number of iterations is chosen equal to 20.

At the first stage of the computational experiments, the effectiveness of applying of each the object selection rule was separately analyzed. In this case was used the simplified packing compaction algorithm which uses only one object selection rule and the transition to the next iteration proceeds only if more dense placement of deleted objects was obtained at the current iteration. The algorithm was applied for placements schemes previously obtained after decoding of placement strings which contain objects in order as they recorded in the considered test problems.

The quality of compacting is estimated by evaluation $\mu = (S_0 - S_1)/S_0 \times 100\%$, where S_0 is a value of the fitness function before compacting, S_1 – a value of the fitness function after compacting.

For each rule of selecting objects, a series of 10 computational experiments was implemented. The averaged results are summarized in table 2.

Table 2. The quality of compacting for different object selection rules.

Rule	Class of test instances										Average
	1	2	3	4	5	6	7	8	9	10	quality, %
MAX_AREA	0.43	2.39	0.10	2.09	0.37	1.12	0.38	1.05	0.00	0.64	0.86
MAX_LAST	0.47	3.96	0.67	3.43	0.45	2.68	0.38	1.41	0.00	0.96	1.44
DOUBLE_AREA	0.56	3.50	0.27	2.24	0.63	1.88	0.60	1.48	0.03	1.11	1.23
DOUBLE_LAST	0.65	4.39	0.74	3.80	0.79	3.31	0.66	1.77	0.09	1.30	1.75
MAXMAX_AREA	1.21	3.37	0.41	2.96	0.76	2.37	0.65	1.91	0.15	1.38	1.52
MAXMAX_LAST	0.94	4.88	1.06	4.29	0.87	4.01	0.67	2.22	0.15	1.62	2.07

The results of the computational experiments have showed that the MAXMAX_LAST rule ensures the greatest efficiency of compacting. This rule provides obtaining of most dense placement schemes in comparison with other rules for 8 test classes, however, when using the MAXMAX_LAST rule, the packing compaction algorithm spends the maximum time. The smallest effect of the applying of the algorithm is achieved after using the MAX_AREA rule, which provides the minimal quality of compacting, while it has the highest time efficiency.

On the basis of the obtained results, was chosen the following order of usage of the object selection rules is set for the packing compaction algorithm: MAXMAX_LAST, DOUBLE_LAST, MAXMAX_AREA, MAX_LAST, DOUBLE_AREA, MAX_AREA (the rules are listed in decreasing order of their quality of compacting). On the second stage of the computational experiments were obtained the averaged results of testing the packing compaction algorithm (table 3).

Table 3. Results of testing the compaction algorithm.

Class of test instances	Average value of the fitness function before compacting	Average value of the fitness function after compacting	Average time, s	Quality of compaction, %
1	203.98	201.76	1.06	1.53
2	69.34	65.56	1.29	6.72
3	577.88	573.20	1.15	1.55
4	229.82	218.98	1.75	6.65
5	1817.86	1804.08	1.19	1.34
6	608.02	579.32	1.97	5.68
7	1685.78	1676.24	1.07	0.78
8	1644.70	1603.48	1.48	3.53
9	3398.78	3394.94	1.04	0.26
10	1052.94	1034.88	1.36	2.17
Average	1128.91	1115.24	1.34	3.02

5. Conclusion

The compaction algorithm for the orthogonal packing problems has been proposed. The work of this algorithm is based on the idea of iterative reallocation of placed objects in a container. The effectiveness of the algorithm for different rules of selection objects has been investigated on the standard two-dimensional packing strip problems. The developed packing compaction algorithm is implemented in a general form, which makes it possible to use it not only for two-dimensional problems as well as for three-dimensional orthogonal packing problems.

6. References

[1] Bortfeldt A and Wascher G 2013 *Eur. J. Oper. Res.* **229(1)** 1–20
[2] Lodi A, Martello S and Monaci M 2002 *Eur. J. Oper. Res.* **141(2)** 241–52
[3] Martello S, Pisinger D and Vigo D 2000 *Oper. Res.* **48(2)** 256–67
[4] Wascher G, Haubner H and Schumann H 2007. *Eur. J. Oper. Res.* **183(3)** 1109–30
[5] Crainic T G, Perboli G and Tadei R 2008 *Informs J. Comput.* **20(3)** 368–84
[6] Johnson D S 2012 *Documenta Mathematica* **Extra Volume ISMP** 359–76
[7] Kierkosz I and Luczak M 2014 *Cent. Eur. J. Oper. Res.* **22(4)** 729–53
[8] Leung S C H, Zhang D F, Zhou C L and Wu T 2012 *Comput. Oper. Res.* **39(1)** 64–73
[9] Chekanin V A and Chekanin A V 2014 *Appl, Mech. Mater.* **598** 377–81
[10] Chekanin A V and Chekanin V A 2013 *Comp. Math. Math. Phys.* **53(10)** 1457–65
[11] Chekanin V A and Chekanin A V 2015 *Adv. Mech. Eng.* **22** 33–8
[12] Chekanin A V and Chekanin V A 2014 *Adv. Mater. Res.* **962–965** 2868–71
[13] Chekanin V A and Chekanin A V 2014 *Adv. Mater. Res.***945-949** 3143–46
[14] Chekanin V A and Chekanin A V 2016 *ARPN J. Eng. Appl. Sci.* **11(13)** 8436–46
[15] Berkey O and Wang P 1987 *Journal of the Operational Research Society* **38(5)** 423–29
[16] Martello S and Vigo D 1998 *Manage. Sci.* **44** 388–99

Acknowledgements

This work was carried out with the financial support of the Ministry of Education and Science of Russian Federation in the framework of the state task in the field of scientific activity of MSTU "STANKIN" (No. 1.7706.2017/8.9).

IOP Conf. Series: Materials Science and Engineering **248** (2017) 012025 doi:10.1088/1757-899X/248/1/012025

Dynamic analysis of suspension cable based on vector form intrinsic finite element method

Jian Qin[1,*], Liang Qiao[1], Jiancheng Wan[1], Ming Jiang[1] and Yongjun Xia[1]

[1]China Electric Power Research Institute, 100055, Beijing, China

*E-mail: qinjian@epri.sgcc.com.cn

Abstract. A vector finite element method is presented for the dynamic analysis of cable structures based on the vector form intrinsic finite element (VFIFE) and mechanical properties of suspension cable. Firstly, the suspension cable is discretized into different elements by space points, the mass and external forces of suspension cable are transformed into space points. The structural form of cable is described by the space points at different time. The equations of motion for the space points are established according to the Newton's second law. Then, the element internal forces between the space points are derived from the flexible truss structure. Finally, the motion equations of space points are solved by the central difference method with reasonable time integration step. The tangential tension of the bearing rope in a test ropeway with the moving concentrated loads is calculated and compared with the experimental data. The results show that the tangential tension of suspension cable with moving loads is consistent with the experimental data. This method has high calculated precision and meets the requirements of engineering application.

1. Preface

The freight ropeway has simple structure and high tensile strength. During working it can fully exert its advantages of small weight, low cost, easy construction and strong transportation ability. It has been widely used in fields such as mining, forestry and aerial transmission line construction, etc.

At present, the dynamic analysis of suspension cable has also been studied. In the papers [1, 2] the discrete method and direct method for nonlinear dynamic analysis are used to analyse the suspension structure. It is found that the result of discrete method may be wrong. In the paper [3] the dynamic model of the cable with large sag is established by multi rigid body system dynamic theory. The results show that the method is accurate. The dynamics of an infinite string on an elastic foundation subjected to a moving load is under investigation in the paper [4]. The free vibration of suspension cable is analysed by using Wavelet-Galerkin method in the paper [5]. In the paper [6, 7] the dynamic study of the fixed load on the suspension cable is carried out. However, the study on suspension cable under moving concentrated loads is rare.

In this paper, the dynamic analysis method of suspension structure under moving concentrated load is established based on the vector form intrinsic finite element (VFIFE) theory [8]. Then the method is applied to calculate the cable tension of an engineering ropeway under the test condition.

2. VFIFE of suspension structure

In the ropeway, the bearing rope can be regarded as the flexible suspension cable, and it follows the assumption of elastic cable.

The truss element is used to analyse the dynamic response of suspension structure by the VFIFE method. The VFIFE method for suspension structure is deduced below.

Content from this work may be used under the terms of the Creative Commons Attribution 3.0 licence. Any further distribution of this work must maintain attribution to the author(s) and the title of the work, journal citation and DOI.
Published under licence by IOP Publishing Ltd

IOP Conf. Series: Materials Science and Engineering **248** (2017) 012025 doi:10.1088/1757-899X/248/1/012025

2.1. The description of space points in suspension structure

In the calculation of VFIFE, the suspension structure is discretized into a set of space points, the points are combined with the truss element. The shape of suspension cable is reflected by the motion trajectory of space points, shown as figure 1. The mass of the structure is shared by the space points, and the elements have no mass. The motion of the space points follows Newton's second law.

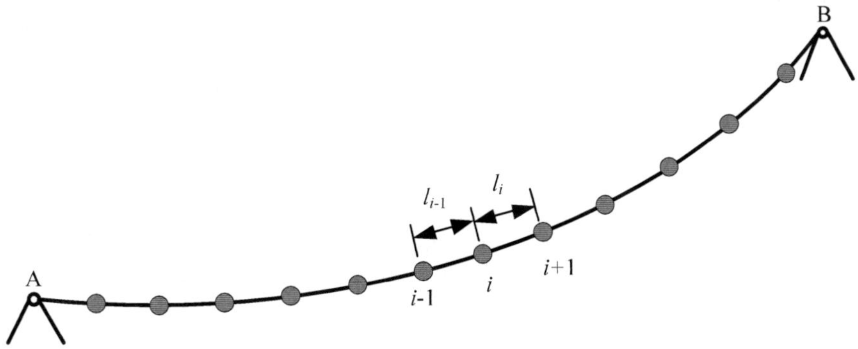

Figure 1. The Description of Space Points in Suspension Structure.

The suspension cable has n-1 elements, and the length of element is l_i. The mass of the suspension cable is evenly distributed on the n space points, so

$$m_i = \frac{1}{2}ql_{i-1} + \frac{1}{2}ql_i \tag{1}$$

Here, m_i is the mass of the space point i, q is the mass of suspension cable per unit length.

Any point on the cable follows Newton's second law in the movement, so its equation of motion can be written as

$$m_i \frac{\mathrm{d}^2\boldsymbol{x}_i}{\mathrm{d}t} = \mathbf{P}_i + \boldsymbol{p}_i + \boldsymbol{f}_i \tag{2}$$

Here, x_i is the position vector function for space point i, $\mathbf{P}_i$ is the external force acting on the space point i directly, $\boldsymbol{p}_i$ is the equivalent external force acting on the space point i which is transformed from the external force acting on the elements connected to the point, $\boldsymbol{f}_i$ is the resultant force acting on the space point i that connected with the truss elements.

2.2. The calculation of element internal force

When the space points are moving, they will be affected by the internal force of the element connected with them. Take one element for force analysis. Define that the cross-sectional area is A, the elastic modulus is E, the length of element is l. In the process of deformation, the cross-sectional area and elastic modulus of the truss element remain unchanged.

The truss element is a kind of axial force element, the axial stress is σ^0, thus the internal force of the element is

$$-\boldsymbol{f}_1^0 = \boldsymbol{f}_2^0 = f^0\boldsymbol{e}^0 = (\sigma^0 A)\boldsymbol{e}^0 \tag{3}$$

Here, $\boldsymbol{f}^0$ is the internal force of the truss element at time t_0, $\boldsymbol{e}^0$ is the unit vector of the truss element at time t_0.

The element moves as a rigid body from time t_0 to t, so it is still straight. The deformation of the element:

$$\Delta\boldsymbol{u} = (l - l_0)\boldsymbol{e}^0 \tag{4}$$

Here, l_0 is the length of the truss element at time t_0, l is the length of the truss element at time t.

So the internal force of the truss element is

$$\boldsymbol{f}_{elem} = \left(f_{elem}^{0} + \frac{EA}{l_0}(l - l_0) \right) \boldsymbol{e}^{t} \tag{5}$$

The suspension cable is absolutely flexible, so it can't bear compressive stress. If $\boldsymbol{f}_{elem}$<0, then the element internal force is defined as zero, $\boldsymbol{f}$=0.

2.3. The external force calculation of the space point
The truss element is axial force element, the external force of the element can be converted into the concentrated force acting on the space point according to the principle of equivalence.

2.4. The process of moving loads
When loads moving along the suspension cable, the gravity forces of loads are constant and the directions of gravity forces are vertical. Assumed that the length of suspension cable changes very small, the loads move from the starting point to the next adjacent space point during the integration step.

3. Solution of the motion equation
In the VFIFE, the central difference formula is used to solve the motion equations. The velocity and acceleration of the space points can be obtained by the central difference formula:

$$\dot{x} = \frac{1}{2h}(x_{n+1} - x_{n-1}) \ (n \geq 1) \tag{6}$$

$$\ddot{x} = \frac{1}{h^2}(x_{n+1} - 2x_n + x_{n-1}) \ (n \geq 1) \tag{7}$$

Here, x_{n+1}, x_n and x_{n-1} is the position vector of space points at step n+1, n and n-1 respectively. h is time increment step. Substituting the formula (6) and (7) into the motion equations of the space points, x_{n+1} can be obtained:

$$\boldsymbol{x}_{n+1} = \frac{\boldsymbol{F}^n}{m} h^2 + 2\boldsymbol{x}_n - \boldsymbol{x}_{n-1} \ (n \geq 1) \tag{8}$$

Here, $\boldsymbol{F}^n$ is the resultant force acting on the space points at the time step n, $\boldsymbol{F}^n = \mathbf{P}^n + \boldsymbol{p}^n + \boldsymbol{f}^n$.

Introducing the energy dissipation mechanism, the structure can reach static balance. In the equation of motion, the virtual damping force is increased, thus the motion equation of the space points is:

$$m\ddot{\boldsymbol{x}} = \boldsymbol{F}^n + \boldsymbol{f}_{dmp} \tag{9}$$

Here, $\boldsymbol{f}_{dmp}$ is the virtual damping force, $\boldsymbol{f}_{dmp} = -\xi m \dot{\boldsymbol{x}}$, $\xi > 0$ is the damping factor.

Substituting the formula (6) and (7) into (9), it can obtain:

$$\boldsymbol{x}_{n+1} = \frac{h^2}{m} c_1 \boldsymbol{F}^n + 2c_1 \boldsymbol{x}_n + c_1 c_2 \boldsymbol{x}_{n-1} \ (n \geq 1) \tag{10}$$

Here, $c_1 = \frac{2}{2 + \zeta h}$, $c_2 = \frac{\zeta h}{2} - 1$.

When n=0, by the formula (6), (7) and (10), the initial position vector can be obtained:

$$\boldsymbol{x}_1 = \frac{\boldsymbol{F}^n}{2m} h^2 + \boldsymbol{x}_0 - c_2 h \dot{\boldsymbol{x}}_0 \tag{11}$$

4. Example

To verify the calculation method, the field test of freight ropeway is finished in Hanzhong, Sichuan in China, shown as figure 2. Here a working case is selected to be calculated by the method and the results are compared with the experimental data.

Figure 2. Testing Site and Concentrated Load.

The test ropeway is a single span ropeway, shown as figure 3. The span is 218.183m, height is 118.668m, diameter of bearing rope is 20mm, the elastic modulus is 80GPa and the mass of suspension cable per unit length is 1.6kg/m. The initial length of bearing rope calculated is 249.06m.

The starting position of the ropeway is marked as A, the end is marked as B. The number of elements of bearing rope is 60, the speed of loads is 0.3m/s.

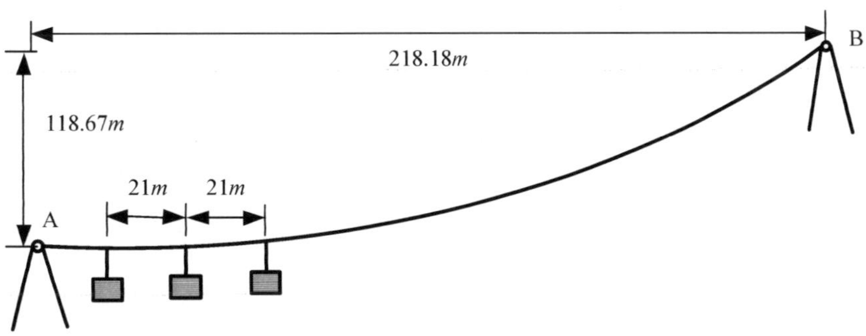

Figure 3. Working Case.

In the test, there are three concentrated loads (640kg, 640kg, 425kg), and the first load is 640kg, the distance between the loads is 21m. The first load starts from the starting position (away A point 53m) and stops at the position away A point 125m. The tangential tension of bearing rope at A and B is calculated by the method presented in this paper during the loads moving.

As shown in figure 4, the tangential tension at position A and B of the ropeway is agreed with the test data. The maximum tangential tension at position A by test is 74.503kN, then the calculation result is 73.106kN, 1.88% smaller than the test result. The maximum tangential tension at position B by test is 75.031kN, then the calculation result is 74.785kN, 0.33% smaller than the test result.

The results show that the relative error between the calculated results and the test results is very small and the method has high precision.

IOP Conf. Series: Materials Science and Engineering **248** (2017) 012025 doi:10.1088/1757-899X/248/1/012025

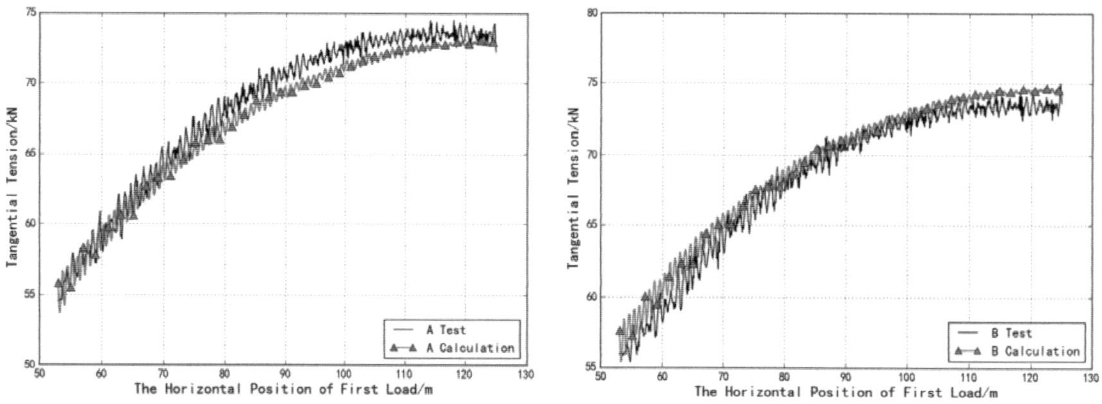

Figure 4. The Tangential Tension at Position A and B during Loads Moving.

5. Conclusions

According to the VFIFE theory and the mechanical characteristics of suspension structure, a vector finite element method for the dynamic analysis of cable structures is presented. A working case of an engineering ropeway verifies the calculation method.

The comparison between the calculated results and experiment data shows that:

1) The calculated tangential tension curves presented in this paper agree with the test curves;

2) The maximum of the relative error is 1.88%.

Therefore, this method has high precision and is easy to be programed in computer, and it is very convenient for engineering application.

6. References

[1] Zhao Y Y and Wang L H, et al 2005 *Acta Mech. Sin.* **37** 329-38.
[2] Cheng D Y 2005 *Exact Element Method for Analysis of Cable Structures* (Beijing: Tsinghua University).
[3] Li B and Li Y H 2000 *Appl. Math. Mech.* **21** 640-6.
[4] Gavrilov S. 1999 *J. Sound Vibr.* **222** 345-61.
[5] Al-Qassab M and Nair S 2004 *J. Sound Vibr.* **270** 191-206.
[6] Cheng S P and Perkins N C 1994 *J. Appl. Mech.* **61** 944-8.
[7] Cheng S P and Perking N C 1992 *J. Acoust. Soc. Am.* A **91** 2654-62.
[8] Ding C X, Duan Y F and Wu D Y 2012 *Vector Mechanics of Structures* (Beijing: Science Press).

Acknowledgment

This research was supported by Science and Technology Research Project of State Grid (Study on the calculation method of multi-bearing cable ropeway and the application in UHVDC).

IOP Conf. Series: Materials Science and Engineering **248** (2017) 012026 doi:10.1088/1757-899X/248/1/012026

Calculation method of multi-span tension stringing in transmission line

Jian Qin[1,*], Jiancheng Wan[1], Liang Feng[1], Fei Peng[1] and Liang Qiao[1]

[1]China Electric Power Research Institute, 100055, Beijing, China

*E-mail: qinjian@epri.sgcc.com.cn

Abstract. The process of multi-span tension stringing in transmission line is analysed based on elastic catenary equation. From the tower condition of transmission line, the conservation equations of span spacing and elevation difference are built for conductor, and the tension balance equations of conductor between different spans are got considering the rolling friction of the pulley. Added the external tension condition, the overall nonlinear equations of multi-span tension stringing are set up which can be solved by Newton iteration method. The iterative process is listed for the calculation of the conductor tension and sags in engineering problems. Compared the results with the numerical example, this method are proved the reliability. The calculation amount, efficiency and accuracy of this method can meet the needs of the practical engineering.

1. Preface

With the development of the Ultra-high voltage transmission line in China, the tension stringing technology has been widely accepted and adopted in the industry. During the whole stringing process of UHV, conductor tension must be kept large enough to raise the conductors off the ground, cross obstacles and facilities. This can avoid conductor damage, ensure the quality of the transmission lines, and minimize the influence for environment. Therefore, the construction technology of tension stringing must be applied in string operation. The selection of construction machinery and equipment is based on the conductor tension and traction, so the tension calculation method for conductor is important for engineering safe [1].

Nowadays, the major way of string construction calculation is the parabolic method [2], but this method has poor precision compared with actual measurement. Other method such as catenary method [3, 4] is more accurate, but the calculation is relatively complex. Therefore, the accurate calculation method of the tension and traction is proposed.

2. Condition of multi-span tension stringing

There are many important conditions for the process of tension stringing construction, such as the span spacing and elevation difference between towers in transmission line, and the material parameters of conductor.

2.1. Main parameter of transmission line

There are n continuous spans in the transmission line, and the span spacing is L_{si}, elevation difference is $h_{si}\,(i=1, \ldots, n)$, shown as figure 1. The unit gravity of conductor is q, the product of cross-sectional area and elasticity modulus is EA_0.

Content from this work may be used under the terms of the Creative Commons Attribution 3.0 licence. Any further distribution of this work must maintain attribution to the author(s) and the title of the work, journal citation and DOI.
Published under licence by IOP Publishing Ltd

IOP Conf. Series: Materials Science and Engineering **248** (2017) 012026 doi:10.1088/1757-899X/248/1/012026

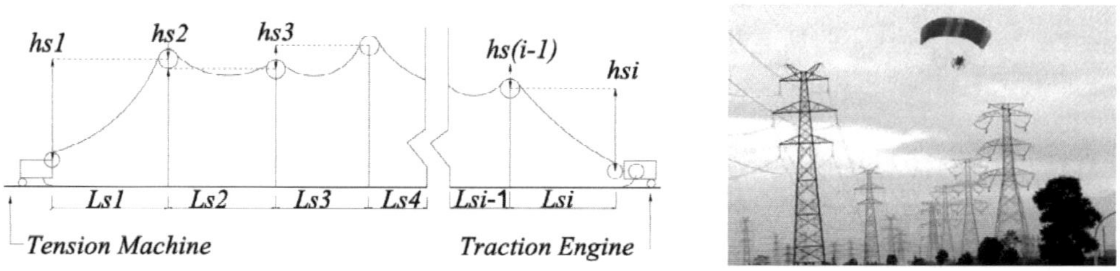

Figure 1. Schematic diagram of the multi-span tension stringing.

The total length s_0 of the conductor can be obtained by measurement before the engineering construction.

Assuming that initial conductor length of each span is s_i, horizontal tension is H_i, vertical tension on point A is $V_{Ai}(i=1,\ldots,n)$, so the vertical tension on point B can be expressed as

$$V_{Bi}=V_{Ai}+qs_i \quad (i=1, \ldots, n). \tag{1}$$

The calculation formula of span spacing and elevation difference of conductor is [5]:

$$\begin{aligned} L_i &= \frac{H_i}{q}\left[\frac{qs_i}{EA_0}+\operatorname{asinh}\frac{V_{Bi}}{H_i}-\operatorname{asinh}\frac{V_{Ai}}{H_i}\right] \\ h_i &= \frac{V_{Bi}^2-V_{Ai}^2}{2qEA_0}+\frac{H_i}{q}\left(\sqrt{1+\frac{V_{Bi}^2}{H_i^2}}-\sqrt{1+\frac{V_{Ai}^2}{H_i^2}}\right) \end{aligned} \quad (i=1,\ldots,n). \tag{2}$$

2.2. *Tension of conductor*

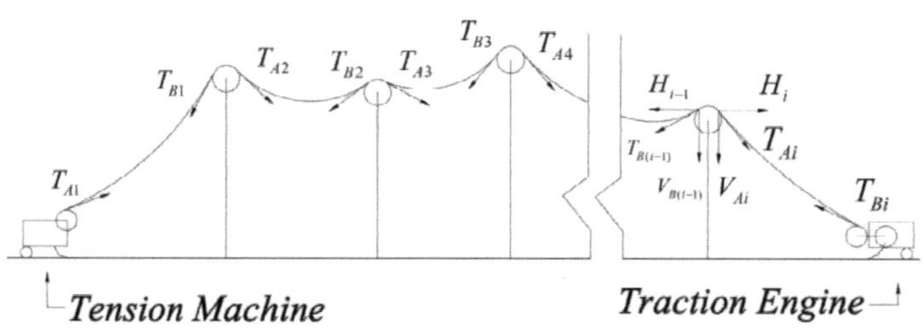

Figure 2. The force diagram of multi-span tension stringing.

One end of the conductor is pulled by traction engine and the other end by the tension machine, shown as figure 2. The tangential tension on the starting point A and end point B of each span are as follow:

$$\begin{aligned} T_{Ai} &= \sqrt{H_i^2+V_{Ai}^2} \\ T_{Bi} &= \sqrt{H_i^2+V_{Bi}^2} \end{aligned} \quad (i=1,\ldots, n). \tag{3}$$

3. Analysis of multi-span tension stringing

The structural characteristics and the force balance of the conductor in the process of multi-span tension stringing are the key points of structural analysis.

3.1. *Constant of spacing and elevation difference of each span*

The spacing and elevation difference of each span is constant no matter how the length of conductor changes, so the equations about the spacing and elevation difference of conductor can be obtained.

$$\begin{aligned}&\frac{H_i}{q}\left[\frac{qs_i}{EA_0}+\operatorname{asinh}\frac{V_{Bi}}{H_i}-\operatorname{asinh}\frac{V_{Ai}}{H_i}\right]-L_{si}=0\\&\frac{V_{Bi}^2-V_{Ai}^2}{2qEA_0}+\frac{H_i}{q}\left(\sqrt{1+\frac{V_{Bi}^2}{H_i^2}}-\sqrt{1+\frac{V_{Ai}^2}{H_i^2}}\right)-h_{si}=0\end{aligned}\quad (i=1,\ldots,n) \tag{4}$$

Therefore, there are $2n$ equations of conductor for spacing and elevation difference of all spans.

3.2. Conductor tension balance besides stringing blocks

The conductor will be influenced by frictional force in tension stringing. The friction is equal to the difference of the tangential tensions of conductor on both sides of the pulley. The tangential tensions of conductor besides of the ith pulley is shown in figure 3.

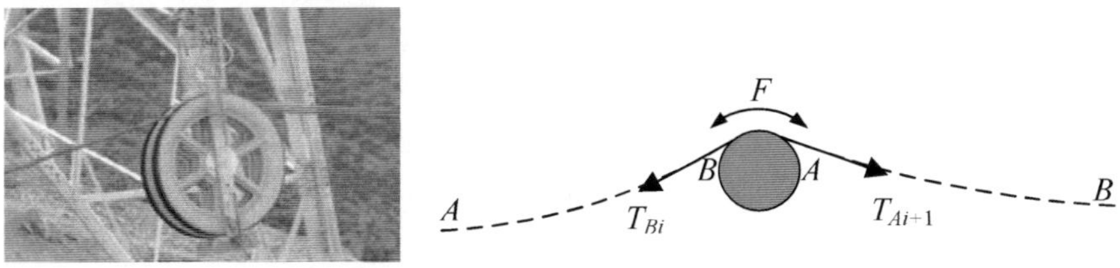

Figure 3. The conductor and pulley and the rolling friction.

The friction is associated with the frictional parameter μ, the envelope angle θ and the conductor tensions on both sides of pulley:

$$F_i = F(\mu, \theta, T_{Ai+1}, T_{Bi}) \tag{5}$$

3.2.1. The envelope angle. Envelope angle is the angle of contact arc of conductor, shown as figure 4.

$$\theta = \tan^{-1}(\frac{V_{Bi}}{H_i}) - \tan^{-1}(\frac{V_{Ai+1}}{H_{i+1}}) \tag{6}$$

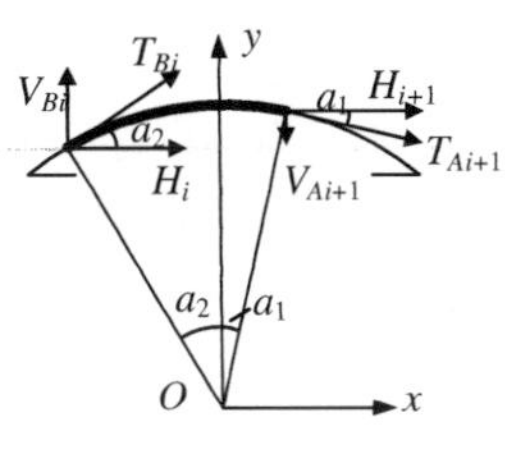

Figure 4. Contact arc of conductor and pulley.

3.2.2. The force of conductor acting on the pulley. Due to small contact part of the arc, the force N_p acting on pulley can be considered as the sum of both sides vertical force, as shown in figure 4.

$$N_p = V_{Ai+1} - V_{Bi} \tag{7}$$

3.2.3. Direction of rolling friction. When T_{Bi}-$T_{Ai+1}>0$, the direction of the rolling friction is along x axis, otherwise the opposite. Therefore, the rolling friction can be expressed as:

$$F_i = \operatorname{sign}(T_{Bi} - T_{Ai+1})\mu F(\theta, N_p) \tag{8}$$

Establish the equilibrium equations of the conductor tangential tensions:

$$T_{Bi} - T_{Ai+1} = F_i = \operatorname{sign}(T_{Bi} - T_{Ai+1})\mu F(\theta, N_p) \quad (i=1,\ldots,n\text{-}1) \tag{9}$$

3.2.4. The condition of traction tension. The end tension of conductor is the traction tension T_c:

$$T_c = T_0 \tag{10}$$

4. Nonlinear equations

Taking the span spacing s_i, horizontal tension H_i, vertical tension V_{Ai} of conductor as the unknown quantities, the nonlinear equations of the conductor for the multi-span tension stringing can be expressed as

$$F(X) = 0 \tag{11}$$

Here $X=[s_1, s_2, \ldots, s_n, H_1, H_2, \ldots, H_n, V_1, \ldots, V_n]^T$.

This nonlinear equations can be solved by Newton iteration method [6].

5. Calculation process

The process of the method for the multi-span tension stringing is shown in figure 5.

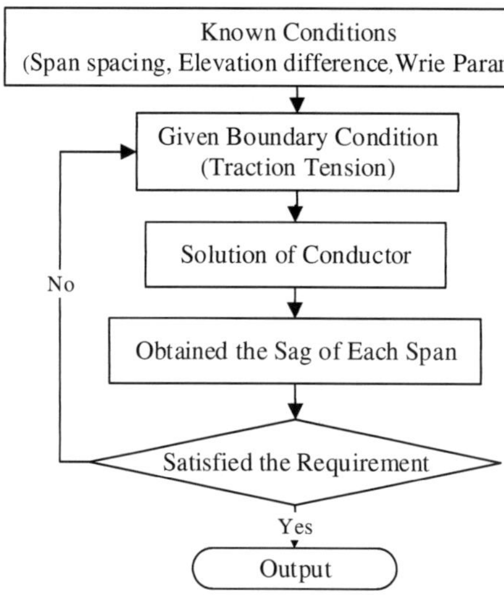

Figure 5. Calculation process.

6. Calculation cases

Taking the ±500kV Caopu-Changkou tension stringing construction project in references [7] as example. The traction tension is 14kN. Conductor parameters and the span information are shown in table 1 and table 2.

Table 1. Conductor parameters.

Type of conductor	LGJ-400/50	**Unit weight (kg/m)**	1.511
Cross-sectional area (m^2)	20	**Gravitational acceleration (kg/N)**	10
Elasticity modulus (Pa)	15	**Coefficient of tigidity**	0.012

Table 2. Information of span.

Tower number		**Span spacing(m)**	**Elevation difference(m)**	**Tower number**		**Span spacing(m)**	**Elevation difference(m)**
N0	N1	114	-16.4	N10	N11	289	31.6
N1	N2	298	32.2	N11	N12	724	87
N2	N3	380	6.5	N12	N13	286	46.7
N3	N4	1035	-162.1	N13	N14	476	73.2
N4	N5	810	149.7	N14	N15	610	127.2
N5	N6	171	21.4	N15	N16	694	56

N6	N7	270	-5.3	N16	N17	520	-37.5
N7	N8	824	-7.6	N17	N18	375	-2.4
N8	N9	522	3.5	N18	N19	206	-5.6
N9	N10	486	39.4	N19	N20	608	-19.7

According to the known condition, the calculated results of the conductor shape and tension are shown in figure 6.

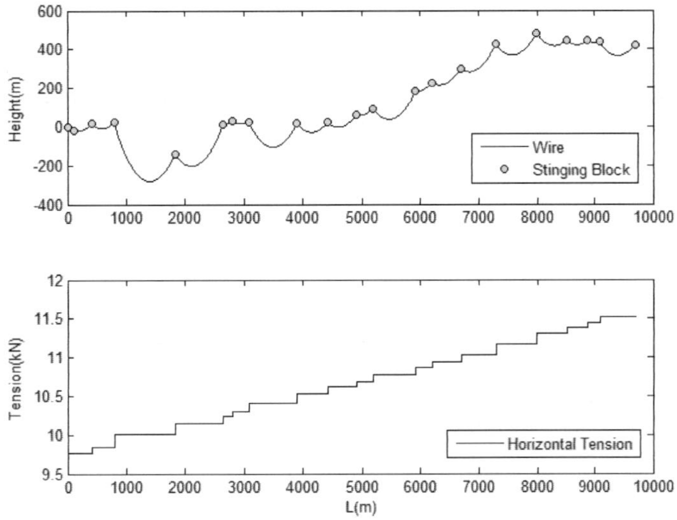

Figure 6. Calculated shape and horizontal tension of conductor.

The comparison of conductor sag and length of each span are shown in table 3 between the results calculated by this method and data in the literature [7].

Table 3. Comparison of conductor sag and length of each span.

Conductor sag of each span			Conductor length of each span		
Calculate result from literature(m)	Calculate result from this method(m)	Error(%)	Calculate result from literature(m)	Calculate result from this method(m)	Error(%)
2.2744	2.407	5.8301	115.2	114	-1.0530
15.5408	17.308	11.3714	300.3	298	-0.7720
25.27	27.9317	10.5330	381.2	380	-0.3160
187.4644	208.8044	11.3835	1070	1035	-3.3820
114.8176	124.587	8.5086	834.4	810	-3.0120
5.1172	5.4627	6.7517	172.4	171	-0.8190
12.7576	13.3717	4.8136	270.5	270	-0.1850
118.8208	126.8477	6.7555	835.5	824	-1.3960
47.6848	49.4378	3.6762	524.9	522	-0.5560
41.3344	42.4979	2.8148	489.9	486	-0.8020
14.6162	14.8857	1.8438	291.2	289	-0.7610
91.7308	93.6205	2.0600	736.9	724	-1.7820
14.3144	14.3106	-0.0270	290.3	286	-1.5030
39.6508	39.3831	-0.6750	483.8	476	-1.6390
65.1176	63.9066	-1.8600	627.7	610	-2.9020
84.2864	82.918	-1.6240	703.1	694	-1.3110
47.32	45.2745	-4.3230	524.2	520	-0.8080
24.6094	23.4323	-4.7830	376.1	375	-0.2930

7.4264	6.9839	-5.9580	206.3	206	-0.1460
64.6912	61.2765	-5.2780	612.9	608	-0.8060

The comparison shows that the results by this method agrees well with the data from literature [7], the maximum error of conductor sag of each span is 11.38%, the maximum error of conductor length of each span is 3.38%. That means the method is valid in calculation of multi-span tension stringing.

7. Summary

The nonlinear equations of multi-span tension stringing are present which include the conservation equations of span spacing and elevation difference and the tension balance equations of conductor. From the method, the traction, length and sag of conductor of each span in transmission line can be calculated to use in the selection of the traction machine.

The proposed method has good convergence and high calculation speed, which can fully meet the needs of engineering.

References

[1] Ye B S 2003 *Hubei Electr. Pow.* **27** 7-8.
[2] Northeastern electric power designing institute of China 1991 *High voltage transmission line engineering design handbook* (Beijing:China Electric Power Press).
[3] Jin M J and Zhang Z G 2004 *China Railway Sci.* **05** 9-10.
[4] Qiu W G 2015 *Phys. Eng.* **25** 11.
[5] Zhang Z G, Jin M J and Zou Z Z 2004 *China Railway Sci.* **25** 67-70.
[6] Qin J and Xia Y J 2013 *Chinese J. Eng. Des.* **20** 404-8.
[7] Shen Z 2010 *Ultra-high voltage overhead transmission line tension stringing construction technology application research* (Kunming : Kunming University of Science and Technology).

IOP Conf. Series: Materials Science and Engineering **248** (2017) 012027 doi:10.1088/1757-899X/248/1/012027

Stress Analysis of Boom of Special Mobile Crane for Plain Region in Transmission Line

Jian Qin[1,*], Tao Shao[2], Jun Chen[1], Jiancheng Wan[1], Zhonghuan Li[2] and Ming Jiang[1]

[1] China Electric Power Research Institute, 100055, Beijing, China
[2] Xuzhou Heavy Machinery Co., Ltd., 221004, Xuzhou, China

[*]E-mail: qinjian@epri.sgcc.com.cn

Abstract. Basis of the boom force analysis of special mobile crane for plain region in transmission line, the load type of boom design is confirmed. According to the different combinations of boom sections, the composite pattern of the different boom length is obtained to suit the actual conditions of boom overlapping. The large deformation model is employed with FEM to simulate the stress distribution of boom, and the calculation results are checked. The performance curves of rated load with different arm length and different working range are obtained, which ensures the lifting capacity of special mobile crane meeting the requirement of tower erection of transmission line. The proposed FEM of boom of mobile crane would provide certain guiding and reference to the boom design.

1. Preface

The mobile crane is the important equipment of tower erection in transmission line, which can reduce the labor intensity, increase the erection efficiency and improve the economic and social benefit [1].

The normal mobile crane of construction industry can't meet the requirement of tower erection. For example, the body size and weight of normal crane is too large to use in small working site [2]. So, it needs a new type mobile crane for tower crane in transmission line with smaller size and weight and enough lifting height.

Therefore, the research of the special mobile crane for plain region in transmission line is carried out. The main technical requirement is proposed as 70m lifting height and 2.7t lifting weight at the maximum height [2].

The boom, as the main load bearing part of crane, should be paid more attention for its design. Its bearing capacity directly affects the lifting level of crane [3-6]. In this paper, through the typological analysis of calculation load, the FEA model of boom of mobile crane for plain region in transmission line is set and simulated. Based of strength and stiffness control criteria, the performance curves of rated load with different arm length and different working range are obtained.

2. Design parameter of boom

The cylinder section of boom is oval shape. The boom head is single-sheet structure, and boom end is vertical plate reinforcement structure. The boom structure is shown as figure 1.

The boom of special mobile crane has 7 sections. The full-retracting length is 13.4m, and the full-extending length is 72m.

Content from this work may be used under the terms of the Creative Commons Attribution 3.0 licence. Any further distribution of this work must maintain attribution to the author(s) and the title of the work, journal citation and DOI.
Published under licence by IOP Publishing Ltd

IOP Conf. Series: Materials Science and Engineering **248** (2017) 012027 doi:10.1088/1757-899X/248/1/012027

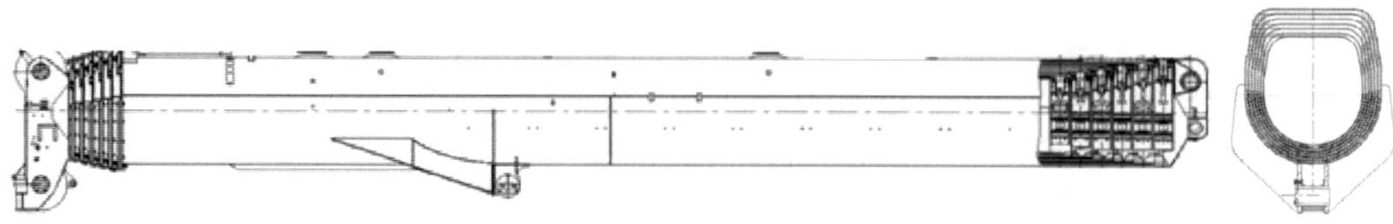

Figure 1. Structure of boom.

3. Calculation loads

The actual loads of boom, such as self-weight, hoisting load, rotary inertia load, deflection load, wind load and the wire tension, are considered in the FEM model.

3.1. Self-weight

Self-weight is the total weight of all sections of boom.

3.2. Hoisting Load

Hoisting load is the gravity of the lifting substances, includes goods, banding device and the hoisting rope. When the goods are lifted off the ground, or braked during descending suddenly, a dynamic load is generated due to inertia. The dynamic load is expressed as the product of lifting quality Q and dynamic coefficient ϕ, which is exceed 1. So the hoisting load P_Q is obtained:

$$P_Q = \phi \cdot Q \cdot g \,. \tag{1}$$

3.3. Rotary Inertia Load

The horizontal tangential inertia force caused by boom dead weight when starting and braking around rotation, is called rotary inertia load. The rotary inertia load P_d is obtained:

$$P_d = 0.002 \cdot R \cdot G_b \,. \tag{2}$$

In which, R is working range; G_b is dead weight of boom.

3.4. Deflection Load

Deflection load refers to the horizontal force caused by lifting quality when rotating or varying amplitude, includes wind force, inertia force and centrifugal force. The deflection load is:

$$P_H = P_Q \cdot \tan\alpha \,. \tag{3}$$

In which, α is swing angle of rope, 3°~6° generally.

3.5. Wind Load

Wind load refers to the wind force worked on the boom:

$$P_W = C \cdot Q_f \cdot \sum A_i \,. \tag{4}$$

In which, C is wind shape coefficient; Q_f is wind pressure (125N/m^2 when boom length less than 50m, 150 N/m^2 when boom length more than 50m); $\sum A_i$ is the sum of areas windward of all sections.

3.6. Wire Tension

During the cargo hoisting, the weight is balanced by the tension of rope. In the FEA model, the tension is replaced by addition load, the value is equal to the weight of cargo.

4. Model of FEA

4.1. Geometrically nonlinear analysis

IOP Conf. Series: Materials Science and Engineering **248** (2017) 012027 doi:10.1088/1757-899X/248/1/012027

4.1.1. Model building.

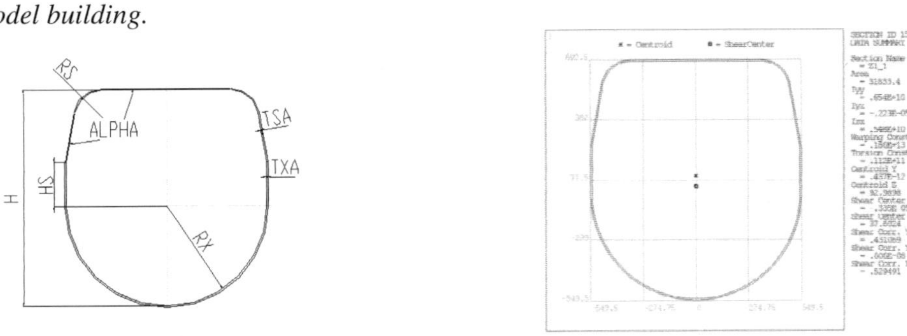

(a) Actual Model (b) FEA Model

Figure 2. Shape of boom section.

The boom of special mobile crane has 7 sections, which can be extendable. The size along extension direction is considerably larger than the size of height and width. The up cover plate of section is arc groove plate, the under plate is elliptical elbow plate. The shape of section is shown as figure 2(a). The section model of boom is established by ANASYS custom section module. The FEA section model is shown in figure 2(b).

4.1.2. Section assembly and load application. Establish the connection between boom sections. The connection of sections is shown as figure 3. Node 1, 2 and 4 simulate the slider of boom head, and node 3, 5 and 6 simulate the slider of boom end. The contact element of boom sections is established by MPC184 slide block unit.

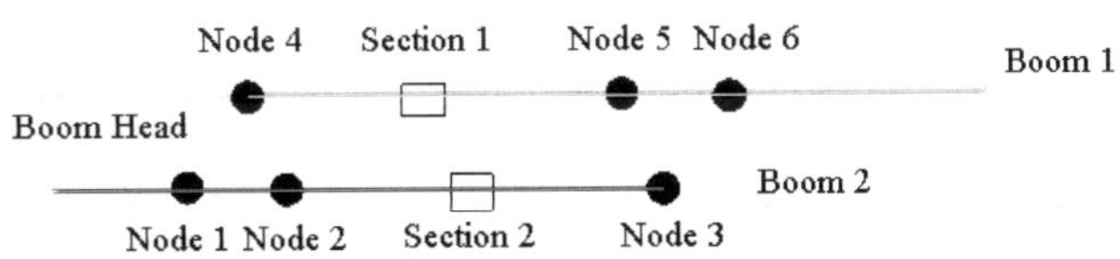

Figure 3. The connection simulated by beam elements.

The geometrically nonlinear analysis is adopted in calculation of large deformation. The model is loaded progressively from 0 to the target weight. Take the case of 72m full-extending length as an example, the model with load is shown as figure 4.

Figure 4. The overall model.

IOP Conf. Series: Materials Science and Engineering **248** (2017) 012027 doi:10.1088/1757-899X/248/1/012027

4.2. Static analysis

4.2.1. Selection of element type. The boom is simulated by shell element basically. All sorts of axle hole are beam elements. The slide block is solid element.

4.2.2. Establishment of connection. The boom sections are connected by slide block. The influence of block at boom head is ignored. The connection of boom sections is shown as figure 5. The slide block decides the boom movement.

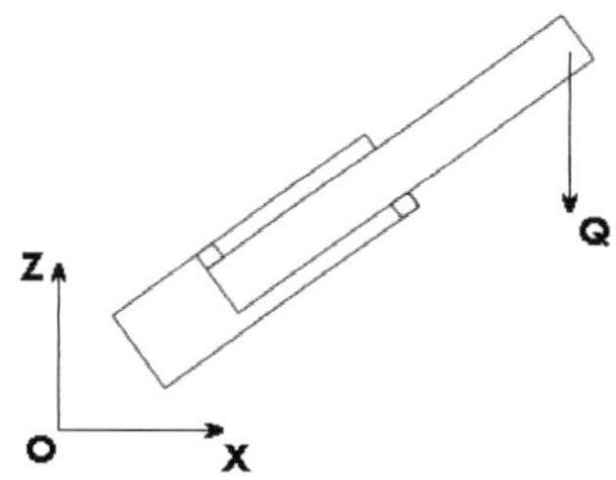

Figure 5. The diagram of main boom connection.

4.2.3. Mesh generation. Mesh generation is a discretization process of structure. If the mesh is too dense, the calculation accuracy is enough, but the computation time is long. Here the triangle and quadrilateral elements about 40mm in size are adopted which has short computation time and high precision.

4.2.4. Boundary conditions. There are two boundary conditions: force boundary condition and displacement condition.

Displacement condition is to constrain the model. The three translational degrees of freedom at later pivot of boom and lower pivot of derricking cylinder are constrained.

Force boundary is to load the model. The pulley at the lifting point has hoisting load, lateral load, rotation inertia load. In the centre of leading sheave is loaded by rope tension. The wind load is applied on the body of boom averagely.

5. Result of calculation

The lifting capacity of boom with different length of arm and different working range is analyzed by geometric large deformation model. And then the partial structure is checked by linear static model. At last, the rated load with different working range is obtained when arm length is 13.4m, 32.6m, 59.1m, 68.2m and 72m respectively.

There are 9 different boom combinations calculated according to the optimization and comparison of working condition, shown in Table 1. Here "0" means full-retracting section, "1" means the section expands 46% in length, "2" means 92%, and "3" means 100%.

Table 1. The combination of expanded length of boom.

Boom length m / Section	13.4	36.2				59.1		68.2	72
2	0	0	1	2	0	1	2	2	3
3	0	0	1	1	2	1	2	2	3
4	0	0	1	1	2	2	2	2	3
5	0	1	1	1	2	2	2	2	3
6	0	2	1	0	2	2	1	2	3
7	0	2	0	0	2	2	1	2	3

The results are shown as figure 6- figure 10.

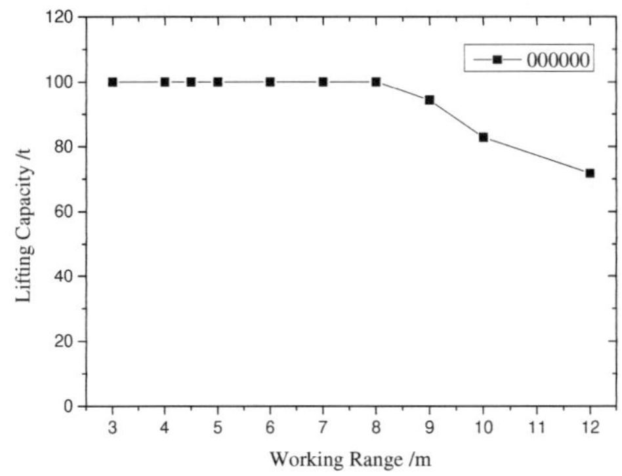

Figure 6. Lifting capacity with arm length 13.4m.

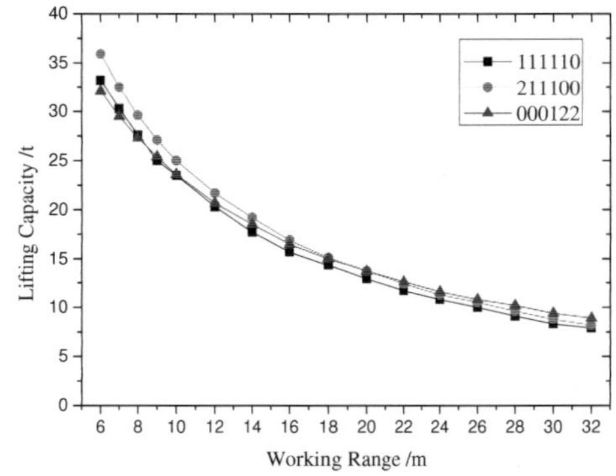

Figure 7. Lifting capacity with arm length 36.2m.

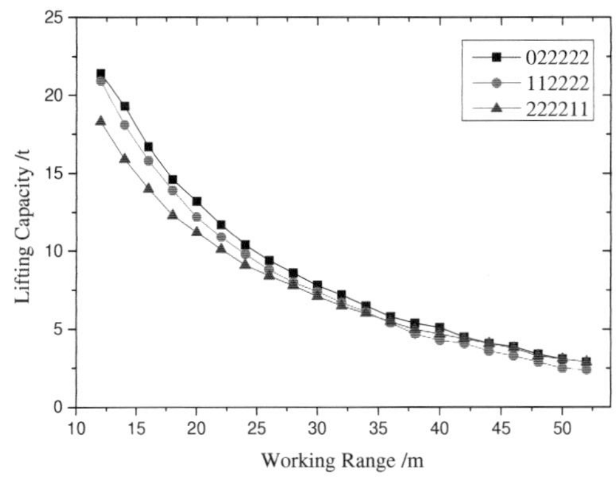

Figure 8. Lifting capacity with arm length 59.1m.

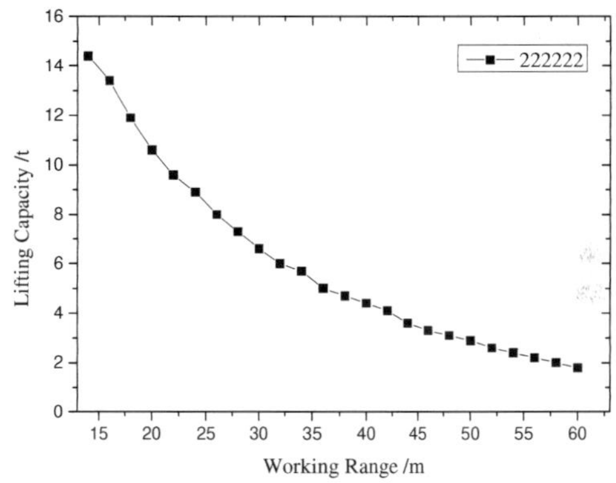

Figure 9. Lifting capacity with arm length 68.2m.

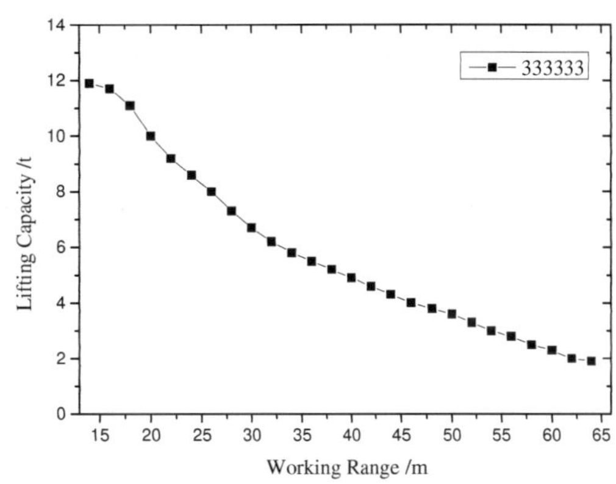

Figure 10. Lifting capacity with arm length 72m.

The results show that the capacity of crane boom with different arm length combinations at different working ranges could meet the requirement of tower erection in transmission line.

6. Summary

The boom is the key part of mobile crane in design and manufacturing. Whether or not its structure is reasonable will affect the working performance and operation stability. In this paper, firstly the load type of mobile crane for plain region in transmission line is analyzed. Then, the FEA model is established by reasonable simplification. Lastly, adopting geometric large deformation model, the working performance with different arm lengths is obtained. The results show that the designed crane can meet the requirement of tower erection in transmission line, and the method can provide enough support for optimization and improvement of crane boom.

7. References

[1] Chen B G and Wang J Q 2004 *Constr. Mach. Tech. Manag.* **12** 53-6.
[2] Qin J and Feng L 2016 *Hoist. Convey. Mach.* **11** 21-5.
[3] Zheng X J and Jia C 2011 *Dev. Innovat. Mach. Electr. Prod.* **24** 125-7.
[4] Zheng Y Q and Wang S M 2013 *J. Mech. Str.* **35** 850-4.
[5] Guo Y S, Zhang X Z and Zhang D W 2009 *Agr. Equip. Vehicle Eng.* **7** 9-11.
[6] Wei S F and Wang S M 2011 *J. Mach. Des.* **28** 92-6.

Acknowledgment

This research was supported by Science and Technology Research Project of State Grid (Study on the key design technologies of mobile crane for tower erection in transmission line).

IOP Conf. Series: Materials Science and Engineering **248** (2017) 012028 doi:10.1088/1757-899X/248/1/012028

Study on Collision of Ship Side Structure by Simplified Plastic Analysis Method

C J Sun[1], J H Zhou[2], W Wu[2]

[1]Military Representative Office in the 426 Shipyard, Dalian 116011, China.
[2]China Ship Development and Design Center, Wuhan 430064, China.

Abstract. During its lifetime, a ship may encounter collision or grounding and sustain permanent damage after these types of accidents. Crashworthiness has been based on two kinds of main methods: simplified plastic analysis and numerical simulation. A simplified plastic analysis method is presented in this paper. Numerical methods using the non-linear finite-element software LS-DYNA are conducted to validate the method. The results show that, as for the accuracy of calculation results, the simplified plasticity analysis are in good agreement with the finite element simulation, which reveals that the simplified plasticity analysis method can quickly and accurately estimate the crashworthiness of the side structure during the collision process and can be used as a reliable risk assessment method.

1.Introduction

For many years, ship engineers have been committed to making ships more safer reduce the casualties, the property damage and the environmental pollution caused by accidents such as collision. In 1912, a ship accident, known as the unsinkable Royal Titanic, has had disastrous consequences. Before the accident, ship design engineers around the world had little or no maritime security cooperation. After the accident, the International Maritime Organization (IMO) developed an International Convention on the safety of life at sea in 1914. To this day, many international conferences, international projects and international studies are conducted to enhance maritime safety and avoid the occurrence of various types of maritime accidents

In the past ten years, with the increase of the number and tonnage of ships, maritime traffic has become more and more busy, and the number of ship collision and contact accidents has increased significantly. Therefore, the research work of modern maritime safety is mainly focused on the ship collision and ship grounding. Ship collision is the focus of most of the research. Risk assessment and safety management of ship collision has become a hot research direction.

To prevent the collision of ships and method of bottom accidents, there are two types: active methods and passive methods such as: the management of shipping traffic and the improvement of hull structure crashworthiness. And the most effective way to reduce the risk of accidents is to use various methods to minimize the possibility of collision with the bottom of the accident. There are many ways to achieve this, for example, the management of shipping traffic, the enhancement of ship traffic service, the use of effective navigation aids, etc.. These methods have been implemented in the shipping industry, but in the long run, these methods do not meet the needs of today's development. So in order to reduce the possibility of accidents, while using the passive method to reduce the possibility of accident, it is necessary to use the passive method to minimize the damage degree of the hull structure after the accident, and reduce the loss to the minimum.

There will be a wide range of structural damage in the event of collision and bottom contact. Many scholars have been trying to analyze the response of ship structure in the course of collision. Ship

Content from this work may be used under the terms of the Creative Commons Attribution 3.0 licence. Any further distribution of this work must maintain attribution to the author(s) and the title of the work, journal citation and DOI.
Published under licence by IOP Publishing Ltd

collision analysis methods can be divided into: simplified analysis method, finite element analysis method, experimental method and empirical formula method. They have their own characteristics, complement each other, and jointly promote the continuous development of ship collision research. The main characteristic of ship collision simplified analysis method is according to the specific characteristics of ship collision, simplifying assumptions are made through a series of calculation model of collision, more concise analytical expression is given to make a reasonable assessment of ship collision. In this paper, a new simplified analytical method called Simplified plastic analysis method is introduced.

2.Simplified plastic analysis method

When the simplified method is used to analysis complex structures, such as ships, the general structure is divided into several basic structures. The collision force of each basic structure is calculated, and then the collision response of the whole structure is obtained. A typical double hull structure is shown in figure 1 a. The structure consists of an outer plate, an inner layer plate and a beam structure between the two layers. The beam structure includes transverse ribs and longitudinal beams. In the simplified plastic analysis method, beam structure is divided into two parts, one part of transverse ribs and longitudinal cross, referred to as the cross structure; the other part is the beam structure in the wide girder structure outside the cross section. In the study of Haris and Amdahl [1], the span of the beam cross section is defined to be equal to half of the height of the beam system, and the rest of the beam structure is a wide truss structure (as shown in figure 1c, d).

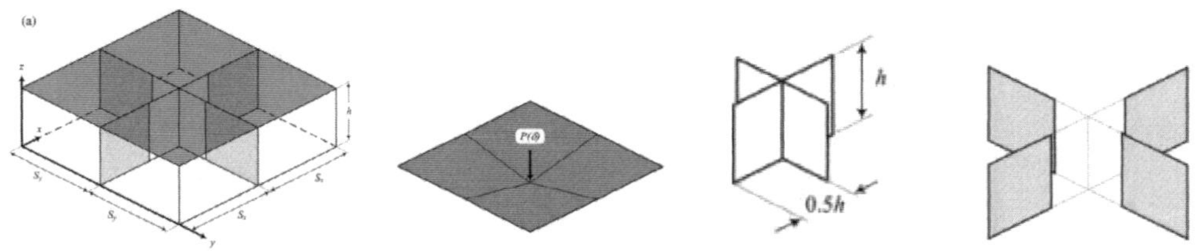

(a) (b) (c) (d)

Figure 1. (a) typical ship side structure, (b) plate, (c) cross structure, (d) 4 wide truss structures.

Many scholars have put forward the analytical formulas for the impact force of the above basic structures, in which the collision force between the cross structure and the wide truss structure is usually assumed to be constant. The collision force increases with the increase of deformation displacement. The collision force formulas of various structures are as follows:

Formula (1) for calculating impact force of cruciform structure (put forward by Hayduk and Wierzbicki in 1984) [2]), Plastic bending moment is defined by formula (2) and (3):

$$P_{cf} = \frac{20.05}{\lambda} M_0 \left(\frac{C}{t_{cf}} \right)^{1/2} \tag{1}$$

$$M_0 = \frac{\sigma_0 \cdot t_{cf}^2}{4} \tag{2}$$

$$\sigma_0 = \frac{\sigma_y + \sigma_u}{2} \tag{3}$$

M_0 is a plastic bending moment of the plate width, C and t_{cf} are the width and thickness of the cross structure. Length factor λ is 0.73.σ_0 is static yield stress of material, σ_y is dynamic yield stress of material, σ_u is Ultimate strength of material.

Formula (4) for calculating impact force of wide truss structure (put forward by Zhang [3]), The formula to calculate the concentrated load impact force of the center of the plate (figure 1 b) is by formula (5) (put forward by Zhang [3]):

$$P_{wg} = \frac{11.26}{\lambda} M_0 \left(\frac{b}{t_{wg}} \right)^{1/3} \tag{4}$$

$$P(\delta) = \frac{8}{3\sqrt{3}} \cdot \sigma_0 \cdot S_x \cdot S_y \left(\frac{t_{px}}{S_x^2} + \frac{t_{py}}{S_y^2} \right) \delta \tag{5}$$

b represents the span of wide truss, t_{wg} is the width of wide truss. Length factor λ is 0.73; The plastic bending moment is the same as the formula (2). S_x and S_y are the dimensions of the plate as shown in figure 1, t_{px} and t_{py} are the equivalent thickness of the plate in the X direction and the Y direction, δ is the displacement of the center position of the plate. But it is not reasonable to calculate the impact load of the bow equivalent to a concentrated force or a uniformly distributed load. So, it is necessary to have a more optimized formula, which can take into account the different shapes of the bow, such as the bow of an Elliptical Parabolic surface. In the recent study, the formula for calculating the impact force of a specific shape of rigid bow has been derived [4]. The geometry of the bow is defined by the following formula:

$$z = \frac{x^2}{\alpha \cdot S_x} + \frac{y^2}{\beta \cdot S_y} \tag{6}$$

α and β are the curvature the X direction and the Y direction. Curvature is not only related to the shape of the bow, but also to the size of the plate. The formula of the impact force of a flat plate subjected to an elliptical paraboloid is as following:

$$p(\delta) = \frac{8}{3\sqrt{3}} \cdot \sigma_0 \left(t_{px} \cdot \frac{S_y}{S_x} \left(1 - \frac{\alpha\delta}{S_x} \right)^{-1/2} + t_{py} \cdot \frac{S_x}{S_y} \left(1 - \frac{\beta\delta}{S_y} \right)^{-1/2} \right) \delta \tag{7}$$

When $\alpha$and $\beta$are equal to 0, the formula (7) is the same as the formula (5). The failure displacement of the plate is affected by the shape of the impact head. The displacement formula for the impact of a flat plate subjected to a circular parabolic head is:

$$\delta_f = \left[1.316 \frac{S_x S_y}{\sqrt{S_x^2 + S_y^2}} \sqrt{\varepsilon_f} \right] \times \sqrt{\alpha} \tag{8}$$

The general range of failure strain is 5-10%. For the beam strengthened plate structure, the contact part between beam and plate is the first failure due to strain concentration. While the formula (8) only considers the membrane strain and neglects the bending strain. When the failure strain is 8%. When the failure strain is 8%, the similarity between the results of finite element analysis and theoretical results is the highest. So, it is suggested that the failure strain of 8% is reasonable. Figure 2 to figure 3 is the comparison of the results of the calculation of the impact of the plate with a rigid impact head with two different curvatures. The equivalent curvature coefficients of the two kinds of impact heads are equal and the shapes are different. Circular parabolic head (α=0.5, β=2), Circular parabolic head (α=β=1), flat plate model shown in figure 1 b, S_x= S_y= 4000mm, t_{px}=t_{py} =20mm. As can be seen from the graph, the simplified calculation results are in good agreement with the finite element results. The peak value of impact force is less than 3%, and the failure displacement is approximately equal.

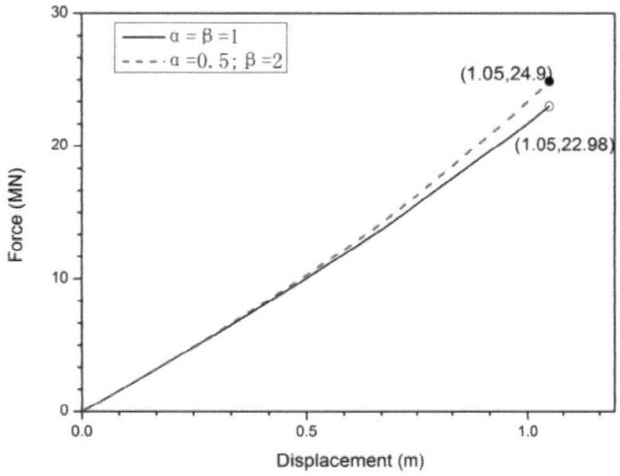

Figure 2. The curve of impact force and displacement (simplified method).

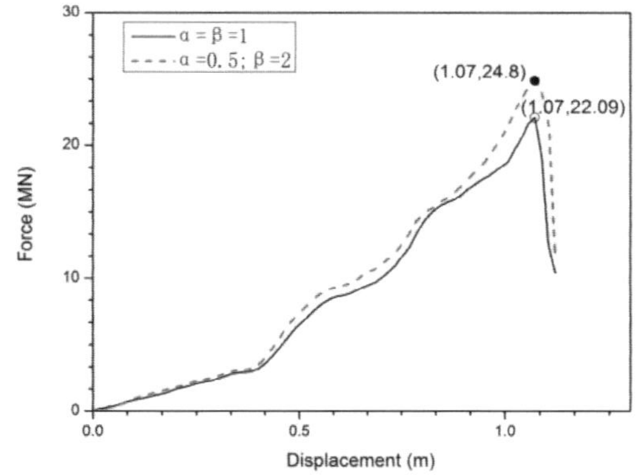

Figure 3. The curve of impact force and displacement (finite element method).

3. Ship side structure collision analysis

The broadside structure of the 120 thousand ton oil tanker is composed of six double hull compartments. The cross section is shown in figure 4.

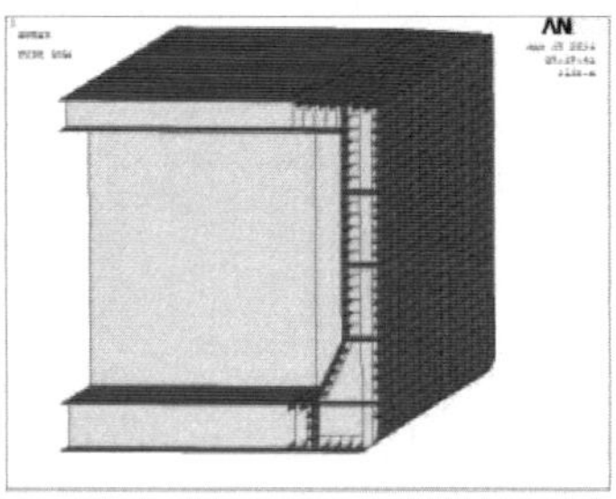

Figure 4. Sketch of broadside structure.

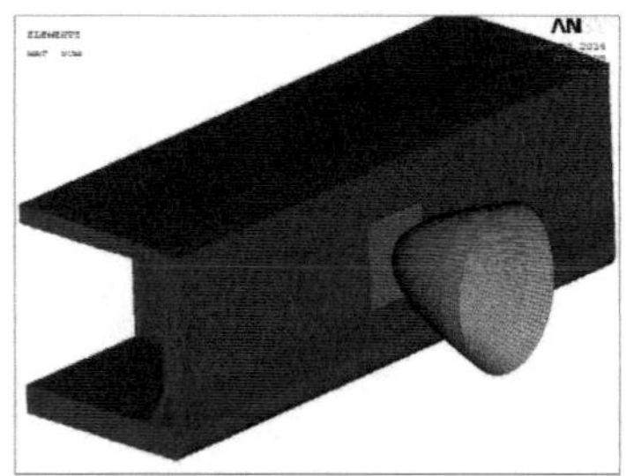

Figure 5. The intent of the bow.

In the hypothetical collision process, the impact bow is rigid and does not absorb energy. We want to calculate the collision response of the rigid bow to the side of an oil tanker. The bow geometry is described by formula (6). The bow's horizontal curvature coefficient α=0.8, vertical curvature coefficient β_1=1.325, β_2=1. The intent of the bow is shown in figure 5. In the model, the material is ordinary steel, density 7.89e-9t/mm^3, Young's modulus 2.07e5MPa, yield stress 235MPa, Poisson's ratio 0.3. RTCL failure criterion for material failure criterion [5]. Material constitutive relationship using power law plasticity. Among them k=740, n=0.24 [6]. In order to simulate the actual ship collision process, it is necessary to define the appropriate boundary conditions and loads. When the actual boundary conditions can not be determined, the appropriate simplified boundary conditions can be used. In the process of calculation, the translational degrees of freedom of the broadside structure should be constrained, fully constrained degrees of freedom of the end of the interior broadside structure. The initial velocity of the 8m/s loaded to the bow. The assumptions made in the process of analysis and calculation: a) the bow is rigid, only the side structure is involved in energy consumption; b) the collision process is quasi-static; c) the collision is a right angle to the heart collision. d) the density of bow increased by 10%. In order to determine the calculation model of the simplified method, a simplified calculation model method to determine the two conditions, need to analyze the deformation mode and load distribution of tanker side structure, the side structure is divided into two parts: the first part called main area, includes the direct collision point and adjacent area; the second part is expansion area, is the adjacent regions of the main area. The regional distribution is shown in figure 6. The deformation and energy variation of the outer shell of the side structure is shown in figure 7, figure 8.

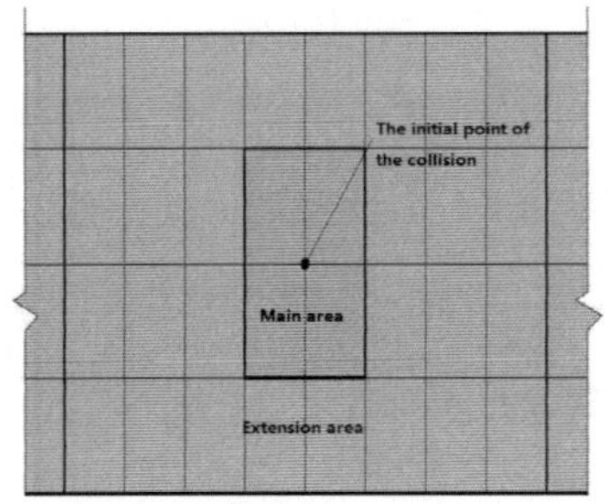

Figure 6. Ship structure zoning map.

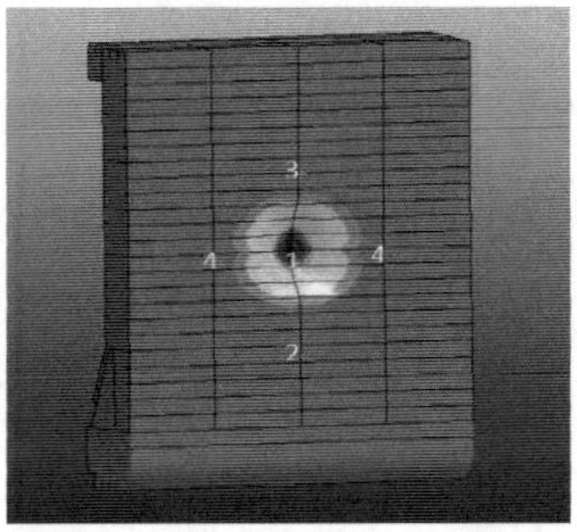

Figure 7. Side structure deformation.

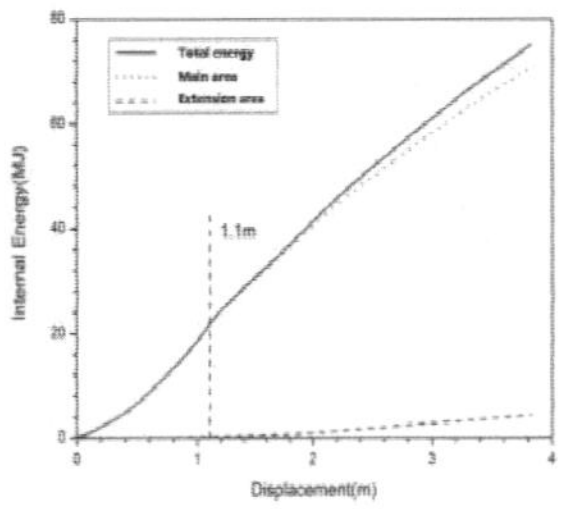

Figure 8. Side structure energy absorption diagram.

It can be seen from the results of finite element calculation, only the deformation of node 1 is obvious, and all the deformed elements are in the main area. The main energy absorption area size is approximately equal to the overall structure of energy absorption size. Therefore, the simplified analysis method is used to calculate only consider the structure of the main area, the calculation model and the related dimensions are shown in figure 9. The dimension parameters are substituted into the formula, and the results are compared with the results of finite element calculation as shown in figure 10. As can be seen from the figure, the results of simplified plastic analysis method are above the finite element curve, the calculated results of failure displacement are the same as those of finite element method, and the ultimate impact force is 14% higher than that of the finite element method.

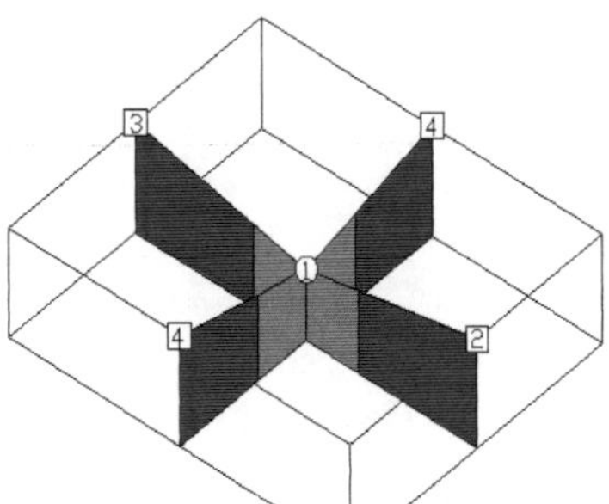

Figure 9. Simplified calculation model.

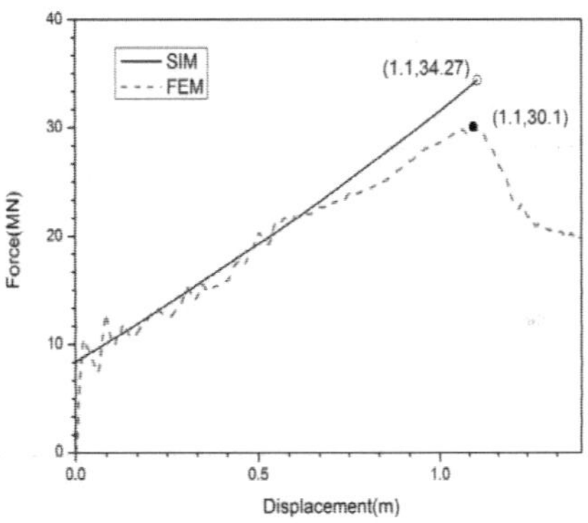

Figure 10. Comparison of impact force displacement curves.

4.Summary

In this paper, a finite element software is used to simulate the process of rigid bowls hitting the side of the oil tanker. The bow shape is defined by an elliptical parabolic formula. The energy consumption of the two sides of the bow side impacted side structure is given in detail. The calculated collision force displacement curve and the failure displacement are used as reference, compared with the simplified method. From the finite element calculation results, it is concluded that the reinforcement structure on the beam structure and the energy absorption of the inner shell in the collision process are negligible and can be neglected. The simplified method divides the side structure into several basic components, including: plate and shell structure, cross structure and wide girder. The collision force of each component is calculated by the formula, and the total collision force of the side structure is obtained. The calculated results are in good agreement with the simulation results. It is revealed that the simplified plasticity analysis method can quickly and accurately estimate the crashworthiness of the side structure during the collision of the ship and can be used as a reliable risk assessment hand.

IOP Conf. Series: Materials Science and Engineering **248** (2017) 012028 doi:10.1088/1757-899X/248/1/012028

5. Reference

[1] Haris S and Amdahl J. 2011 *Ships Offshore Struct.* **7(2)** 185-95.
[2] Hayduk R and Wierzbicki T. 1984 *Comput. Struct.* **18(3)** 447-58.
[3] Zhang S 1999 *The mechanics of ship collisions*. Technical University of Denmark.
[4] Haris S and Amdahl J 2012 *Ships Offshore Struct.* **7(4)** 431-48.
[5] Tornqvist R and Simonsen B 2004 *Safety and structural crashworthiness of ship structures;Modelling tools and application in design*. Lzu, Japan: 285-94.
[6] Alsos H, Amdahl J and Hopperstad O 2009 *Impact Eng.* **36(7)** 875-87.

IOP Conf. Series: Materials Science and Engineering **248** (2017) 012029 doi:10.1088/1757-899X/248/1/012029

Kinematic responses and injuries of pedestrian in car-pedestrian collisions

T L Teng[1], C C Liang[2], C Y Hsu[3] and S F Tai[2]

[1]Department of Mechanical Engineering, Hsiuping University of Science and Technology, Taichung, 412, Taiwan
[2]Department of Mechanical and Automation Engineering, Da-Yeh University, Zhanghua, 515, Taiwan
[3]Department of Marine Mechanical Engineering, ROC Naval Academy, Kaohsiung, 813, Taiwan

Abstract. How to protect pedestrians and reduce the collision injury has gradually become the new field of automotive safety research and focus in the world. Many engineering studies have appeared and their purpose is trying to reduce the pedestrian injuries caused by traffic accident. The physical model involving impactor model and full scale pedestrian model are costly when taking the impact test. This study constructs a vehicle-pedestrian collision model by using the MADYMO. To verify the accuracy of the proposed vehicle-pedestrian collision model, the experimental data are used in the pedestrian model test. The proposed model also will be applied to analyze the kinematic responses and injuries of pedestrian in collisions in this study. The modeled results can help assess the pedestrian friendliness of vehicles and assist in the future development of pedestrian friendliness vehicle technologies.

1. Introduction

Considering the traffic safety request for most people, automakers and related research institutions put a lot of manpower, material, and financial resources to develop and improve the car body structure and safety equipment, such as airbags, and energy absorbing steering columns. The purpose of these equipment are to reduce the severity of injury to the occupant during a crash. According to the statistics of fatal traffic accidents shows these safety structures and equipment have obvious protective effects on the passengers, so that the casualty rate of passengers in the vehicle's collision accident is greatly reduced. However, pedestrians hit by cars are still very serious [1]. For the traffic accident casualties, the number of pedestrian casualties are less than the number of casualties in the passengers and ranked second. Therefore, how to protect pedestrians and reduce the collision injury has gradually become the new field of automotive safety research and focus in the world.

According to statistical data, injury of pedestrian concentrates on some regions such as head, thorax, upper limbs, and lower limbs [2]. Injury distribution of each body region for fatalities and injuries has different severity level. To effectively reduce pedestrian fatalities and injuries, designing a pedestrian friendly vehicle and protection system in relation to considerations of pedestrian safety is a relevant and urgent task. Since the 1970s, many scientists in Europe, Japan or American have started researching to find out study for pedestrian protection. Many engineering studies have appeared and their purpose is trying to reduce the pedestrian injuries caused by traffic accident. The EEVC Pedestrian Working Groups (WGs 7, 10, 17) have worked to produce the test methods and criteria [3, 4].

Content from this work may be used under the terms of the Creative Commons Attribution 3.0 licence. Any further distribution of this work must maintain attribution to the author(s) and the title of the work, journal citation and DOI.
Published under licence by IOP Publishing Ltd

To facilitate the evaluation and redesign the vehicle model for pedestrian protection, Working Group 17 of the European Enhance Vehicle Safety Committee (EEVC/WG17) has developed components system tests for evaluating the pedestrian friendliness vehicles, including full–scale dummy testing and impactor models testing [5]. These are the anthropomorphic test devices that simulate the dimensions, weight proportions and articulation of some testing human body parts or whole human body.

The physical model involving impactor model and full scale pedestrian model are costly when taking the impact test. Thus, to save money and time for taking impact test, computer simulation model is an effective, alternative way to replace the physical model. Nowadays, the impact between vehicle and pedestrian studies were simulated by computer program [6-9]. For instance there were some studies built model by MADYMO (Mathematical Dynamic Model) 2D and 3D. The data obtained was used to evaluate the kinematic in reconstructions of pedestrian crash and using its injury prediction ability to find impact area on vehicle.

To effectively assess pedestrian injuries resulting from impacts with vehicles, a computer simulation model must be developed for vehicle-pedestrian collision analysis. This study constructs a vehicle-pedestrian collision model by using the MADYMO. To verify the accuracy of the proposed vehicle-pedestrian collision model, the experimental data (Post-Mortem Human Subject (PMHS)) are used for the validation of pedestrian model [10]. These data provide the corridor trajectories for the head, pelvis, knee, and foot used to validate the pedestrian model. The proposed model also will be applied to analyze the kinematic responses and injuries of pedestrian in collisions in this study. The modeled results can help assess the pedestrian friendliness of vehicles and assist in the future development of pedestrian friendliness vehicle technologies.

2. Multi-body model

2.1 Pedestrian model

MADYMO offers a powerful and extensive human model database. A pedestrian model with a height of 175 cm and a weight of 77 kg was used in this study. In order to simultaneously get the femoral and tibial injury values at the same time, this study modified the original femoral and tibial models from a single ellipsoid to two same size of ellipsoids, and established a joint in between. The pedestrian model is described by 37 ellipsoids to replicate the human bodies. The ellipsoids of the model act as several individual body parts such as head, torso or upper leg, where the stiffness, shape, size, mass and inertia are predefined based on the MADYMO. 16 joints exists to connect these ellipsoids together. The contact response of the ellipsoids is measured by 24 accelerometers mounted in the ellipsoid surface. Seven chains connect all segments of pedestrian model. As mostly collision accident statistics show that pedestrians are in the process of walking and the vehicle impact on the side. In this study, the relative position between pedestrian and vehicle is set to the side impact mode.

2.2 Vehicle model

Figure 1 shows the frontal shape of the car model used in this study. Moreover, LEH denotes the height of the leading edge of the hood, BL represents the bumper lead (longitudinal distance from front of bumper to front of hood), BCH is the bumper center height, HL is hood length and α is hood angle. The vehicle model is constructed with multiple rigid ellipsoids representing the windshield, hood, hood front, bumper and tires, respectively.

2.3 Vehicle - pedestrian collision model

Figure 2 displays the vehicle-pedestrian collision model in present study. A vehicle impacts the left side of pedestrian with a velocity of 40 km/hr. In this model, the friction coefficients between foot and ground is 0.67, the friction coefficients between pedestrian model and car is 0.25 and the friction coefficients between car and ground is 0.8.

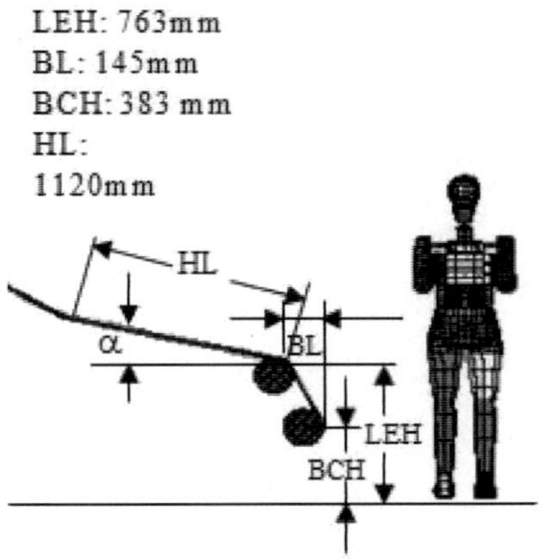

Figure 1. The frontal shape of the car model.

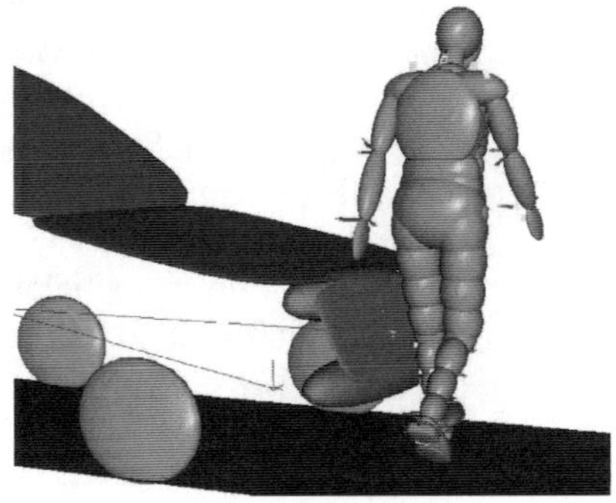

Figure 2. The vehicle-pedestrian collision model.

3. Analysis of pedestrian collision trajectory

Figure 3 presents a simulated impact sequence in a vehicle-pedestrian collision simulation model at an impact speed of 40 km/h. As figure 3 presented, the shank hit against the bumper and the thigh also hit the edge of the hood at 20 ms. The shank and thigh is still in contact with the vehicle at 40 ms. The pelvis shows a clockwise rotation owing to step forward with left thigh. The wrist contacted with the hood at 60 ms, and the pelvic parts also began to collide with the hood. The shank part from the bumper with the curvature action, then the thigh has left the front of the bumper and turned on the hood surface at 80 ms. The thigh with the body away from the vehicle, but the pelvis, arm maintain contact with hood surface at 100 ms. The left hand arm completely contact with the hood at 120 ms, the torso also began to hit the hood. The head hit the windshield at 140 ms, the torso also collided with the hood.

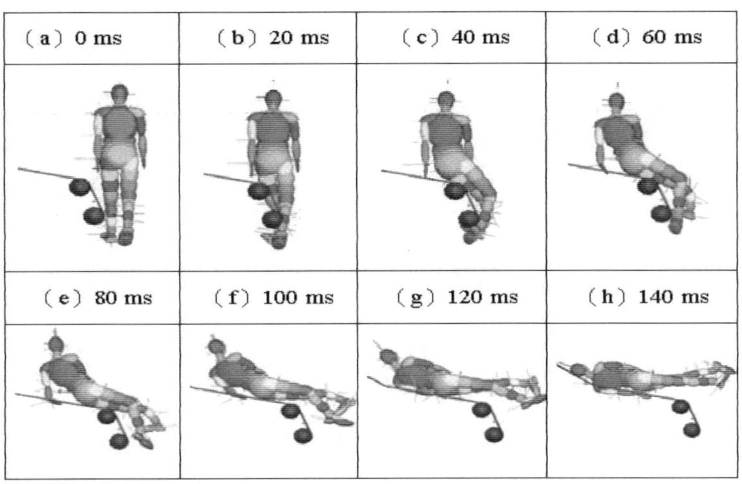

Figure 3. Impact sequence in a vehicle-pedestrian collision simulation model.

The vehicle-pedestrian collision model was validated using post-mortem human subject (PMHS) corridor for the head, pelvis, knee and foot [10]. Figure 4 to figure 7 list the head, pelvis, knee, and foot trajectories of pedestrian. The head displacement separates a little from the limitation, particularly at the end of the curve (figure 4). Because the slope angle of hood surface is 12° relative to the ground that will force the pelvis moves upward when it contacts and slides on the hood. The increase in upward pelvis movement is accumulated during the sliding process so that according to the pelvis

IOP Conf. Series: Materials Science and Engineering **248** (2017) 012029 doi:10.1088/1757-899X/248/1/012029

curve, the final position is higher than the upper limitation of PMHS data. Because of the relation between the pelvis and the head, together with upward movement of the pelvis, the head displacement also exceeds the upper limitation. The pelvis and knee trajectories fall entirely within the corridor (figure 5, 6). The good agreement of knee displacement with experimental data clearly indicates that the hip joint in the pedestrian model is very suitable for free impact. The end of the foot trajectory curve does not fall within the PMHS corridor owing to the deformation of the leg upon impact with the bumper (figure 7). It is presumed that the ankle is made up of multiple ellipsoids in the pedestrian model. The thighs and calves are also joined by two ellipsoids, and more than one set of joints. It is more likely to make the pedestrian model have excessive bending, resulting in an ankle trajectory moving to the left of the situation.

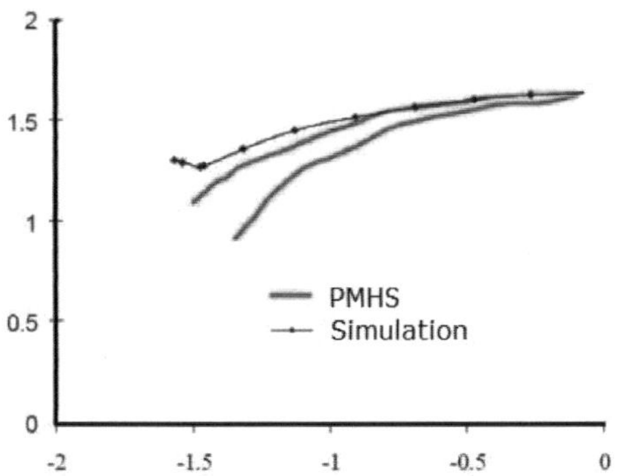

Figure 4. Head trajectory of pedestrian model together with PMHS corridor.

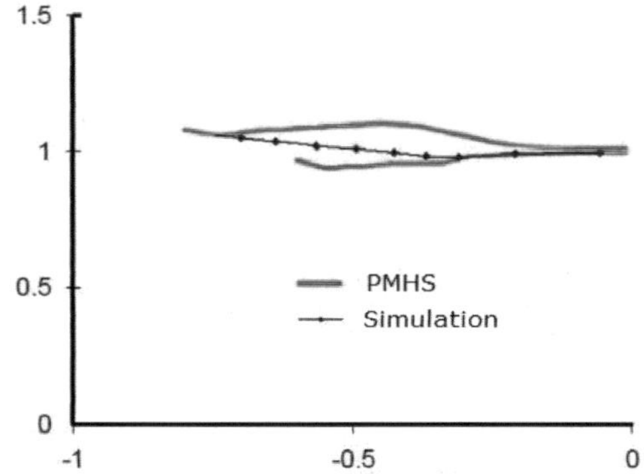

Figure 5. Pelvis trajectory of pedestrian model together with PMHS corridor.

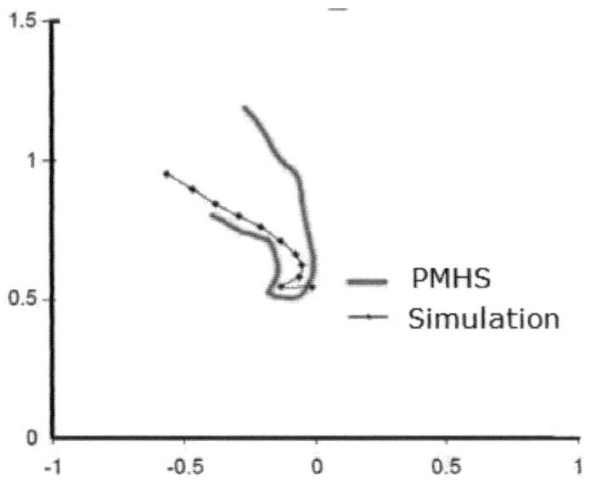

Figure 6. Knee trajectory of pedestrian model together with PMHS corridor.

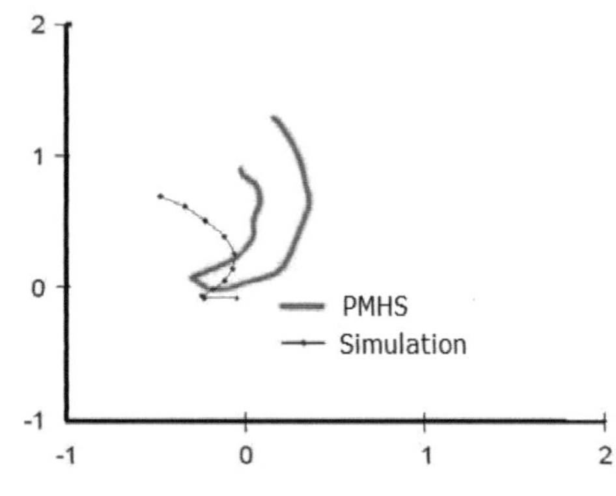

Figure 7. Foot trajectory of pedestrian model together with PMHS corridor.

4. Injury analysis of pedestrian

In the MADYMO simulation analysis, the head, pelvis, thigh and shank were used to analyze the pedestrian injuries. In the injury of various parts of the human body, the head HIC value of 1255 exceeds the safety threshold of 1000. To translate HIC value into AIS severity scores, an AIS 4 head injury was produced. It may cause skull rupture and nerve damage which can lead to the risk of death. The pelvic injury reached 623.75 N which is lower than the criteria of pelvic fracture. Femoral and tibial injury is a common injury in pedestrian suffering impacts with vehicles. These injury are evaluated based on femur and tibia fracture resulting from bending moment of the femur on impact with the bumper and the resultant acceleration of tibia. The bending moment of the femur is 392.67N.m, and the tibia acceleration is 216.5 g. Limitations to avoid serious femur and tibia injury are 200N.m for bending moment and 150g for tibia acceleration, respectively. The maximum bending moment and acceleration values for the femur and tibia exceed the threshold value. These results demonstrate that the femur and tibia suffers fracture on impact with the bumper.

In order to confirm the rigid multi-body model for the injury analysis, the injury results of are compared with the numerical simulation results of Rooij [8]. The injury data are shown in Table 1. According to the analysis model of Rooij and the injury model in this study, it can be found that the difference of analysis results are less than 15%. The frontal shape of the finite element vehicle model is more fine than that of the rigid multi-body model. It is the reason why Rooij result causes a high injury value of femur. The analysis of the head and pelvic injury values are converted to the AIS, as shown in Table 2. The table indicate that two models of the head and the pelvis of the AIS are in the same injury level.

Table 1. Injury risk comparison between the Rooij and present study.

	Rooij [8]	Present study
Head Injury (HIC)	1420	1255
Pelvis Injury (N)	675	623.75
Femur Injury (N.m)	99.56	94.95
Tibia Injury (g)	237	216.5

Table 2. The head and the pelvis of the AIS

	Rooij [8]	Present study
Head	AIS 4 ~ AIS 5	AIS 4 ~ AIS 5
Pelvis	AIS 0 ~ AIS 1	AIS 0 ~ AIS 1

5. Conclusion

In this study, a vehicle-pedestrian collision model was constructed by using the MADYMO to analyze the kinematic responses and injuries of pedestrian. The current study shows the vehicle-pedestrian collision model has a reasonable correlation with Post-Mortem Human Subject experimental data. And the model is more visible in assessing injuries of pedestrian. The modeled results can help assess the pedestrian friendliness of vehicles and assist in the future development of pedestrian friendliness vehicle technologies.

6. References

[1] Information on http://www-fars.nhtsa.dot.gov/Main/index.aspx
[2] Chidester A and Isenberg R 2001 *The 17th ESV conference* (Amsterdam).
[3] Information on http://www.ergo-safety.ca/text/about.html
[4] Information on http://eevc.net/history/wg10/wg10index.html
[5] EEVC 1998 *EEVC Working Group 17 Report*: Improved test methods to evaluate pedestrian protection afforded by passenger cars.
[6] Twigg D W and Tocher D L 1977 U.S. Department of Transportation Report No. DOT HS-902 419 vol 1
[7] Yoshida S and Matsuhashi T 1998 *The 16th ESV conference* (Windsor, Ontario, Canada).
[8] Van Rooij L, Bhalla K, Meissner M, Ivarsson J, Crandall J, Longhitano D, Takahashi Y, Dokko Y and Kikuchi Y 2003 *The 18th ESV conference* (Nagoya, Japan).
[9] Yao J, Yang J, Otte D and Fredriksson R 2005 *Proceedings of the 2005 International IRCOBI Conference on the Biomechanics of Impact* (Czech Republic).
[10] Stammen J and Barsan-Anelli A 2001 *The 4th MADYMO User's Meeting of The America's* (Detroit, Michigan, USA).

IOP Conf. Series: Materials Science and Engineering **248** (2017) 012030 doi:10.1088/1757-899X/248/1/012030

Experimental study on water content detection of traditional masonry based on infrared thermal image

Baoqing Zhang[1, 3], Zukang Lei[2, b*]

[1]PhD students, Department of Huazhong University of Science and Technology, China
[2]Associate professor, Department of Huazhong University of Science and Technology, China
[3]Lecturer, College of Water & Architectural Engineering, Shihezi University, China

Abstract. Based on infrared thermal imaging technology for seepage test of two kinds of brick masonry, find out the relationship between the distribution of one-dimensional two brick surface temperature distribution and one-dimensional surface moisture content were determined after seepage brick masonry minimum temperature zone and water content determination method of the highest point of the regression equation, the relationship between temperature and moisture content of the brick masonry reflected the quantitative and establish the initial wet masonry building disease analysis method, then the infrared technology is applied to the protection of historic buildings in.

1. Introduction

Historical building protection, for its main purpose, is to preserve and extend the information and value of legacies furthest [1]. Researches on modern historical buildings are helpful for us to get a good understanding of cultural inheritance and transformation, as well as technological innovation and development during modernization of traditional buildings [2]. However, historical buildings, due to particularity of their materials, texture and structure, may give rise to some disease problems. When a building disease occurs, mechanical performance and thermal performance of the building materials will be lowered, thus accelerating the speed of building damage and leading to unrecoverable losses [3].

In China, modern historical buildings are dominated by masonry buildings, most of which are constructed by traditional masonries. For such buildings, most of the building diseases are caused by moisture, which not only threatens effective preservation of the building and relevant cultural relics, but also induces other diseases, such as biological damage and corrosion etc. Hence, timely diagnosis and effective countermeasures against moisture of masonry buildings are of great significance [4]. Masonry buildings exposed to moisture and the subsequently induced saltpetering is a complicated process [5]. However, most of literatures at present focus on detection of mechanical performance of masonry buildings [6], and some literatures pay attention to qualitative analysis on moisture-based disease of masonry buildings [7], without enough attention to their quantitative analysis. Hence, it is more urgent to carry out quantitative analysis on the disease of masonry buildings exposed to moisture.

The application of infrared thermal imaging inspection in construction works dates back to the 1980s [8], and at present, such technology is mainly applied to non-destructive detection of buildings [9, 10]. Defect detection in materials and components, as well as evaluation and thermal performance detection, are, at present, main research contents of non-destructive detection [11]. In addition,

Content from this work may be used under the terms of the Creative Commons Attribution 3.0 licence. Any further distribution of this work must maintain attribution to the author(s) and the title of the work, journal citation and DOI.
Published under licence by IOP Publishing Ltd

quantitative analysis method for thermal performance detection has been put into the exploration list [12], without extensive application in detection of moisture-based building diseases but with sound research prospect in how to quantitatively analyze such moisture-based diseases.

In this paper, infrared thermal imaging technology is adopted to find out the correlation between one-dimensional distribution of surface temperature and surface moisture content of traditional brick masonry (sintered clay brick and autoclaved sand-lime brick) commonly used by protective historical buildings, and find out the method for quantitative analysis on moisture-based masonry diseases, thus extending the application scope of infrared thermal imaging technology in evaluation and treatment of building diseases.

2. Basic information on water penetration experiment in the masonry

2.1. Experimental design

Masonry samples for the experiment include a block of sintered clay brick and a block of autoclaved sand-lime brick. In which, the sintered clay brick, in accordance with the national standard for Sintered Common Brick (GB5101-2003), is sintered and formed from clay, with a solid core or void ratio no greater than 15%; while the sand-lime brick, in accordance with the national standards for Autoclaved Sand-lime Brick (GB11945-1999), is fabricated in three stages, including blank preparation, compression molding and autoclaved curing, from sand (approximately 90%), lime (approximately 10%) and color blending materials. Dimension of the two bricks are the same, namely 240×115×53mm. Since the two bricks have different fabrication process and void ratio, thus getting their temperature distributed presented in different characteristics, research on temperature and moisture distribution is separately carried out in the two bricks to obtain the correlation between the temperature and moisture distribution suitable for the two bricks.

Main measurement parameters for the experiment include time, visually measured dry-wet boundary height, surface temperature and moisture distribution parallel to the long side etc (table 1). Both of the two bricks are put into a water tank for water absorption, and then a parameter measurement is carried out to find out the relationship between surface temperature and moisture distribution along the long side, thus working out the regression equation for surface temperature and one-dimensional moisture distribution of the two bricks.

Table 1. Experimental measurement parameters.

Measurement parameters	Symbols	Unit	Unit symbol
Time (experiment)	*T*	Hour	H
Visually measured dry-wet boundary height (distance from the top)	*H*	Millimeter	Mm
Brick surface temperature	*ST*	Centigrade	℃
Brick moisture content	*MC*	Percentage	%

2.2. Experimental method

Key instruments for the experiment include FLIR E60 infrared thermal imager (accuracy: 0.1%), FD-100 blank moisture meter (accuracy: ±5%) and ST60+infrared thermometer (accuracy: ±1%), in which infrared thermal imager is used to measure the brick surface temperature. However, infrared thermal imager does not merely carry out a direct temperature measurement, but also include radiation of the object itself and reflected radiation towards the surrounding. Such radiation is attenuated through the atmosphere and finally arrives at the detector. Transmission of the atmosphere itself and radiation inside the thermal imager will also get involved, as is shown in figure 1.

In figure 1, ε serves as emissivity of the object, τ as transmittance of the atmosphere, tobj as temperature of the tested object, tsur as the environment temperature, tatm as the atmospheric temperature, (1 -ε)τWsur as reflected radiation energy of the surrounding, and (1 -τ)Watm as the atmospheric radiation energy [13].

To obtain a clear and accurate image, a preprocessing is necessary for the infrared image to effectively control the temporal noise and spatial noise [14]. In this experiment, the calculation formula for surface temperature error of the graybody target in general conditions is used to verify the temperature measured by infrared thermal imager [15]. Blank moisture meter is used for measurement of moisture content on the block surface, and infrared thermometer is used for point depiction.

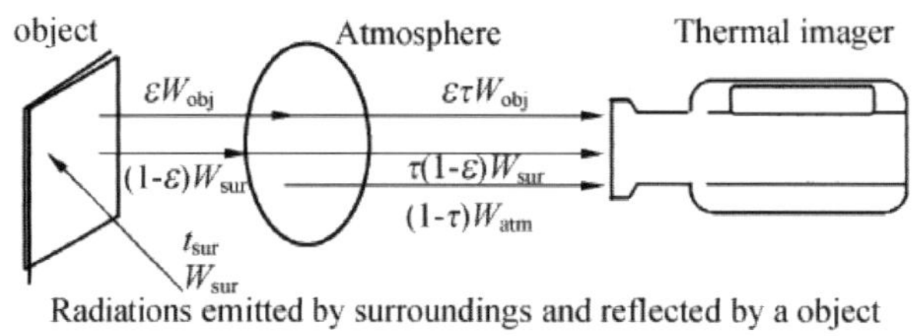

Figure 1. Principle of thermal radiation.

2.3. Experimental environment

The experiment is proposed to begin at 11 on November 26, 2016 in a closed laboratory of Wuhan (lab size: 6.5×6×3.6m), and the test block is placed on plane center to protect the experimental data from impact of temperature and humidity inside the lab wall. For experiment at time zero, air temperature and water temperature in the laboratory is 13.3°C and 13.5°C, with a brick surface temperature of 13.3°C and relative air humidity of 56%, and an air conditioning system is adopted to maintain such temperature/humidity for about 26h (see figure 2).

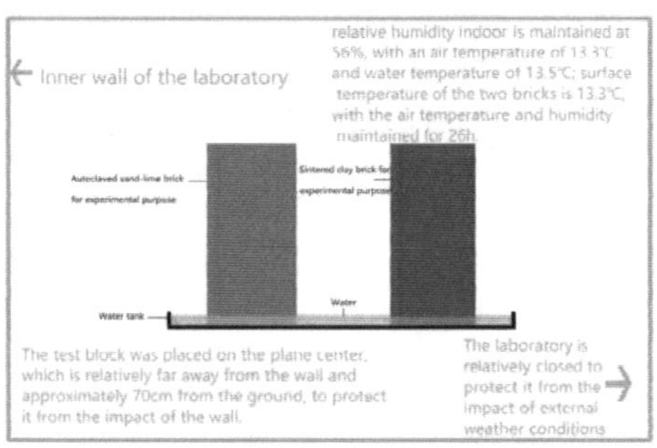

Figure 2. Experimental setting.

Figure 3. Experimental process photo.

2.4. Experimental method

In this experiment, the FLIR E60 infrared thermal imager is used to shoot infrared thermal images of the post-experiment sintered clay brick and autoclaved sand-lime brick, and then these images are imported into FLIR tools 5.9 for analysis, thus exporting the temperature distribution along center connection line (**hereinafter referred to as centerline**) of at broadside (115m long) of the two bricks; then 49 points are identified in the centerline, with a spacing of 5mm, and the FD-100 moisture meter is used to measure moisture content of the two bricks point by point; finally, infrared thermometer is used to, in the infrared thermal images, identify the surface temperature corresponding to the 49 points, and then the two groups of data are imported into relevant software for regression analysis, thus working out the regression equation for surface temperature and moisture distribution of the two bricks along the centerline.

IOP Conf. Series: Materials Science and Engineering **248** (2017) 012030 doi:10.1088/1757-899X/248/1/012030

3 Experimental process and results

3.1. Experimental process

Before the experiment, infrared thermometer was used to measure surface centerline temperature of the two bricks, both at 13.3°C; at beginning of the experiment, the two bricks were simultaneously put into a water tank and infrared thermal imager was used to capture the infrared thermal images for the brick temperature. Since there is a difference between water penetration rate of the clay brick and sand-lime brick, the two bricks were provided with different experimental periods. Water penetration test for both the bricks were carried out at 11:05AM, in which the test for sintered clay brick ended at 12:48PM on the very day, with an experimental period of 1h43min; test for autoclaved sand-lime brick ended at 12:57PM the next day, with an experimental period of 25h52min.

After the experiment, FLIR E60 infrared thermal imager was used to shoot infrared thermal images (front view and vertical view) of the two bricks, and FD-100 moisture meter was used to measure moisture content in the 49 measuring points; then these captured infrared thermal images (front view and vertical view) of clay brick and sand-lime brick were imported into FLIR tools 5.9, thus exporting the centerline surface temperature distribution by using line temperature measuring instrument in the software; finally, an observation was carried out in temperature characteristics of these infrared thermal images, brick surface temperature corresponding to the 49 moisture measuring points was identified under the help of infrared thermometer, and then the two groups of data were imported into IBM SPSS Statistics 22 for regression analysis.

3.2. Experimental results

Through an analysis by the software FLIR tools 5.9 and verification by infrared thermometer, surface temperature and moisture content of the two bricks after experiment were presented as follows:

3.2.1. Temperature variation characteristics of sintered clay brick along the centerline (figure 4c). From the top to a point at 60mm from the top (Point A in the figure), the temperature (from top to bottom) was in a trend of steady drop, with a relatively narrow drop range; from Point A to the point at 90mm from the top (Point B in the figure), the temperature (from top to bottom) was presented in a drop of the widest range, with fluctuations at Point B; from Point B to the point at 190mm from the top (Point C in the figure), the temperature (from top to bottom) was in a state of fluctuations. The minimum temperature occurred in the area between the point at 155mm from the top and the point at 185mm from the top, with the lowest average temperature and a top-to-bottom distance of approximately 30mm. Minimum temperature (13.2℃) along the centerline also occurred in the area, the visually measured dry-wet boundary (Point D in the figure) was approximately 20mm above the minimum temperature zone; From Point C to the bottom, temperature of the area, under impact of water temperature in lower part, enjoyed a narrow-range recovery. Since the bottom was immersed in water, the water temperature (13.5℃) was identified as its temperature.

3.2.2. Moisture content variation characteristics of sintered clay brick along the centerline (figure 4d). From top to the point at approximately 30mm from the top (Point A in the figure), the area was presented with stable moisture content, while the area from top to the point at about 10mm from the top, under the impact of evaporation, was measured with relatively low moisture content; from Point A to the point at approximately 100mm from the top (Point B in the figure), the area (from top to bottom) enjoyed a relatively fast rise in moisture content, with fluctuations when reaching Point B; while from Point B to the point at approximately 210mm from the top (Point C in the figure), the area, from top to bottom, fluctuated obviously in moisture content compared with sand lime brick. The moisture content reached its maximum value (16.7%) in the point at approximately 125mm from the top (Point E in the figure). Point E was about 10mm above the dry-wet boundary (Point D in the figure); from Point C to bottom, the area, under the impact of temperature recovery, had a narrow-range drop in moisture content.

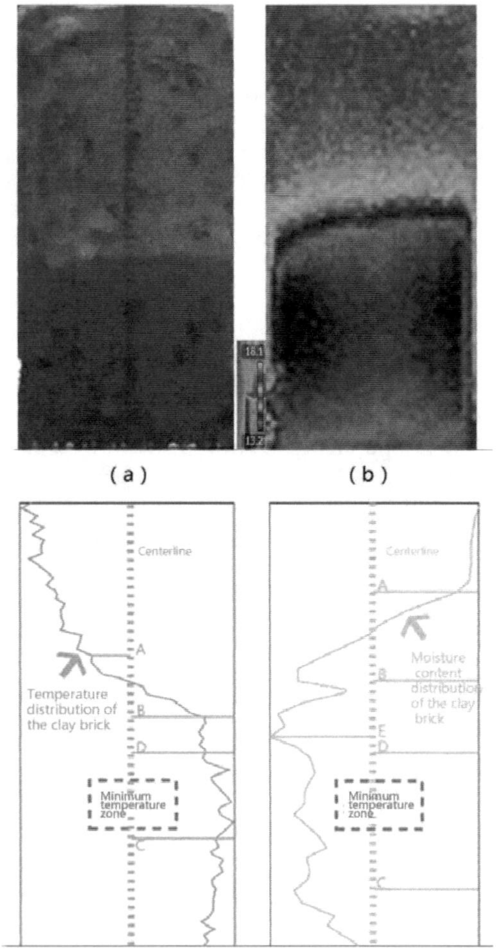

Figure 4. Temperature and moisture distribution of clay brick.

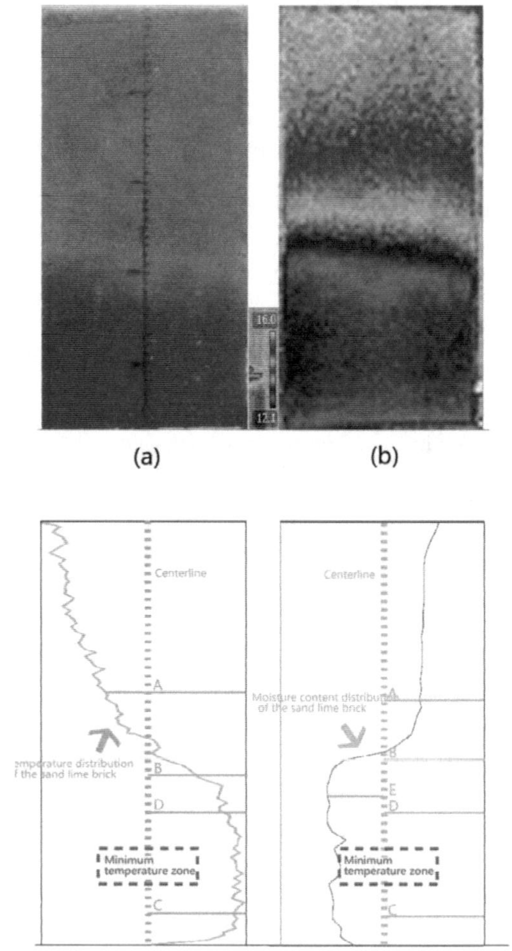

Figure 5. Temperature and moisture distribution of sand lime brick (drawn by the Author).

3.2.3. Temperature variation characteristics of sand lime brick along the centerline (figure 5c). From the top to the point at approximately 90mm from the top (Point A in the figure), the area, from top to bottom, was presented in a trend of slow temperature drop, with a narrow drop range; from Point A to the point at approximately 145mm from the top (Point B in the figure), there was the widest range of temperature drop in the area, which slowed down at Point B; from Point B to the point at about 230mm from the top (Point C in the figure), the area (from top to bottom) was presented in a trend of temperature stabilization. The minimum temperature occurred between 200mm and 225mm, with a range of approximately 25mm along the long side; the average temperature reached its maximum value in the area, in which the minimum temperature of sand lime brick (12.7℃) occurred. From Point C to the bottom, the area, under the impact of water temperature in lower part, enjoyed a narrow-range temperature recovery (from top to bottom). Since the bottom was immersed in water, the water temperature was identified as its temperature, namely 13.5℃. The visually measured water line was about 20mm above the minimum temperature zone.

3.2.4. Moisture content variation characteristics of sand lime brick along the centerline (figure 5d). From top to the point at 11mm from the top (Point A in the figure), the area was presented with stable moisture content, while the area from top to the point at 20mm from the top, under the impact of evaporation, was measured with relatively low moisture content;

from Point A to the point at 140mm from the top (Point B in the figure), the area (from top to bottom) enjoyed a relatively fast rise in moisture content. While from Point B to the point at 225mm from the top (Point C in the figure), the area, from top to bottom, fluctuated slightly in moisture content, which was maintained at 17%, with a relatively lower fluctuation level compared with clay brick; and in this area, the moisture content reached its maximum value (17.8%) in the point at approximately 155mm from the top (Point E in the figure). Point E was about 5mm above the dry-wet boundary (Point D in the figure); from Point C to bottom, the area, under the impact of temperature recovery, had a narrow-range drop in moisture content.

4 Analysis of experimental results

In the experiment, minimum temperature zone showed up in both of the two bricks. Minimum temperature zone refers to the temperature occurring at the centerline, with a difference of no greater than 0.2℃ between the minimum centerline temperature, and gathering with each point. The experimental results show that if a regression analysis is conducted in the minimum temperature zone and temperature value above/below such temperature zone, its distribution bears a negative correlation to the moisture content of corresponding points.

4.1. Analysis on correlation between surface temperature and moisture content distribution of sintered clay brick

4.1.1. Minimum temperature zone of sintered clay brick. The results show that after the clay brick is immersed in water for 1h43min, the minimum temperature zone occurs on the surface, with a distance of approximately 155mm~185mm from the top and a length of about 30mm along the centerline; and as is shown in the measurement by FD-100 moisture meter, moisture content of the clay brick reaches its maximum value at 20mm above the dry-wet boundary along the centerline (see figure 4).

4.1.2. Regression analysis on temperature and moisture content above the minimum temperature zone. Based on the determined minimum temperature zone, infrared thermometer was used to identify infrared thermal imaging temperature corresponding to the 49 moisture measuring points, and temperature and moisture content distribution at long side of the clay brick were divided into upper part and lower part, both of which included the minimum temperature zone (namely that temperature and moisture content in the minimum temperature zone are used twice). Then the temperature and moisture content data in upper part were imported into IBM SPSS Statistics22 for linear regression analysis, which indicated that there was a negative correlation between the temperature and moisture content in the upper part, with a correlation coefficient of -0.950. At the same time, a linear regression equation for temperature and moisture above the minimum temperature zone (inclusive) was worked out as:

$$\omega_u = -5.866t_u + 91.650 \tag{1}$$

(N=31, S=1.6755, F=261.765, R2=0.900, Sig=0.000)

In which, ωu represents the moisture content of each point above minimum temperature zone of the clay brick; While tu represents the temperature of each point (along the long side) above minimum temperature zone of the clay brick.

4.1.3. Regression analysis on temperature and moisture content below the minimum temperature zone. Temperature and moisture content data in the lower part were imported into IBM SPSS Statistics22 for linear regression analysis, which indicated that there was a negative correlation between temperature and moisture content in the upper part, with a correlation coefficient of -0.753. At the same time, a linear regression equation for temperature and moisture below the minimum temperature zone (inclusive) was worked out as:

$$\omega_D = -5.838t_D + 92.253$$
$$(N=24, S=1.0514, F=25.051, R2=0.546, Sig=0.000) \quad (2)$$

In which, ωD represents the moisture content of each point below minimum temperature zone of the clay brick; While tD represents the temperature of each point (along the long side) below minimum temperature zone of the clay brick

4.1.4. Regression analysis on centerline temperature and moisture content of the sintered clay brick. All the 49 centerline temperature and moisture content data were imported into BM SPSS Statistics22 for an analysis, which indicated that there was still a linear negative correlation between the overall temperature and moisture content, with a correlation coefficient of -0.950. At the same time, a linear regression equation for the temperature and moisture content was worked out as:

$$\omega = -6.123t + 95.657$$
$$(N=49, S=1.5953, F=414.039, R2=0.950, Sig=0,000) \quad (3)$$

In which, ω represents the moisture content of each point along centerline of the clay brick; While t represents the temperature of each point along centerline of the clay brick

4.2. Analysis on correlation between surface temperature and moisture content of the autoclaved sand-lime brick

4.2.1. Minimum temperature zone of the autoclaved sand-lime brick. The results show that after the sand lime brick is immersed in water for 25h52min, the minimum temperature zone occurs on the surface, with a distance of approximately 200mm~225mm from the top and a length of about 25mm along the direction parallel to long side of the clay brick; the visually measured dry-wet boundary (Point D in the figure) was 20mm above the minimum temperature zone, and as is shown in the measurement by FD-100 moisture meter, moisture content of the sand lime brick reaches its maximum value at 5mm above Point D(see figure 5).

4.2.2. Regression analysis on temperature and moisture content above the minimum temperature zone. Based on the determined minimum temperature zone, temperature and moisture content distribution at long side of the sand lime brick were divided into upper part and lower part, both of which included the minimum temperature zone. Then the temperature and moisture content data in upper part were imported into IBM SPSS Statistics22 for linear regression analysis, which indicated that there was a negative correlation between the temperature and moisture content in the upper part, with a correlation coefficient of -0.824. At the same time, a linear regression equation for temperature and moisture above the minimum temperature zone (inclusive) was worked out as:

$$\omega_U = -6.338t_U + 94.506$$
$$(N=46, S=2.3825, F=63.593, R2=0.669, Sig=0.000) \quad (4)$$

In which, ωU represents the moisture content above minimum temperature zone of the sand lime brick; While tU represents the centerline temperature (along the long side) above minimum temperature zone of the sand lime brick

4.2.3. Regression analysis on temperature and moisture content below the minimum temperature zone. Temperature and moisture content data in the lower part were imported into IBM SPSS Statistics22 for linear regression analysis, which indicated that there was a negative correlation between temperature and moisture content in the upper part, with a correlation coefficient of -0.707. At the same time, a linear regression equation for temperature and moisture below the minimum temperature zone (inclusive) was worked out as:

$$\omega_D = -1.942t_D + 41.967$$
$$(N=8, S=0.5994, F=18.027, R2=0.473, Sig=0.000) \quad (5)$$

In which, ωD represents the moisture content below minimum temperature zone of the sand lime brick; While tD represents the centerline temperature (along the long side) below minimum temperature zone of the sand lime brick

4.2.4. Regression analysis on centerline temperature and moisture content of the autoclaved sand-lime brick. All the 49 centerline temperature and moisture content data were imported into BM SPSS Statistics22 for an analysis, which indicated that there was still a linear negative correlation between the overall temperature and moisture content, with a correlation coefficient of -0.819. At the same time, a linear regression equation for the temperature and moisture content was worked out as:

$$\omega = -6.123t + 95.657$$

$$(N=49,\ S=2.9111,\ F=95.997,\ R2=0.664,\ Sig=0.000) \tag{6}$$

In which, ω represents the moisture content of each point along centerline of the sand lime brick; While t represents the temperature of each point along centerline of the sand lime brick

5. Conclusion

5.1. Range determination for the minimum temperature zone

The experimental results show that both the clay brick and sand lime brick, after a period of water penetration test, are respectively presented with a minimum temperature zone, both of which are approximately 20mm below the dry-wet boundary. In addition, height of such minimum temperature zone is between 25mm and 30mm.

5.2. Method for measurement of moisture content distribution (one-dimensional) of the two bricks by infrared thermal imager

After an analysis based on the experimental results, the method for measurement of moisture distribution (one-dimensional) by infrared thermal imager was concluded. Step 1: shoot the infrared thermal images for surface temperature of the water-penetrated bricks, and then, based on these images, use analysis software of the infrared thermal imager to identify the one-dimensionally distributed minimum temperature zone, which is about 20mm below the visually measured dry-wet boundary, with a height between 25mm and 30mm; step 2: identify the point with maximum moisture content along the long side at 5mm~10mm above the visually measured wet-dry boundary; step 3: considering that the regression equations obtained from SPSS analysis are all of statistical significance, they can be used to calculate the moisture content of each point along the centerline after identifying the above-mentioned points. Since there is a high correlation between formula (3) and formula (6), both of the two formulas can be adopted to determine one-dimensional moisture distribution of the two bricks under the help of infrared thermal imager. For moisture content variation at the visually measured dry-wet boundary, appropriate formula can be selected based on the actual conditions.

5.3. Practical examples

As is shown in figure 6, there is a protective historical building made from ordinary sintered clay bricks suffering from moisture-based diseases and requiring timely repair. However, the disease degree in each part of the wall, before such repair, should be identified to carry out a disease-oriented analysis on the entire wall.

Based on research achievements in this paper, vertical moisture distribution (measurement direction shown in figure 5) on the wall surface can be obtained. First, the visually-measured dry-wet boundary is identified, and then infrared thermal images of the wall are shot to determine the minimum temperature zone and point with the maximum moisture content along the measuring line; then formula (1) and (2) are used to verify whether there is a negative correlation between upper/lower points and temperature value in the minimum temperature zone; finally, corresponding adjustment is made in the included points and formula (3) is adopted to obtain the moisture content distribution in this direction.

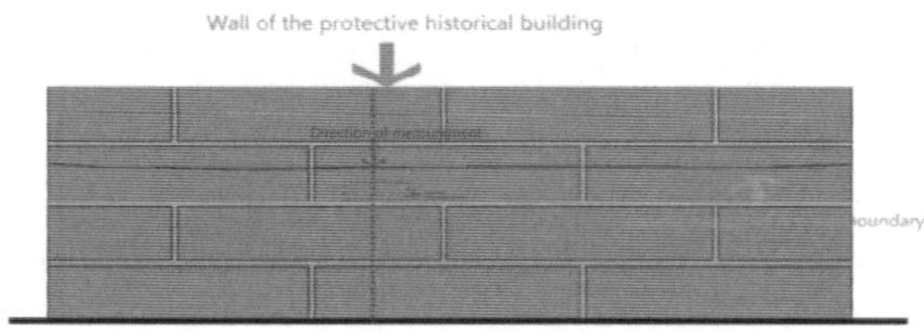

Figure 6. Practical examples.

5.4. Limitation and prospect

In this paper, a preliminary one-dimensional quantitative analysis is conducted in one-dimensional surface temperature and moisture content of sintered clay brick and autoclaved sand-lime brick. Future research shall, based on this research, seek for the correlation between two-dimensional/ three-dimensional surface temperature and moisture content of the two bricks, and the research scope shall, at the same time, be extended towards bricks of different materials/mix ratio and their external decorative finish, to obtain the quantitative analysis method for moisture-based diseases of traditional masonry buildings and extend the infrared technology to protection of modern historical buildings.

6. References

[1] Xiao J L 2009 *Research on Methodology and Establishment of Scientific System for Historical Building Protection in China*. Tsinghua University.

[2] Ru J H 2007 *Research on Guide Rules and Technologies for Historical Building Protection* Tianjin University, 2007.

[3] Sun P 2012 *Research on Prevention and Treatment against Moisture-based Diseases of Traditional Wooden Architecture in Tujia of Southwest Hubei* Huazhong University of Science and Technology.

[4] Zheng H J 2013 *Experimental Research on Moisture-based Diseases of Masonry Buildings* Huazhong University of Science and Technology, 2013.

[5] Delgado J M P Q, Guimarães A S and Freitas V P D, et al. 2016 *Adv. Mater. Sci. Eng.* **2016** 1280894.

[6] Cheng S 2011 *On Technological Support for Maintenance and Repair of Modern Historical Buildings* Tianjin University.

[7] Balaras C A and Argiriou A A 2002 *Energ. Build.* **34(2)**171-83.

[8] Tang Y F, Zhang Z Y and Tang C H et al. 2008 *Chem. Build. Mater.* **06** 26.

[9] Wang K, Yan B G, Zhang H Y and Cheng Z C.2015 *Concrete*, **05** 154.

[10] Chiang L W 2017 *Adv. Tech. Innov.* **02** 56.

[11] Li G H, Wu L X, Wu M and Qu J X 2004 Infr. Laser Eng. **03** 227.

[12] Zhang W, Li A. 2007 *Infr. Tech.* **06** 311-5.

[13] Li Y, Sun X, and Yuan G 2007 *Opt. Prec. Eng.* **09** 1336-41.

[14] Li X, Ni G. 2008 *Infr. Las. Eng.* **S2** 523-6.

[15] Lu Z. 2010 *Error Analysis on Radiometric Calibration and Temperature Measurement by Infrared Thermal Imager*. Graduate University of Chinese Academy of Sciences (Changchun Institute of Optics, Fine Mechanics and Physics).

IOP Conf. Series: Materials Science and Engineering **248** (2017) 012031 doi:10.1088/1757-899X/248/1/012031

SDRE controller for motion design of cable-suspended robot with uncertainties and moving obstacles

Ahad Behboodi[1] and Seyedmohammad Salehi[2]
[1]Biomechanics and Movement Sciences University of Delaware, USA
[2]Computer and Information Sciences University of Delaware, USA

Abstract. In this paper an optimal control approach for nonlinear dynamical systems was proposed based on State Dependent Riccati Equation (SDRE) and its robustness against uncertainties is shown by simulation results. The proposed method was applied on a spatial six-cable suspended robot, which was designed to carry loads or perform different tasks in huge workspaces. Motion planning for cable-suspended robots in such a big workspace is subjected to uncertainties and obstacles. First, we emphasized the ability of SDRE to construct a systematic basis and efficient design of controller for wide variety of nonlinear dynamical systems. Then we showed how this systematic design improved the robustness of the system and facilitated the integration of motion planning techniques with the controller. In particular, obstacle avoidance technique based on artificial potential field (APF) can be easily combined with SDRE controller with efficient performance. Due to difficulties of exact solution for SDRE, an approximation method was used based on power series expansion. The efficiency and robustness of the SDRE controller was illustrated on a six-cable suspended robot with proper simulations.

1. Introduction

A new class of highly applicable crane-like robots, driven by cables, is designed with advantages including convenient assembly, simple structure, lightness, acceptable accuracy, huge workspace, and relatively high Dynamic Load Carrying Capacity (DLCC) [1]. Using cables as actuators in this kind of robot makes the dynamic equations more complicated and nonlinear. Therefore, a controller is needed to capture nonlinearities as well as to guarantee optimal performance in keeping DLCC at a maximum level. Another important feature of cable-suspended robots is their huge workspace, which might contain moving or fixed obstacles. Thus, an appropriate controller must be capable of capturing supplementary techniques for motion design in the presence of obstacles in a convenient and efficient manner.

Controlling and regulating the performance of dynamical systems based on the desired task is the key objective in almost all mechanical engineering research fields not only in vibrations [2], dynamics and robotics [3-5] as the most common ones, it is also extended to fluid mechanics [6, 7] and boundary layer analysis [8, 9]. For cable-suspended robots, many control methods have been proposed; open-loop controllers like Hamilton-Jacobi-bellman (HJB) [10] (which is weak against noises and uncertainties), closed-loop controllers like feedback linearization [11] and Sliding Mode Control (SMC) [12] (which are not optimal), Optimal linear and closed-loop controllers like Linear Quadratic Regulator (LQR) with feedback linearization basis [13] (which is unable to capture nonlinearities properly).

SDRE method provides an optimal controller that can be applied to complicated nonlinear dynamic systems easily and systematically [14]. Authors in [15] has done a thorough survey of SDRE that contains details like formulation and structure, optimality concept, state dependent parameterization

Content from this work may be used under the terms of the Creative Commons Attribution 3.0 licence. Any further distribution of this work must maintain attribution to the author(s) and the title of the work, journal citation and DOI.
Published under licence by IOP Publishing Ltd

and stability conditions. In comparison with LQR, SDRE is proved to be more robust [16]. Difficulties of SDRE's exact solution has given rise to numerical techniques like power series [17], which are developed for the most general cases [14].

Motion design for cable-suspended robots that perform in huge workspace might be problematic and are usually subjected to different constraints like obstacles [13] and moving boundaries [18]. In this paper, obstacles were considered based on their more commonalities over other constraints. Most of the existing obstacle avoidance techniques are based on the APF concept [19]. The potential term created by this method is usually nonlinear which is problematic for linear based control methods like LQR and makes them more complex and restricted [13]. While SDRE is shown to be capable of being combined with this technique very conveniently and perform efficiently in the presence of fixed obstacles [14], many other potentials of such controllers are not yet studied. In this paper, by proper simulations, we show that the SDRE combined with APF techniques is able to avoid moving obstacles in motion design. Robustness of this controller against uncertainties is also illustrated with simulation results on six-cable-suspended robot.

2. Controller formulation

A nonlinear dynamical system equation can be always factorized and be expressed in the form below:

$$\dot{x}(t) = A(x)x(t) + B(x)u(t) \tag{1}$$

where $x \in R^n$ and $u \in R^m$ are state and input vectors respectively. The input $u(t)$ is designed to minimize the objective function:

$$W = \int_0^{\infty} [x^T(t)Q(x)x(t) + u(t)^T R(x)u(t)]dt \tag{2}$$

SDRE method suggests the optimal control law as:

$$u(x) = -R^{-1}(x)B^T(x)K(x)x \tag{3}$$

where $K(x)$ is unique symmetric and positive definite solution of the Riccati equation below:

$$K(x)A(x) + A^T(x)K(x) - K(x)B(x)R^{-1}(x)B^T(x)K(x) + Q(x) = 0 \tag{4}$$

Solving state dependent Riccati equation is very difficult and even impractical when system has complicated dynamical equation. Consequently, an approximate solution approach is needed. Power series method has been developed for dynamic systems for the case where A, B and even Q matrix is state dependent [14]. The same method has been used in this paper.

APF was used as obstacle avoidance technique in this paper. The blow potential term was added to the objective function.

$$\sum_{i=1}^{n} w_i' \left(\frac{1}{r_{o_i} + r_m + S_p - d_i}\right)^2 \tag{5}$$

where n is the number of the obstacles, r_{o_i} is the radius of circumscribed sphere to the i-th obstacle, r_m is the radius of safe zone sphere for the end-effector, S_p is the safety factor, w_i' is weight of the potential term and d_i is the distance between end-effector and the i-th obstacle. After adding this term to the objective function, since Q can be state dependent, it will be resolved to the Q matrix to keep the standard format of SDRE objective function. In this way, all the same formulation of SDRE can be used for motion design with obstacle avoidance. This is a very convenient capability of SDRE method that can be easily combined with complementary techniques like APF.

3. Simulations

The schematic of spatial cable-suspended robot is presented in figure 1. Dynamic equations of the robot were derived in [20] and dynamic equations that were adapted to use in SDRE formulation are presented in [14]. There were six motors in A, B, C (two in each one) and six cables to control the end-effector DEF.

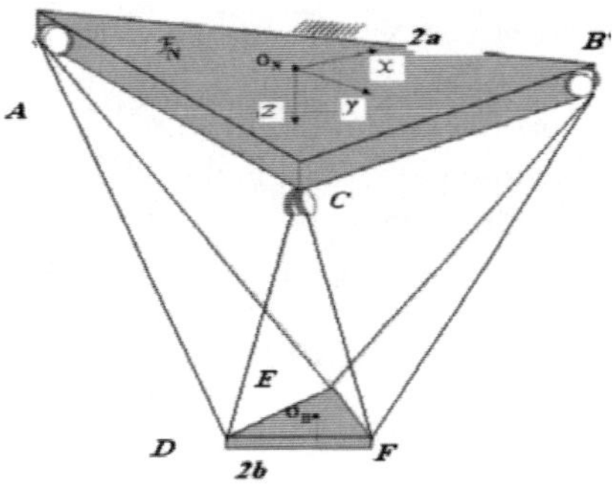

Figure 1. Schematic of six-cable-suspended robot [14].

3.1. Trajectory tracking

In addition to SDRE controller's capability in acceptable trajectory tracking, the parameters of the controller can be tuned to compensate for the uncertainties in the initial position. In the conducted simulation, end-effector should follow a predefined path while at the beginning it was not on the desired path. Thus it should first go to the desired path as quickly as possible and then follow it. The predefined trajectory is given by:

$$x = 0.035\cos(t) - 0.035\sin(t)\,;\; y = 0.035\sin(t)\,; z = 1 + 0.025\cos(t) + 0.025\sin(t) \qquad (6)$$

where $0 \leq t \leq 2\pi$. Comparing the results with LQR method demonstrated how capturing nonlinearities of the system helps the controller to be more precise. Trajectories of SDRE and LQR are shown in figure 2 and motor torques for the first three motors is demonstrated in figure 3.

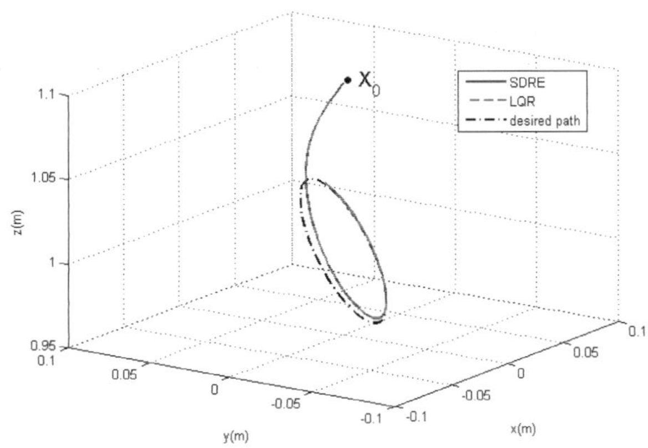

Figure 2. Trajectories of the end-effector.

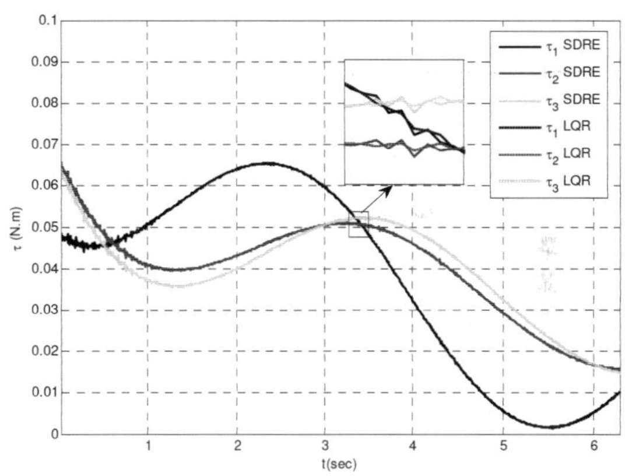

Figure 3. Motor torques as control inputs.

The results show that how the proposed controller successfully overcomes the uncertainties at initial position and makes end-effector follow the desired trajectory precisely. Errors from real trajectories to the desired one are shown in figure 4. Clearly SDRE performed more accurate than LQR since it captures nonlinear behavior of the dynamic system.

3.2. Moving obstacle

To make the situation more difficult and realistic and to assess the performance of the proposed controller in combination with APF obstacle avoidance technique, a moving obstacle was considered during the robot performance. The motion of the obstacle of radius 2 cm is characterized by:

$$x_{obs} = (0.0332)t; \quad y_{obs} = (0.0332)t - 0.81; \quad z_{obs} = 0.7595 \qquad (7)$$

Simulations were conducted for three cases; first, no obstacle avoidance technique, in second and third case the obstacle avoidance technique was added but the weight of the potential term was different and set to 0.05 and 0.15 respectively in order to show the effect of this parameter.

Trajectories and the distance between end-effector and obstacle during motion are shown in figure 5 and figure 6 respectively. The results illustrated how a complicated technique like APF can be easily combined with SDRE formulation to successfully design an optimal motion for the robot to avoid moving obstacles.

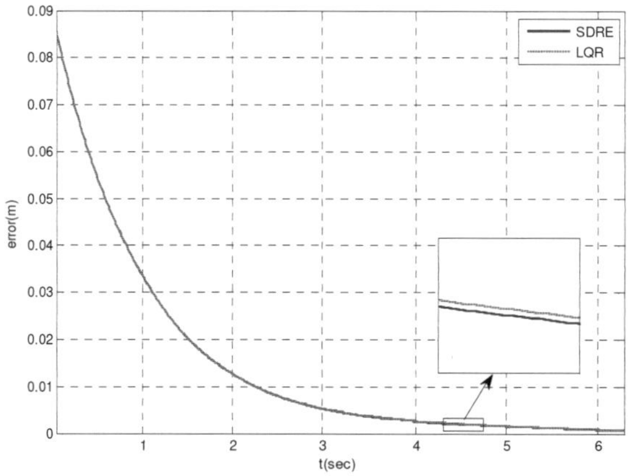

Figure 4. Comparison of trajectory errors of SDRE and LQR.

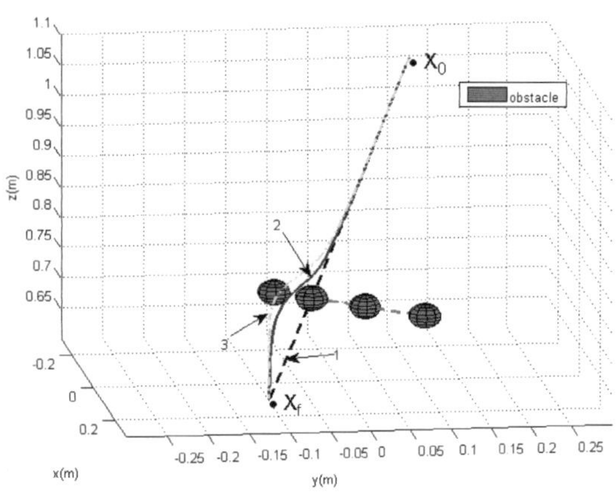

Figure 5. Obstacle avoidance of the end-effector.

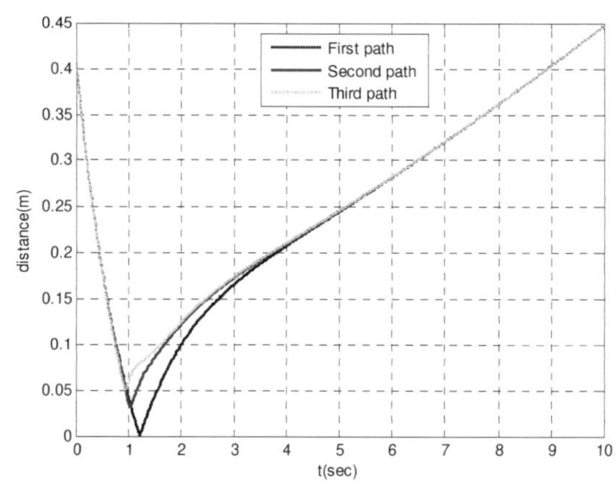

Figure 6. End-effector and obstacle distance.

4. Conclusion

In this work, capabilities of SDRE controller in handling uncertainties were investigated. The simulation results on trajectory tracking of cable-suspended robot with uncertainty in initial position illustrated that SDRE method can provide a robust and systematic basis for controlling nonlinear dynamical systems. One advantage of SDRE method is the ability to be easily combined with supplementary techniques for motion design such as APF-based obstacle avoidance techniques. Results in difficult and realistic situation i.e. moving obstacles in the robot workspace showed the efficiency of the controller in designing an optimal motion for the robot in the presence of obstacles.

5. References

[1] Albus J S, Bostelman R and Dagalakis N G 1993 *J. Robot. Syst.* **10** 709-24.
[2] Motahar M S and Ahmadian M T 2011 *Proc. ASME Int. Des. Eng. Tech. Conf. Comput. Inform.*

Eng. Conf. **37** 1-7.
[3] Motahar M S, Veer S and Poulakakis I 2015 *Proc. IEEE/RSJ Int. Conf. Intell. Robot. Syst.* 1004-10.
[4] Veer S, Motahar M S and Poulakakis I 2015 *Proc. IEEE/RSJ Int. Conf. Intell. Robot. Syst.* 997-1003.
[5] Motahar M S, Veer S and Poulakakis I 2016 *Proc. IEEE Conf. Dec. Contr.* 6368-74.
[6] Yousefi K and Saleh R 2015 *Meccanica* **50**1481-94.
[7] Yousefi K, Saleh R and Zahedi P 2014 *J. Mech. Sci. Tech.* **28** 1297-310.
[8] Yousefi K and Saleh R 2014 *J. Theoret. Appl. Mech.* **52** 165-79.
[9] Yousefi K, Saleh R and Zahedi P 2013 *Int. J. Eng.* **7** 10-24.
[10] Korayem M H, Davarzani, E and Bamdad, M 2010 *Sci. Iran.* **17** 315–26.
[11] Alp A B and Agrawal S K 2002 *Proc. IEEE Int. Conf. Robot. Autom.* **4** 315-26.
[12] Oh S R, Mankala K K, Agrawal S K and Albus J S 2005 *Multibody Syst. Dyn.* **13** 385-99.
[13] Korayem M H, Tourajizadeh H, Zehfroosh A and Korayem A H 2015 *Robotica* **33** 933-52.
[14] Korayem M H, Zehfroosh A, Tourajizadeh H and Manteghi S 2014 *Nonlin. Dyn.* **76** 1423-41.
[15] Çimen T 2008 *Proc. 17th World Congr. Int. Federat. Autom. Contr.* **41** 3761-75.
[16] Abdelrahman M, Chang I and Park S Y 2011 *Int. J. Non-Lin. Mech.* **46** 758-71.
[17] Banks H T, Lewis B M and Tran H T 2007 *Comput. Opt. Appl.* **37** 177-218.
[18] Korayem M H, Tourajizadeh H, Zehfroosh A and Korayem A H 2014 *Nonlin. Dyn.* **78** 1515-43.
[19] Khatib O 1986 *Int. J. Robot. Res.* **5** 90-8.
[20] Alp A B 2001 Cable suspended parallel robots *Thesis for the degree of Master of Science in Mechanical Engineering in University of Delware.*

IOP Conf. Series: Materials Science and Engineering **248** (2017) 012032 doi:10.1088/1757-899X/248/1/012032

Analysis on Influence of Metro Shield on Ground Subsidence

X Q Liu[1], Y H Yang[1], W B Pei[1]

[1]Tianjin Research Institute for Water Transport Engineering, M.O.T Tianjin, 300456, China

Abstract. The shield tunnel construction inevitably disturbs the surrounding strata, which in turn causes ground subsidence, and the excessive sedimentation will endanger the normal use and structural safety of adjacent buildings. Based on the analysis of the ground subsidence during the construction of the shield tunnel, the settlement trough of the surface cross section in the process of metro shield is similar to the normal distribution curve. The maximum settling volume usually occurs within 5 meters from the center of the tunnel. The ground subsidence arising from shield construction can be roughly divided into three stages: firstly, before the shield machine reaches the observation section, the subsidence (-30m-0m) occurs, and the settling volume is about 10% to 20% of the total settlement; secondly, when the shield machine passes through the observation section, the subsidence occurs (0-30m), and the settling volume is about 40% to 70% of the total settlement; and finally, after the shield machine passes through 30m of the observation section, the later consolidation settlement of the section occurs, and the settling volume is 10% or less of the total settlement. The transverse influenced range of the ground subsidence is about 30 meters or less and the main influence area of shield construction is around 5 meters from the left or right of the tunnel axis. In the case of the shield pass-through or the shield tail pull-away, the ground subsidence is relatively larger, accounting for 30% to 50% of the maximum settlement value.

1. Introduction

The shield thrust technology is widely used in the urban metro construction, but the shield tunnel construction inevitably disturbs the surrounding soil and causes the soil to move, which may lead to the tilt or even cracking and collapse of houses, viaducts and other buildings and destroy the underground pipelines, so the influence of the shield construction process on the surrounding environment has become a matter of great concern. Many scholars at home and abroad have carried out extensive and in-depth research on the ground subsidence caused by the shield tunnel construction. Japan has accumulated rich experience in the construction of long-term urban tunnels, so it is very concerned about the problem of surface subsidence caused by tunnel construction. Based on the measured data, Hanya put forward the prediction method of the maximum ground settlement. Fujita carried out the research on the influence of different shield construction methods on the strata in soft soil strata, and the practical formulas were also put forward by Mori and Shimada. Lin Zhenhong studied the influence of the expansion construction of the existing tunnel on the tunnel structure mechanics behavior and the stability of the surrounding rock; Wu Zhangzhong derived the specific form of the complex stress function in the case of the lateral enlarging excavation of the existing tunnel. The engineering practice shows that the shield tunnel construction technology has been developed very maturely, but it inevitably produces disturbance to the soil and causes different levels of formation displacement and deformation, which is usually expressed as ground subsidence. In this paper, the metro shield construction of one project under construction in Tianjin was taken as an example to study the ground subsidence law of horizontal and vertical sections along the subway

Content from this work may be used under the terms of the Creative Commons Attribution 3.0 licence. Any further distribution of this work must maintain attribution to the author(s) and the title of the work, journal citation and DOI.
Published under licence by IOP Publishing Ltd

tunnel in the process of Tianjin subway tunnel shield construction, and to analyze the influence of metro shield on the surrounding environment.

2. Field experiment

2.1 Arrangement of measuring points

The test section for the shield tunnel construction was about 1,000 meters long, the four test sections were selected, and eight observation points were arranged at each observation section (the specific locations are shown in figure 1 and figure 2), so as to monitor and master the changing situation of the vertical displacement of the ground point during the shield construction and the characteristics of the subsidence of the soil of the measuring section perpendicular to the direction of the shield axis. The Leica DNA03 digital level was chosen as the observation instrument.

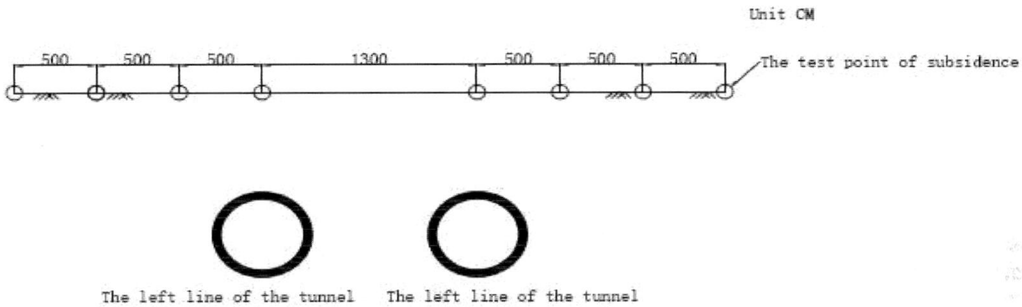

Figure 1. Arrangement diagram of the measuring points of cross-section for ground subsidence in shield zone.

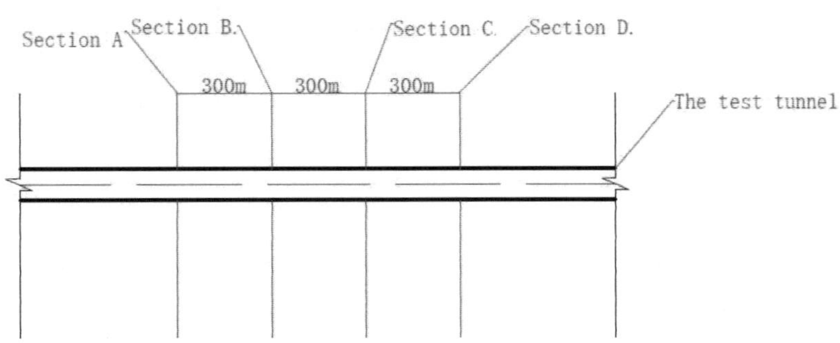

Figure 2. Arrangement diagram of the measuring Transverse.

2.2 Project overview

According to the exploration, the depth range can be divided into Quaternary Holocene artificial fill layer (Q4ml), beach alluvium (Q42al) Tertiary (N) sandstone and conglomerate. The geology of the region consists of a set of settled layers varying from the fine particles gradually to thick ones, usually expressed as clay, silt, medium sand, coarse sand, gravelly sand and thick pellet soil, and the silt and fine sand are rarely seen, with part strata of the local section missing. The permeability coefficient is generally between 50 to 100 m/d, hydraulic slope is around 1 ‰.

3. Experimental data analysis

The transverse subsidence data of four monitoring sections were selected from the right line of the test section for analysis. According to the shield construction process, the continuous measurement was applied to measure the four monitoring sections firstly with 30 meters from the shield, until 40 meters of the section were passed by the shield machine and the surface subsidence stabilized. The transverse subsidence of ground surface of these four monitoring sections is shown in figure 3 to figure 6. The longitudinal surface subsidence can help analyze the influence range and size of front and rear soils during shield construction. In this paper, four monitoring sections were selected, and figure 7 to figure 10 were drawn based on the surface subsidence values of the measuring points distributed from about 30 meters from the sections to 40 meters of the sections.

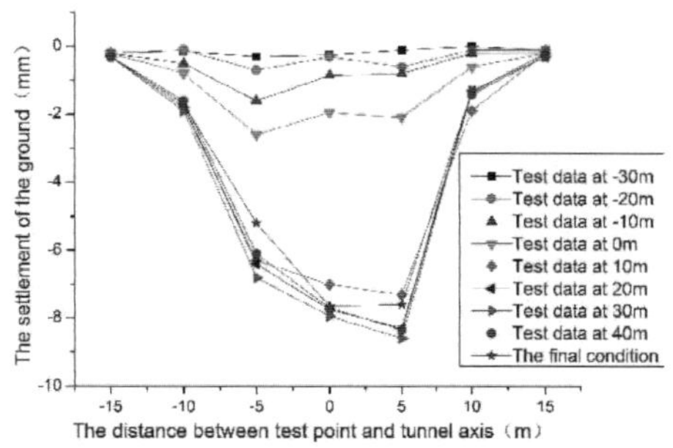

Figure 3. The Transverse surface settlement process chart of section A.

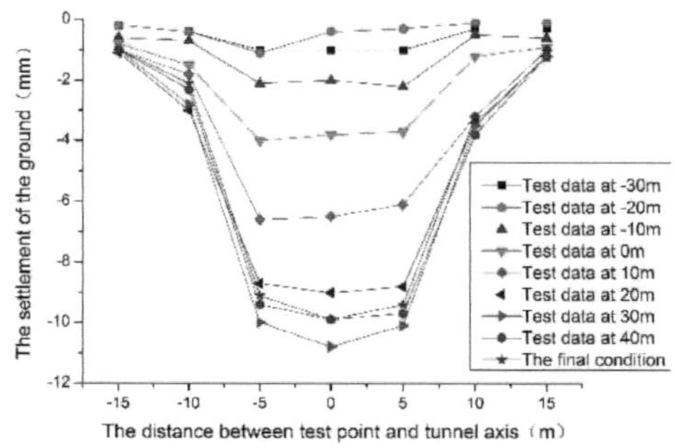

Figure 4. The Transverse surface settlement process chart of section B.

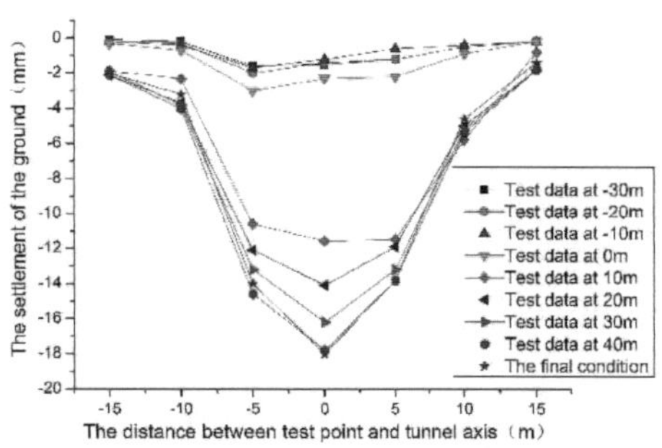

Figure 5. The Transverse surface settlement process chart of section C.

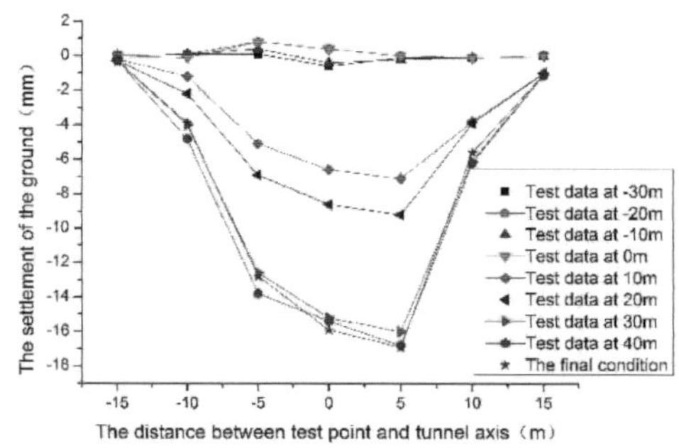

Figure 6. The Transverse surface settlement process chart of section D.

It can be seen from the ground subsidence curve that:

(1) The settlement trough of the surface cross section is similar to the normal distribution curve. When the shield machine is on the flat surface with no ground load, the maximum settling volume occurs at the center of the tunnel.

(2) The ground subsidence caused by shield construction can be divided into five stages:

① As to the ground subsidence before the arrival of the shield, the shield machine is around -30 to 10 m away from the observation section, and the surface subsidence value is 1 to 2 mm at this stage, as small as 5% to 10% of the maximum settlement value.

② As to the ground subsidence when the shield reaches, the shield machine is -15 to 0 m away from the observation section, and if the earth pressure of chamber is higher, the upheaval of the earth surface may occur at this stage, like Section DK6 +608.680. According to the general data, the ground subsidence value is 2 to 4 mm at this stage, equal to 10% to 20% of the maximum settlement value.

③ As to the ground subsidence when the shield passes, the shield machine is 0 to 10 m from the observation section, and due to the friction shear stress between the shield and soil and changes in stress of the sandy soil resulting from the shield's "head-up" and "kowtow", the subsidence at this stage generally has a sudden change, up to 5 to 8 mm, equal to 30% to 50% of the maximum settlement value.

④ As to the ground subsidence when the segment pulls away from the shield tail after the shield passes through, the shield machine is 10 to 40 m from the observation section, and mainly due to "shield tail gap" and stress release, the subsidence value at this stage is 5 to 8 mm, as large as 30% to 50% of the maximum settlement value.

⑤ As the soil is disturbed by the shield excavation to lead to the re-consolidation of the soil, the long-term consolidation settlement occurs after the shield passes. The sedimentation value at this stage is not large, which generally accounts for 10% or less of the maximum settlement value.

(3) The transverse influenced range of the ground subsidence is about within 30 meters and the main influence area of the shield construction is within 5 meters from the left or right of the tunnel axis. The ground subsidence turning points of the four monitoring sections are located within 5 meters from the left or right of the center of the shield tunnel axis. In this area, the settlement trough's volume accounts for 60% to 70% of the total volume, and the average settlement value is 60% to 80% of the maximum settlement value.

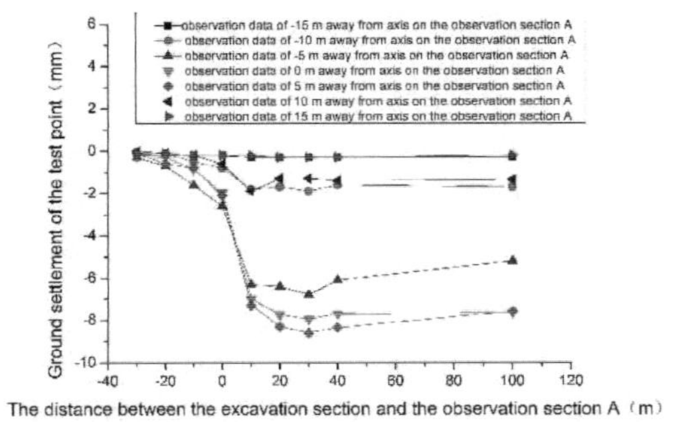

Figure 7. The relationship between the surface subsidence and excavation distance of the axis of section A.

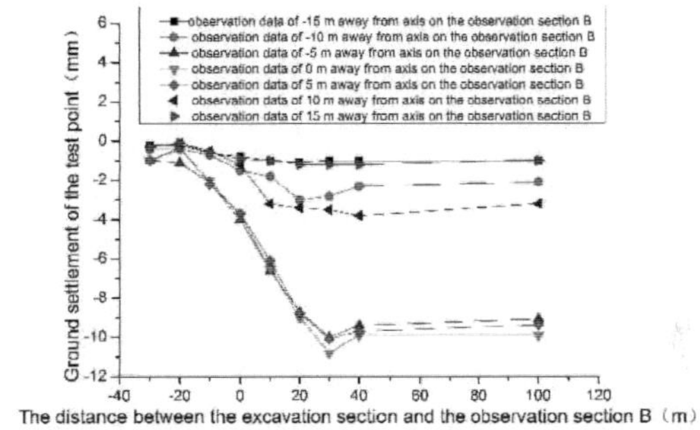

Figure 8. The relationship between the surface subsidence and excavation distance of the axis of section B.

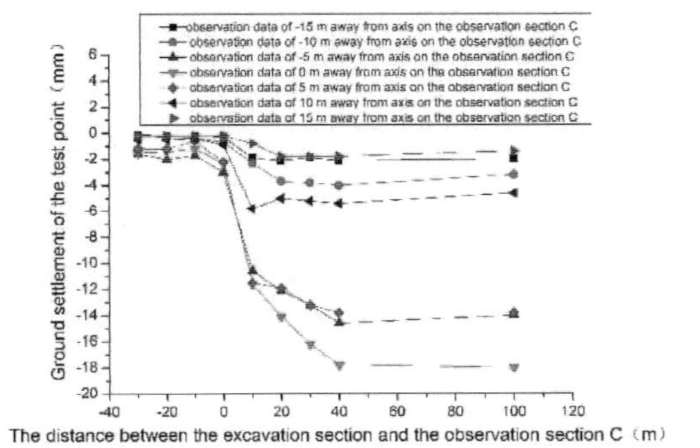

Figure 9. The relationship between the surface subsidence and excavation distance of the axis of section C.

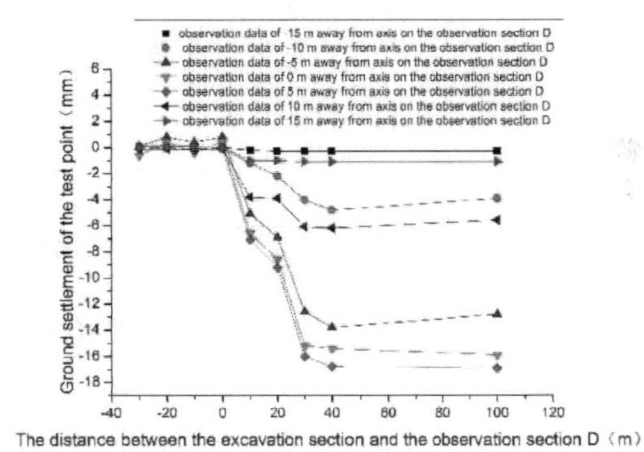

Figure 10. The relationship between the surface subsidence and excavation distance of the axis of section D.

The ground settlement in longitudinal surface indicates that:

(1) The influence range of the shield thrust on the front soil is within 30 m. As can be seen from the figure, the settling volume for the front soil is generally not more than 5 mm, but the duration is longer. The subsidence is done in a slow process and there is a phenomenon of surface upheaval.

(2) The influence range of shield thrust on the disturbance and settlement of the rear soil mainly ranges from 0 to 10 m, and the sand strata are different from the clay strata in such range. The shear strength and cohesion of the sandy soil are very weak, so the deformed settlement occurs quickly after the soil is disturbed, without the delay-time effect of the sticky soil. Therefore, the settlement

generally has a sudden change at this stage, with the maximum rate of change of 5 to 8 mm, accounting for 30% to 50% of the maximum settlement value.

(3) The ground subsidence is also larger at the stage of shield pass-through or shield-tail pull-away, accounting for 30% to 50% of the maximum settlement value.

4. Conclusion

Based on the analysis of the ground subsidence during the construction of the shield tunnel, the following conclusion is obtained:

(1) The settlement trough of the surface cross section in the process of metro shield is similar to the normal distribution curve.

(2) The maximum ground settlement which caused by shield construction is occurs between the shield machine reaches the observation section and leave the observation section 30m, and the settling volume is about 40% to 70% of the total settlement.

(3) The influence of shield construction is within 15m on both sides of the subway line.

5. References

[1] Lin Z H 2009 J. Railway Sci. Eng. **6(4)** 46-50.

[2] Mashimo H, Ishimura T and Isago N, et al. 2005 *Behavior of support structure of existing tunnels during enlargement of cross section* ITA-AITES 2005 World Tunnel Congress. Rotterdam, 133-8.

[3] Huang M L, Fan H Z and Tan Z S. 2010 *Study of construction technique of metro station driven with enlarging existing shallow air defense tunnel* Proceedings of the 11th National Academic Conference of Rock Mechanics and Engineering. Beijing: China Electric Power Press, 307-12.

[4] Alba J and Castro E 2002 *Analysis of settlements caused by the underground construction of the Tren Urbano using the New Austrian Tunneling Method (Puerto Rico).* 2002, A Dissertation Submitted to university of Purto rici mayaue campus for Ph.D.

[5] Duan Z Y 2001 *Ground movement associated with micro tunnelling* A Dissertation Submitted to college of engineering and science Louisiana Tech university for Ph.D.

[6] Leca E 1989 *Analysis of NATM and Shield Tunneling in Soft Ground* A Dissertation Submitted to Virginia Polytechnic institute and state university for Ph.D.

Acknowledgements

This paper is funded supported by Tianjin Research Program of Application Foundation and Advanced Technology (15JCQNJC07900), Central Public Research Institutes Fundamental Research (TKS160103), Key Research and Development Program of Tianjin (16YFXTSF00280), The Science and technology plan projects of Tianjin port, the views expressed are authors alone.

Application of Foldcore Sandwich Structures in Helicopter Subfloor Energy Absorption Structure

H Z Zhou[1] and Z J Wang[1]

[1]Key Laboratory of Fundamental Science for National Defence -Advanced Design Technology of Flight Vehicles, Nanjing University of Aeronautics and Astronautics, 29 YuDao Street. Nanjing 210016, China

Abstract. The intersection element is an important part of the helicopter subfloor structure. The numerical simulation model of the intersection element is established and the crush simulation is conducted. The simulation results agree well with the experiment results. In order to improve the buffering capacity and energy-absorbing capacity, the intersection element is redesigned. The skin and the floor in the intersection element are replaced with foldcore sandwich structures. The new intersection element is studied using the same simulation method as the typical intersection element. The analysis result shows that foldcore can improve the buffering capacity and the energy-absorbing capacity, and reduce the structure mass.

1. Introduction

When considering the ability to absorb kinetic energy, the subfloor structure is one of the most important structural components in helicopter. Consequently, it has to be designed to limit the deceleration forces by structural deformation and provide post-crash structural integrity of the cabin floor [1]. Figure 1 shows a typical subfloor structure. In such a structure, the structural intersections of beams and bulkheads play an important part in the overall crash response of the subfloor structure. The intersections have higher stiffness and strength, so they create high deceleration peak loads and cause dangerous inputs to the occupant system. In order to design a subfloor structure which is able to fulfil the crashworthiness requirements, it's first of all to design a good structural intersections.

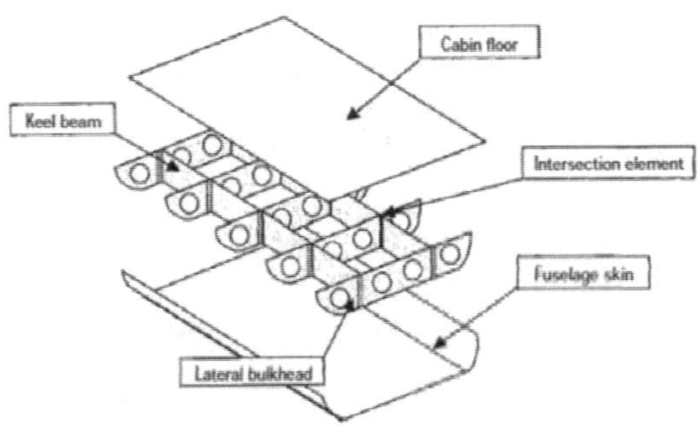

Figure 1. Typical subfloor structure.

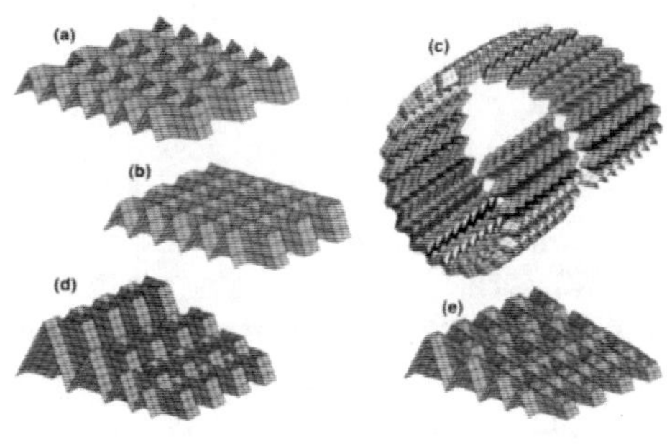

Figure 2. Some kinds of foldcore.

Experimental and numerical studies of structural intersections are conducted by Bisagni C [1-5]. A fast reanalysis methodology based on neural network is proposed with the intent to reproduce the

Content from this work may be used under the terms of the Creative Commons Attribution 3.0 licence. Any further distribution of this work must maintain attribution to the author(s) and the title of the work, journal citation and DOI.
Published under licence by IOP Publishing Ltd

crash behaviour of structural intersections. Based on the method, size and topological optimization are conducted.

The foldcore structures, as shown in Figure 2, are geometric structures formed by folding plates or foils according to regular repeated lines. According to different lines, M-type (Figure 2 (a),(b)), V-type and other types of foldcore structures can be formed, even shells with large curvature (Figure 2(c)), stepped plates (Figure 2 (d)) and plates with variable thickness (Figure 2 (e)) can be manufactured using foldcore. Foldcore sandwich structures have a lot of cavities, which provides the possibility of large plastic deformation. So the foldcore has the common feature of energy absorbing structures and may have good energy absorbing capacity.

The geometric design method and the mechanics model of foldcore have been studied by Wang Z J and her team [6, 7]. Methods of designing different kinds of foldcore structures are introduced. The relations of elastic constants and geometric parameters of a foldcore are derived.

The energy absorption capacity of foldcore sandwich structures is studied by Zhang Y C. Due to his study, foldcore is proved to be a good anti-crash element [8, 9]. The basic mechanical properties of foldcore is studied by Ren Y F [10]. Experimental and numerical studies are conducted on the bending capacity of foldcore composite sandwich beams by Sun Y G [11].

Many experiments and simulations have also been conducted on the foldcore by S. Heimbs and his team [12-20]. Compression and shear experiments have been conducted on foldcore from CFRP and aramid paper. Besides, a numerical simulation method has been carried out to study the mechanical properties in all period from starting loading to the densification stage of the core. The residual strength and the impact property of honeycomb core and foldcore sandwich structures have been studied with the help of experiments and numerical simulations. In order to model foldcore with higher accuracy, a new way to model the material imperfections and geometrical defects of foldcore has been set up by E. Baranger, P. A. Guidault and C. Cluzel [21, 22]. The issue of the relation between the strength of wedge-shape foldcore sandwich structures and the geometrical parameters has been considered by Falk Hähnel and Klaus Wolf [23].

Foldcore sandwich structures have been proved to have good energy absorption capacity. In order to reduce the deceleration peak loads and improve the energy-absorption capacity of structural intersection elements of subfloor structures, foldcore sandwich structures are considered to be applied in the intersection elements. In this paper, an intersection element containing foldcore sandwich structures is designed and studied.

2. Numerical simulation model of a typical intersection element

A typical intersection element is shown in Figure 3. The engineering drawing of the intersection element experiment sample is shown in Figure 4. Comparison of the experiment sample before and after crush test is shown in Figure 5.

The finite-element analysis software ABAQUS is used to simulate the crush procedure of the intersection element. The numerical simulation model is established according to the crush test in paper [1]. The impact velocity is 7.4m/s. The geometric parameters of the intersection element sample are shown in Figure 4. The thickness of the cabin floor and the skin are 1.27mm, and the thickness of the rest panels are 0.81mm. The material used is aluminium alloy 2024 T3.

In the numerical simulation model, the element type is chosen as 4-node shell reduced integration element S4R. The impact mass is applied on a reference point. The reference point locates on the middle of the floor top surface, and is coupled with the floor top surface. Contact in the structure are modelled using ABAQUS general contact controls. The contact property is defined as "Hard" contact in the normal direction and "penalty" in the tangential direction. The friction coefficient is 0.17. The rivets are simulated by connecting nodes around the rivets of two panels using tie constraint.

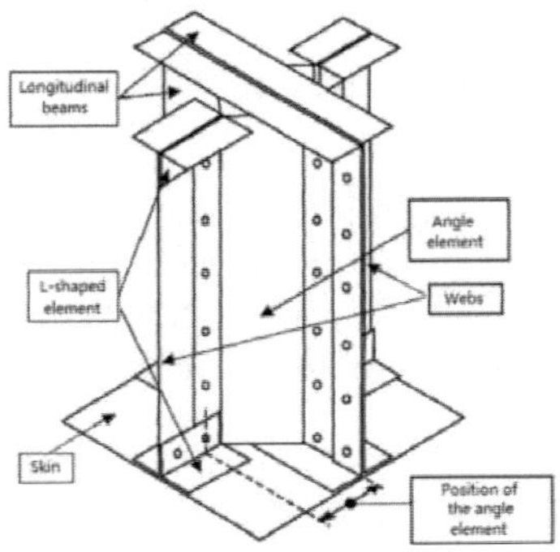

Figure 3. A typical subfloor intersection element.

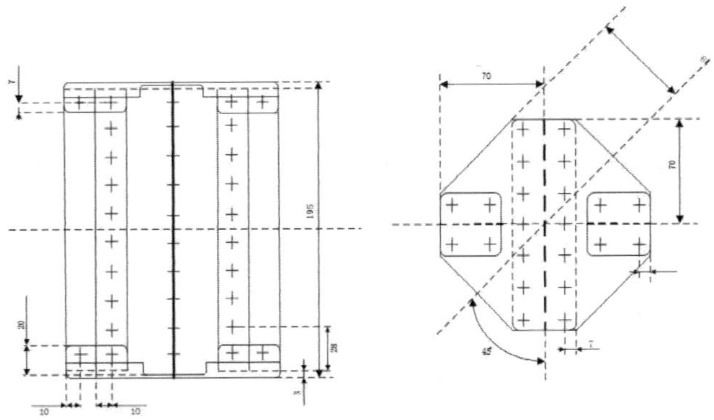

Figure 4. Engineering graphic of a typical subfloor intersection element sample.

Comparison of the simulation results and the experiment results in paper [1] is shown in table 1. The force-shortening diagrams obtained from simulation and experiments are shown in Figure 6. The deformed structure after experiment and the corresponding simulation results are shown in Figure 7.

Figure 5. Comparison of the experimental sample before and after crush test.

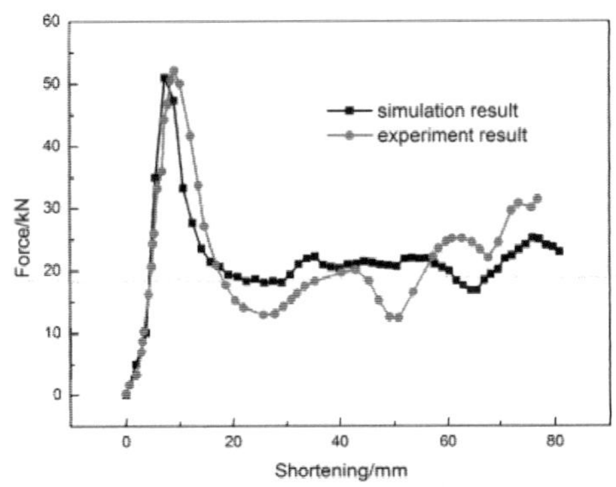

Figure 6. Comparison of the force-shortening diagrams of the simulation and experiment results.

Table 1. Comparison of the simulation results and the experiment results of the intersection element.

	Experiment result	Simulation result	Error/%
Peak load/kN	52.0	50.1	3.65
Average load/kN	25.5-28.0	22.8	
Mass/kg	0.481	0.481	0.00
Specific energy absorption/KJ/kg	6.5	6.29	3.23

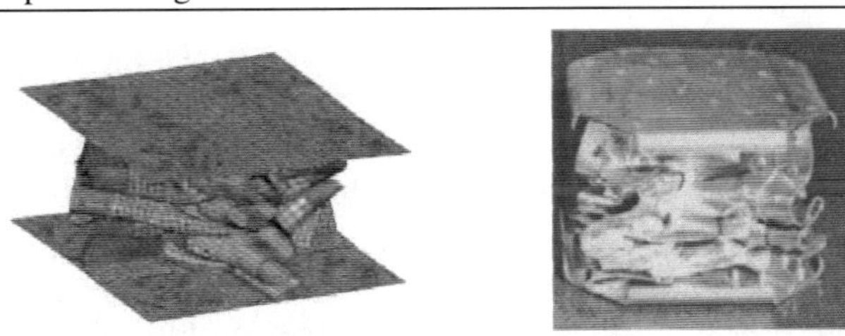

(a) Simulation result. (b) Experiment result.

Figure 7. Comparison of the experimental sample before and after crush test.

It can be concluded from the comparison above that the simulation method can correctly reflect the crush procedure of the intersection element.

3. Intersection element with foldcore sandwich structures

The intersection element with foldcore sandwich structures can be got by replacing the skin and the floor of the intersection element with foldcore sandwich structures. The floor is replaced with V-type foldcore structure and the skin is replaced with M-type foldcore structure. The two kinds of foldcore are shown in Figure 8. The simulation method of the foldcore sandwich structures are introduced in paper [24, 25].

The height of the folded core sandwich structures is set as 20mm. In order to maintain the bending stiffness of the skin and the floor, the thickness of the foldcore sandwich structure plates can be calculated as follows:

The bending stiffness of a panel can be calculated as follows:

$$D = Et^3 / 12(1-\mu) \tag{1}$$

In formula (1), D is the bending stiffness, t is the thickness of the panel, E and μ are the elastic model and the Poisson ratio of the material of the panel.

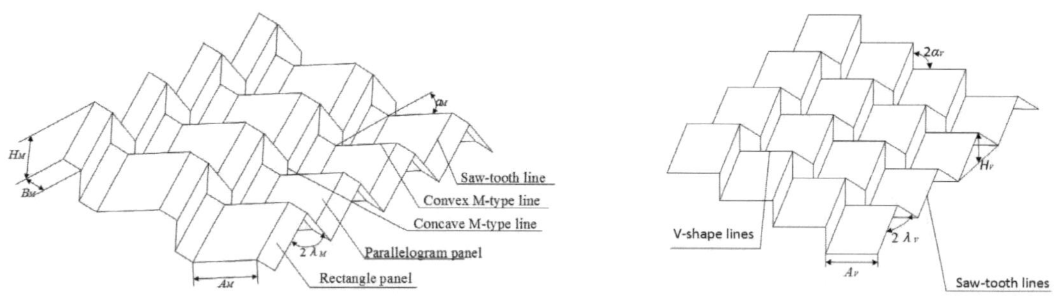

(a) M-type foldcore. (b) V-type foldcore.

Figure 8. M-type and V-type foldcore and their geometric parameters.

For a sandwich structure, when calculating the bending stiffness, stiffness of the core can be ignored. According to the definition of the bending stiffness, the bending stiffness of a sandwich structure can be expressed as:

$$D_s = \left(E_s t_s / 6\left(1-\mu_s\right)\right)\left(H^2 - 2Ht_s + 4t_s^2\right) \tag{2}$$

In formula (2), D_s is the bending stiffness of the sandwich structure, t_s is the thickness of the plate, E_s and μ_s are the elastic model and the Poisson ratio of the material of the plate, and H is the height of the folded core.

Let $D \le D_s$, H=20mm, t=1.37mm, it can be calculated that:

$$t_s \ge 0.002\text{mm} \tag{3}$$

Considering the manufacturing technology and the ability to resistant impact, t_s is set as 0.3mm.

The height of the intersection element is 195mm, which is the distance of the middle surface of the skin and the floor. The height of the foldcore sandwich structures is 20mm, so the height of the beams and the angle elements are 175mm. The geometric parameters of other panels don't change. The parameters of the foldcore are set as: $H_M=H_V$=20mm, $A_M=A_V$=15mm, $a_M=a_V=30°$, $\lambda_M=\lambda_V=20°$, B_M=5mm, the material of the foldcore is also aluminium alloy 2024 T3, and the thickness is 0.15mm. The intersection element structure with foldcore sandwich structures is shown in Figure 9.

Crush simulation of the structure is conducted. The impact velocity, impact mass and the boundary conditions are the same as the typical intersection element model. The deformation process of the structure with foldcore sandwich structures is shown in Figure 10. The force-shortening diagrams are shown in Figure 11 and the acceleration of the impact mass-shortening diagrams is shown in Figure 12.

Figure 9. The intersection element structure with foldcore sandwich structures.

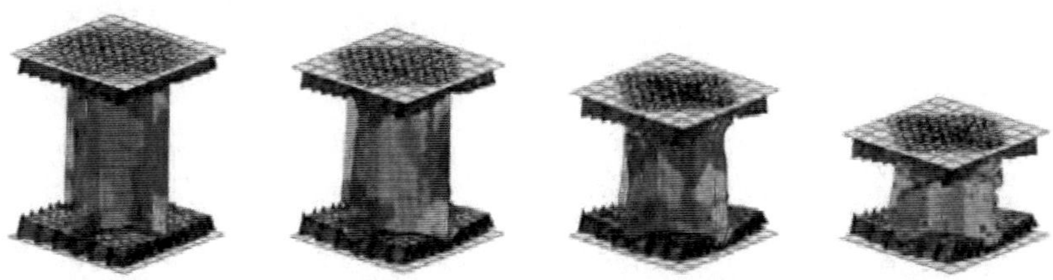

(a) 1.35ms (b) 3.15ms (c) 8.1ms (d) 31.5ms

Figure 10. The deformation process of the intersection element structure with foldcore sandwich structures.

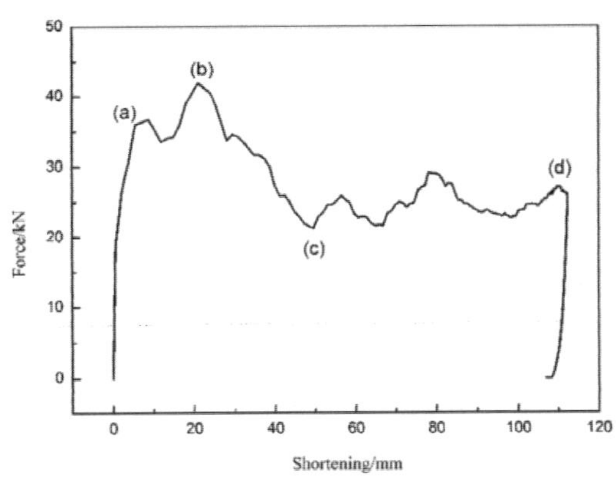

Figure 11. The force-shortening diagrams of the intersection element structure with foldcore sandwich structures.

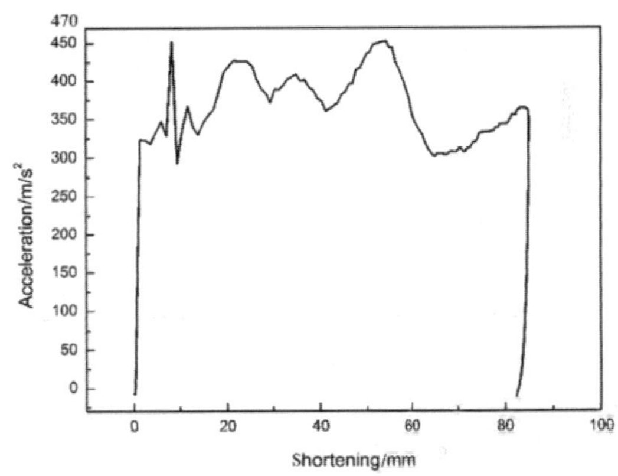

Figure 12. The acceleration-shortening diagrams of the intersection element structure with foldcore sandwich structures.

It can be seen from Figure 10 that the damage first appears at the foldcore near the beam webs. The deformation of the rafters produce small disturbance to the webs, so the webs buckle easier, which decreases the peak load and improves the buffering capacity of the structure. The deformation of the foldcore form concave zones near the webs. In the deformation process of the webs and the angle elements, a series of plastic hinges are generated. The concave zones can affect the position of the plastic hinges near the foldcore. Besides, the height of the webs and the angle elements are also changed. Thus the distance between two plastic hinges are changed, which improves the crush load of the structure.

The force-shorting diagrams of the intersection element structure with foldcore sandwich structures and typical intersection element structure are shown in figure 13.The crush analysis results of the two structures are shown in table 2. It can be seen that the structure with foldcore sandwich structures has small peak force and bigger crush force. It means that the structure with foldcore sandwich structures has better buffering capacity and energy-absorption capacity. Besides, the mass of the structure with foldcore sandwich structures is smaller, so it’s better compared to the typical intersection element.

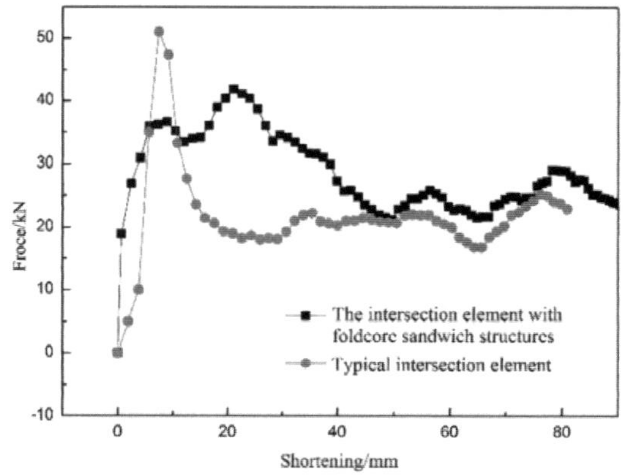

Figure 13. Comparison of the force-shortening diagrams of the structure with foldcore sandwich structures and typical intersection element.

Table 2. Comparison of the simulation results and the experiment results of the intersection element.

	Typical structure	Structure with foldcore sandwich structures	Variable ratio /%
Peak load/kN	52.0	41.9	-19.4
Average load/kN	25.5-28.0	26.8	
Mass/kg	0.481	0.480	-0.2

4. Conclusions

(1) The intersection element is an important part of the helicopter subfloor structures. A numerical simulation method is established to model the crush process of the intersection element. The analysis result is compared with the experiment result and they match well.

(2) The foldcore sandwich structures are used as buffering and energy-absorbing structures in helicopter subfloor structures and the intersection element is redesigned.

(3) Compared with typical intersection element structure, structure with foldcore sandwich structures have less mass, smaller peak force and bigger crush force. It means that foldcore can improve the buffering capacity and energy- absorbing capacity of the intersection element structure.

References

[1] Bisagni C 1999 *Aircraft Eng. Aeros. Tech.* **71(1)** 6-11.
[2] Bisagni C 1999 Int. J. Crashworth. **4(2)** 199-212.
[3] Bisagni C, Lanzi L and Ricci S. 2013 *Size and Topological Optimization for Crashworthiness Design of Helicopter Subfloor* Aiaa/issmo Symp Multidisciplinary Anal Optimization Atlanta p 1-11.
[4] Lanzi L, Bisagni C and Ricci S 2004 *Comput. Struct.* **82(1)** 93-108.
[5] Bisagni C, Lanzi L and Ricci S 2002 *J. Aircraft* 296-304.
[6] Wang Z J and Khaliulin V 2003 *J. Nanjing Univ. Aeronaut. Astronaut.* **34(1)** 6-11.
[7] Wang Z J and Xu Q H 2004 *J. Nanjing Univ. Aeronaut. Astronaut.* **36(4)** 449-53.
[8] Zhang Y C, Yu J M and Zhang S L, et al 2014 *J. Vib. Shock* **33(1)** 113-8.
[9] Zhang Y C, Wang Z L and Zhang S L 2010 *J. Ship Mech.* **14(Z1)** 114-20.
[10] Ren Y F 2014 *Study on Basic Mechanical Properties of Flodcore Sandwich Structures* Harbin Harbin Ins of Tech.
[11] Cong L, Sun Y and Wang X 2015 *Comp. Mater. Continua.* **47(1)** 1-14.
[12] Heimbs S, Middendorf P and Hampf C, et al 2008 *Aircraft Sandwich Structures with Folded Core under Impact Load* Int. Conf. Sandwich Struct. Porto. pp. 169-380.
[13] Heimbs S, Middendorf P, Kilchert S, et al 2007 *Appl. Compos. Mater.* **14(5-6)** 363-77.

[14] Heimbs S, Middendorf P, Kilchert S, et al 2007 *Numerical simulation of advanced folded core materials for structural sandwich applications* Ceas. Eur. Air Space Conf. Berlin, pp. 1-8.
[15] Heimbs S 2009 *Comput. Mater. Sci.* **45** 205-16.
[16] Heimbs S, Mehrens T, Middendorf P, et al 2007 *Numerical Determination of the Nonlinear Effective Mechanical Properties of Folded Core Structures for Aircraft Sandwich Panels.* 6th LS-DYNA Users Conf. Sweden Gothenburg p 29-30.
[17] Heimbs S, Cichosz J, Kilchert S, et al 2009 *Sandwich Panels with Cellular Cores Made of Folded Composite Material: Mechanical Behaviour and Impact Performance* Int. Conf. Compos. Mater. Edinburgh p 1485–97.
[18] Fischer S, Heimbs S, Kilchert S, et al 2009 *Sandwich Structures with Folded Core: Manufacturing and Mechanical Behavior Sampe* Eur. Int. Conf. Paris pp. 256-63.
[19] Heimbs S, Cichosz J, Klaus M, et al 2010 *Compos. Struct.* 92(6) 1485-97.
[20] Klaus M, Reimerdes H G and Gupta N K 2012 *Int. J. Impact. Eng.* **44** 50–8.
[21] Baranger E, Guidault P A and Cluzel C 2011 *Compos. Struct.* **93(10)** 2504–10.
[22] Baranger E, Cluzel C and Guidault P A 2011 *Strain* 47(s2) 170–8.
[23] Hähnel F, Wolf K, Hauffe A, et al 2011 *CEAS Aeronaut. J.* **2(1)** 203-12.
[24] Zhou H Z and Wang Z J 2016 Acta *Aeronaut. Astronaut. Sin.* **37(2)** 579-87.
[25] Zhou H Z and Wang Z J 2017 *Acta. Mater. Compos. Sin. Preprint* (DOI: 10.13801/j.cnki.fhclxb.20170321.001).

IOP Conf. Series: Materials Science and Engineering **248** (2017) 012034 doi:10.1088/1757-899X/248/1/012034

Test and Analysis on Peeling and Cracking on Surface of 45 Steel Shafts

Xinghui Li[1]

[1]Department of Architecturl Engineering,Jiangsu Prvince Xuzhou Technician Institute Xuzhou, Jiangsu, China

Abstract. Aiming at the peeling and cracking appear on the surface of a 45 steel shaft after surface quenching and grinding, we've conducted chemical composition analysis, hardness test and metallographic microstructure analysis for the cracked shaft sample. The results shows that the main reason for the peeling and cracking of 45 steel shafts is the stress generated due to the uneven structures produced on the surface during quenching and tempering after forging and surface quenching.

1. Problem

As a kind of quality carbon structural steel used for manufacturing shaft parts, 45 steel has high strength, certain degree of plasticity and toughness and good machinability, and can get good comprehensive mechanical properties after quenching and tempering. While in actual manufacturing process, steel parts are prone to form various cracks after quenching and processing due to multiple factors, and the condition is difficult to control and will lead to increased rejection rate. A shaft, with diameter of 130mm, manufactured by XXX company and made of 45 steel, has the following processing steps: casting – forging – (rough machining) – quenching and tempering – finish machining – surface quenching – low temperature tempering – grinding. In actual manufacturing, cracking occurs on the shaft during the storage after grinding, with the macro appearance of cracks shown in figure 1. It can be concluded from observations that most cracks appear on the thin layer on the surface, with thickness of about 0.3mm-0.5mm, and cracks are generally in parallel to the surface of parts and have large area, acting as the direct reason for surface peeling. Some workpieces have secondary cracks along axial direction, some even have tertiary cracks along tangential direction. To probe into the reason for shaft cracking, we've conducted in-depth and detailed tests on material of parts, microstructure, heat treatment process, etc. as well as research works such as chemical composition analysis, hardness test and metallographic microstructure analysis, aiming to provide effective ways to deal with failures of shafts.

Content from this work may be used under the terms of the Creative Commons Attribution 3.0 licence. Any further distribution of this work must maintain attribution to the author(s) and the title of the work, journal citation and DOI.
Published under licence by IOP Publishing Ltd

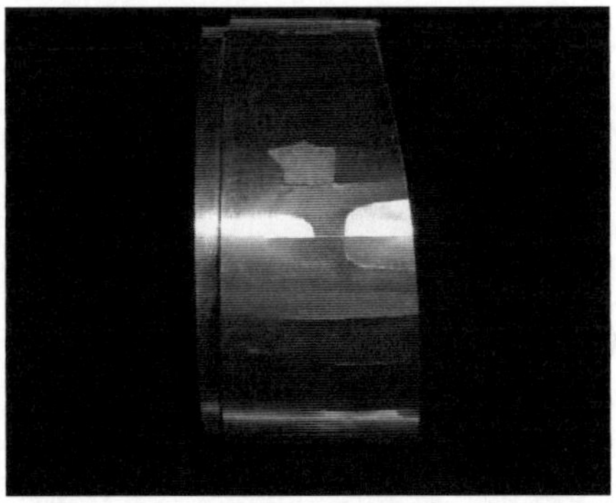

Figure 1. Macro appearance of cracks.

Figure 2. Picture for section of the sample.

2. Methods and steps for test research

2.1 Sampling
Take samples with length of 50mm-60mm from shafts and conduct material analysis and original structure analysis.

2.2 Test research
Conduct chemical composition analysis, hardness test and metallographic structure analysis, etc. on the sample and research into the influence of shafts' material, structures and mechanical property on generation of cracks based on characteristics of cracks.

2.3 Results analysis
Analysis results, conclude reasons for generation of shaft cracks and its spreading rules, and propose methods and measured to prevent cracks.

2.4 Major instruments and equipment used
Spectrum analyzer, Rockwell hardness tester, P-2 type metallographic microscope and S-530 type scanning electron microscope (SEM).

3. Test research results and analysis

3.1. Chemical composition analysis
By applying the spectrum analyzer, we've conducted chemical composition analysis for the shaft samples with cracks, determined the major chemical compositions of shafts, analyzed the materials of shaft materials as well as possible influencing factors. The analysis result is shown in table 1. The result shows that the chemical composition of the steel used for manufacturing the shaft meets the technical requirements for the 45 steel in accordance with GB699-88 Quality Carbon Structure Steel – Technical Requirements.

Table 1. Chemical composition of steel (mass fraction).

Element	C	Si	Mn	P	S	Cr
Content (%)	0.47	0.26	0.65	0.027	0.021	0.16
Content stipulated in GB699-88 (%)	0.42 ~ 0.50	0.17 ~ 0.37	0.50 ~ 0.80	≤0.035	≤0.035	≤0.25

IOP Conf. Series: Materials Science and Engineering **248** (2017) 012034 doi:10.1088/1757-899X/248/1/012034

3.2 Hardness test

The picture for section of the sample is shown in figure 2. The white highlighted area on the periphery is the hardened layer. The shaft has gone through surface quenching with a hardened layer thickness of about 3mm. We measured hardness of the surface and core of shaft by conducting Rockwell hardness test, with test results shown in table 2.

Table 2. Hardness test results.

Location	Quench hardened layer on surface	(The core is located at about 20mm from the surface)
Measured hardness value (HRC)	55.5, 56.3, 55.5, 51.7	11.2, 13.4, 12.1, 10.3, 13.0, 12.6, 11.6, 12.6, 12.1
Average hardness value (HRC)	54.7	12.1 (≈192HB)

After surface quenching of the shaft, the surface is the quenching structure, while the core goes through quenching and tempering treatment, so the surface and the core have different microstructure and thus definitely have different hardness values. It can be concluded from table 2 that the average hardness value of the quench hardened layer on surface is HRC54.7, the core hardness is HRC12.1 (20mm from the surface), which is about HB192 in terms of Brinell hardness. The surface hardness is relatively high, and the core hardness is relatively low, demonstrating huge difference between the two values.

4. Metallographic structure and analysis

4.1 Core structure

According to documents introduction [1], the cooling period of quenching for work pieces with diameter over 100mm should be as follows: for water quenching 3-4 min, and for oil quenching 15-25 min, to ensure the temperature of core cools down to around the Ms point (about 300℃). According to GB699-88, the hardness of carbon steel (of 35 steel, 45 steel or 50 steel and with diameter generally not more than 200mm) after quenching and tempering can reach HB240-280[2], and the hardness after annealing generally stays below HB197 (with blank dimension of 25mm). But the hardness of the core of shaft is only HB192. Although differences exist in sizes, the hardness after quenching and tempering is much lower than the hardness after annealing. This is absolutely abnormal. To identify the structure of the shaft, take sample from the core 10mm-20mm from the surface and determine the core microstructure by using scanning electron microscope, as is shown is figure 3. It can be concluded from figure 3 that the microstructure for the core of shaft is pearlite+ ferrite, of which the ferrite is distributed in network structure as shown in figure 3(a) due to low cooling speed and short cooling period during the quenching and tempering process after forging, causing proeutectoid ferrite separates out along the grain boundary and form network structure ferrite and eventually leading to low hardness of the core.

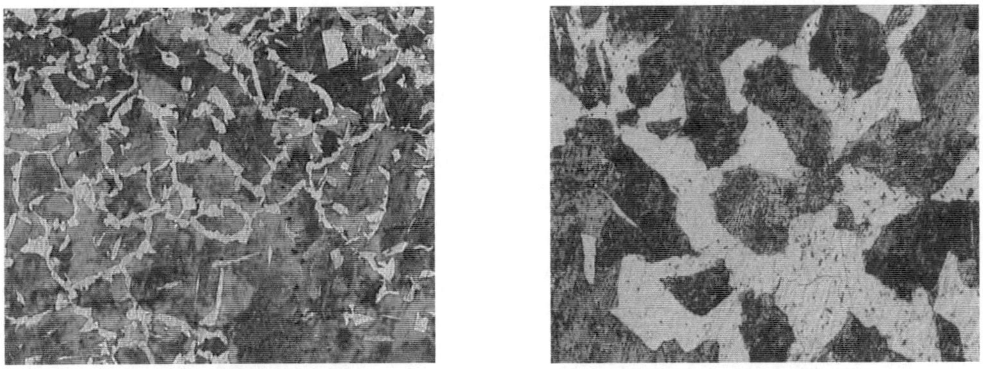

(a) ×100 **(b) ×500**

Figure 3. Metallographic structure of the core of shaft.

4.2 Surface structure

The metallographic structure of the surface and around the cracks is shown in figure 4. We can see that the metallographic structure around the cracks has no obvious oxide layer and decarburized layer, which means that the cracks do not occur during the processes of casting, forging and quenching and tempering after forging, but during the surface quenching as well as later manufacturing processes.

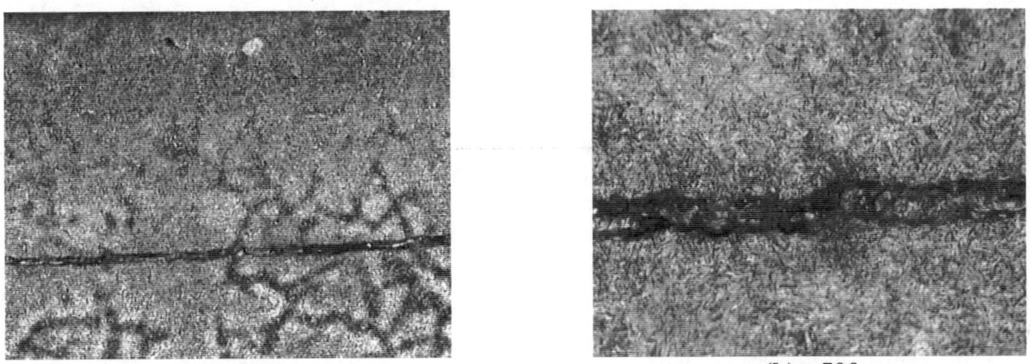

(a) ×100 **(b) ×500**

Figure 4. Metallographic structure of the surface of shaft.

It can also be concluded from figure 4 that the lath and acicular martensite exist in the quenched layer on surface [3], which may be caused by insufficient low temperature tempering after surface quenching. It also produces a relatively high hardness of surface – HRC54.8.

It can be concluded from figure 4(a) that obscure and black network structure exists in the quench hardened layer on surface, which is the low carbon martensite formed by network ferrite structure that goes through quenching and tempering during surface quenching. Low carbon martensite itself is prone to produce self-tempering phenomenon, and therefore make the sample susceptible to corrosion. High carbon martensite is formed by pearlite structure that goes through quenching and tempering, which is of large specific volume and is prone to generate volume expansion, thus make the surface layer produce radial tensile stress. This is the root cause for the peeling and cracking on the shaft [4].

Figure 5 is the subsurface structure under the quench hardened layer. It can be concluded that the subsurface structure is basically the same as core structure (observation can be conducted through comparison between figure 3 and figure 5).

Because the structure in subsurface is network structure ferrite [5], and the surface structure is the quenched high carbon martensite structure, coefficient of linear expansion of subsurface and surface structure is different, large tensile stress will produce in cooling and contraction, which is one of the reasons for the peeling and cracking of the metal on the surface.

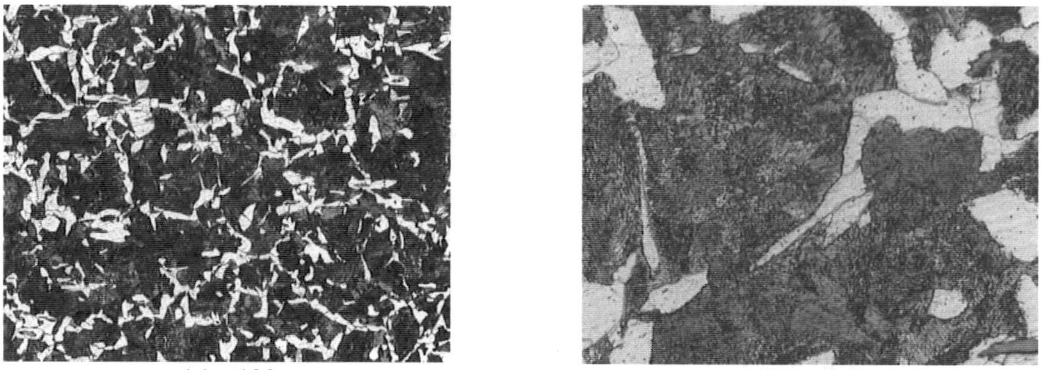

(a) ×100 **(b) ×500**

Figure 5. Metallographic structure of subsurface.

5. Conclusions

Reasons for the peeling and cracking on the surface of the shaft manufactured of 45 steel with a diameter of 130mm are as follows: the low cooling speed and short cooling period during the quenching and cooling process after forging leads to generation of network structure ferrite (and low hardness of the core); low carbon martensite (prone to erosion) is generated from network structure ferrite during surface quenching, and the pearlite structure surrounded by network structure ferrite, of which the carbon contents is higher than that of steel, generates high carbon martensite during surface quenching. And because the high carbon martensite has large specific volume and is prone to generate volume expansion, it makes the surface layer produce high radial tensile stress. The thickness of surface layer reduces after grinding and thus the peeling and cracking is produced under the tensile stress. According to documents, the peeling and cracking belongs to one of the 4 typical quenching cracks of steel pieces. Uneven structure along the hardened layer during surface quenching is easy to produce this typical peeling and cracking on surface.

Meanwhile, it shall be pointed out that delayed tempering or insufficient tempering of the shaft after surface quenching which leads to failure to timely eliminate the quenching stress on surface (and a relative high hardness of surface– HRC54.8) may also be the reason for the peeling and cracking on surface of the shaft.

Reference

[1] Zhejiang University, University of Shanghai for Science and Technology, Hefei University of Technology 1978 Ferrous Materials and Heat Treatment Technology (Shanghai: Shanghai Scientific & Technical Publishers).

[2] Gao W G, Wang G S & Li D F 2012 *Formation and Spreading of Quenching Cracks of 45 Steel Gears* (Natural Science Edition).

[3] Wang C J 2017 *Failure Analysis for the Quenching Cracks on 45 Steel Taper Sleeve* (Beijing:Heat Treatment of Metals)

[4] Zhang L W, Liu S F & Wang F 2015 *Analysis on Causes for Quenching Cracks on 50CrV Leaf Spring* (Heat Treatment of Metal).

[5] Wang G S, Shi K C and Zhou J G 2007 *Analysis Defect of Metal Heat Treatment and Cases* (Beijing: Machinery Industry Press) 13-51.